GEOMECHANICAL PROCESSES DURING UNDERGROUND MINING

PROCEEDINGS OF THE SCHOOL OF UNDERGROUND MINING, DNIPROPETROVS'K/YALTA, UKRAINE, 24-28 SEPTEMBER 2012

Geomechanical Processes During Underground Mining

Editors

Genadiy Pivnyak
Rector of National Mining University, Ukraine

Volodymyr Bondarenko
Department of Underground Mining, National Mining University, Ukraine

Iryna Kovalevs'ka
Department of Underground Mining, National Mining University, Ukraine

Mykhaylo Illiashov
PJSC "Donetsksteel", Ukraine

CRC Press is an imprint of the
Taylor & Francis Group, an **informa** business
A BALKEMA BOOK

CRC Press
Taylor & Francis Group
6000 Broken Sound Parkway NW, Suite 300
Boca Raton, FL 33487-2742

First issued in paperback 2019

Typeset by Olga Malova & Kostiantyn Ganushevych, Department of Underground Mining, National Mining University, Dnipropetrovs'k, Ukraine

ISBN-13: 978-0-415-66174-4 (hbk)
ISBN-13: 978-0-367-38091-5 (pbk)

Visit the Taylor & Francis Web site at
http://www.taylorandfrancis.com

and the CRC Press Web site at
http://www.crcpress.com

© 2012 Taylor & Francis Group, London, ISBN 978-0-415-66174-4

Table of contents

Geomechanical Processes During Underground Mining – Pivnyak, Bondarenko, Kovalevs'ka & Illiashov (eds)
© 2012 Taylor & Francis Group, London, ISBN 978-0-415-66174-4

Preface

Present collection of scientific papers is dedicated to mining engineers, engineering technicians, designers, scientific and research personnel, students, postgraduates and all mining-related professionals working in coal and ore industry.

The authors of this book have contributed their articles that cover economic aspects of mining companies' development strategies, peculiarities of various mineral deposits development techniques, imitational modeling of mine workings with rock massif, methane extraction technologies during coal mining, geomechanical processes during plow mining, mining transport importance for mineral extraction, massif strain-stress state management using non-explosive destructing materials, surface mining negative influence on the environment.

The alternative ways of mining such as borehole underground coal gasification for extraction of hardly accessible coal and development of gasification plant is paid a large amount of attention to. Also development and use of alternative sources of energy such as gas hydrates and sun energy are given consideration in this book.

Genadiy Pivnyak
Volodymyr Bondarenko
Iryna Kovalevs'ka
Mykhaylo Illiashov
Dnipropetrovs'k
September 2012

DEDICATED TO G.G. PIVNYAK, ACADEMICIAN OF THE NATIONAL ACADEMY OF SCIENCES OF UKRAINE, WHO HAS BEEN THE RECTOR OF THE NATIONAL MINING UNIVERSITY FOR 30 YEARS

The year of 2012 is a momentous year for the history of the National Mining University. Thirty years ago Gennadiy Grygorovych Pivnyak, the youngest rector at that time, headed one of the leading technical higher institutions of Ukraine. Today he is a talented organizer, recognized scientist and Academician of the National Academy of Sciences of Ukraine. His life and scientific work are a unique phenomenon not only in Ukraine but also abroad.

Since that time the National Mining University has developed rapidly and there have been qualitative changes in all spheres of its activities.

The National Mining University, founded in 1899, is one of the oldest state higher technical educational institutions of Ukraine. Its dynamic development has always favoured engineering education. Internationally recognized scientific schools of mining, geology, geomechanics, power engineering, geophysics, and mineral dressing continued their development. Nowadays this helps achieve the best scientific and practical results necessary for the innovative development of Ukraine's economy.

The history of University reflects the economic situation of the country and region, the level of scientific and technological progress as well as social, political and cultural life. It is the Mining University that became the basis for establishing twenty higher educational institutions and nine scientific and research institutes of the National Academy of Sciences of Ukraine.

The National Mining University has been continuously developing, its scientific achievements constantly expanding. It is a modern scientific and educational centre, autonomous research national university characterized by fundamentality and consistency of knowledge, combination of education, science and innovations, and many-sidedness of international cooperation and collaboration. According to the UNESCO, the National Mining University is reckoned among three principal technical universities of Ukraine. Being in compliance with the time has always been the priority of the University activities.

The whole life of Gennadiy Grygorovych Pivnyak is inseparably connected with the National Mining University.

G.G. Pivnyak was born on the 23rd of October, 1940 in the town of Oleksandria in Kirovograd region in the family of teachers. In 1958 he entered the Dnipropetrovsk Mining Institute graduated with distinction in 1963. He was qualified as a mining electrician engineer, and he cast in his lot with Alma Mater.

Educational work is dominant in the life of G.G. Pivnyak. He heads the Department of Electric Power Supply Systems, delivers lectures, supervises projects, and encourages his students to carry out research. In 2005 the textbook "Transients in the Systems of Electric Power Supply", published under the editorship of Academician G.G. Pivnyak in Ukrainian, Russian, and English, was awarded second State Prize of Ukraine in the field of science and engineering. That was the recognition of high level of his pedagogical art. He is the author of a number of textbooks and monographs in mining and metallurgical power engineering, systems of electric power supply of enterprises, electric drive, electrical safety, and automated control systems. Some of his books were published abroad, and they are popular with specialists and students of different countries.

Academician G.G. Pivnyak is the founder of scientific school of mining and metallurgical power engineering. He enriched modern science with the research having the paramount importance for the development of electric power engineering and creation of new electrotechnical complexes and equipment. His scientific activity is devoted to solving physicotechnical problems of transforming and regulating parameters of electromagnetic energy, development of scientific bases for innovative technologies, equipment and control facilities providing efficiency and safety of mining and metallurgical production.

G.G. Pivnyak and his disciples elaborated the newest systems for automation of modern power engineering complexes and efficient power supply of complex structure electromechanical systems, and automatic electric drive of processing stations.

G.G. Pivnyak is well-known for 3 scientific discoveries, over 500 scientific publications including 37 monographs, 13 textbooks, 22 tutorials, and 107 inventions and patents. During the last 35 years he continued and cherished the traditions of Dnipropetrovsk school of electrical power engineers. 18 PhD theses and 29 MPhil theses were defended under his supervision. Academician G.G. Pivnyak, a famous scientist, generator of ingenious ideas, organizer of large-scale and urgent research, gifted lecturer, always shares his knowledge and experience with his colleagues. Students, postgraduates, and doctoral students are always working with him. They are his disciples and followers. The main credo of their teacher and adviser is: education and science must serve the national interests of Ukraine.

Research activity of G.G. Pivnyak is aimed to support the innovative model for the development of key sectors of Ukrainian economy. In 1996 G.G. Pivnyak got the Academician Lebedev Award of the National Academy of Sciences of Ukraine for the development in the field of science-intensive production control at the enterprises of mining and metallurgical complex. In 1998 he was awarded the State Prize of Ukraine in the field of science and engineering for the development and implementation of methods and technologies of geological environment conservation for providing energy-efficient coal mining. That Prize helped Gennadiy Grygorovych establish a charitable foundation to support young scientists, which is still functioning at the University.

G.G. Pivnyak was awarded the order of Badge of Honour (1981), the Red Banner of Labour order (1986), Prince Yaroslav the Wise order of the 5th (1999) and 4th (2004) degrees. He has the Certificate of Merit of the Cabinet of Ministers of Ukraine (1999) and Certificate of Merit of Verkhovna Rada of Ukraine (2003). Besides, he was awarded Merits of Education of Youth medal (1998, Poland), Merits of Scientific Achievements by the Ministry of Education and Science of Ukraine Badge (2007) and Merits of Scientific Achievements by the National Academy of Sciences of Ukraine (2009).

Long-term foreign relations of G.G. Pivnyak (Germany, Austria, Poland, Great Britain, the USA, Canada, Russia, Moldova, Switzerland, France, Spain, China, and Japan) helped him to gain great experience in implementing international scientific and research projects, in developing integration relations in science, education, and innovations. This creates necessary conditions for his successful mission as an expert of the Committee on sustainable energy of ECE UN testifying that Academician G.G. Pivnyak is recognized as a scientist world-wide.

Academician G.G. Pivnyak is Honourable Professor of such technical universities as Krakov Mining and Metallurgical Academy, Freiberg Mining Academy, and Moscow State Mining University; honoured worker of science and technology of Ukraine (1990), honoured worker of oil and gas industry of Poland (1994), honoured Professor of the National Mining University, the NMU award winner in the field of education and science (2002) – these are the main titles which testify his high authority in Ukraine and in the world.

These very thirty years were marked by the events which livened up the NMU activities in the all spheres; they approved its international recognition, they influenced further progress of the institution of higher education, and determined its position in education, science, and economy of Ukraine.

Geomechanical Processes During Underground Mining – Pivnyak, Bondarenko, Kovalevs'ka & Illiashov (eds)
 ISBN 978-0-415-66174-4

Development of coal industry of Ukraine in the context of contemporary challenges

O. Vivcharenko
Department of Coal Industry Ministry of Energy and Coal Industry of Ukraine

ABSTRACT: Having analyzed accepted in the Euro Union ways of energy safety problems solving, comparative analysis of approaches to optimization of solution in Ukraine and other countries has been conducted. Measures that are powerful counterforce to modern external and internal challenges and threats of the country's energy safety are marked out. Measures necessary for coal extraction volumes increase and the industry functioning efficiency increase are substantiated.

Coal industry – one of the most important and integral components of the fuel and energy complex (FEC) and economy of Ukraine, which plays a significant role in ensuring of energy security of the country. At the same time, the energy security has been one of the weakest links in the economy and national security of Ukraine for last twenty years.

Modern external challenges that were faced before Ukraine in the energy sector and that are common to most countries is the need to overcome the risks associated with the instability and unpredictability of the situation on world energy markets, rising prices for all kinds of energy at the same time. Under such circumstances, guaranteeing of needs of Ukrainian economy in fuel and energy resources (FER) at reasonable prices for consumers extremely difficult.

Modern external challenges that were faced before Ukraine in energy sector dependence of on energy imports (about 55% of the total consumption of energy resources FER, taking into account conditional – primary nuclear energy). Close to such a level of energy dependence on most European countries: Germany – 61.4%, France – 50%, Austria – 64.7%. However, unlike the EU, where imports of energy diversified, energy dependence of Ukraine has monopolistic character: by the import of petroleum and petroleum products economy of Ukraine depends on Russia for 65% of natural gas – by 72%, nuclear fuel – 100%, which is critical.

For Ukraine, effective response to external challenges in the coming years could be measures to activation of energy-safety, maximum use of all possible and available at the moment of its own energy resources and the substitution of natural gas (where it is technologically possible and economically reasonable), accelerating the modernization of the Extractive Industries FEC to increase the volume of fuel production, modernization of energy-intensive industries with the introduction of energy-saving technologies and optimization of fuel supply, the implementation of structural changes in the economy as well.

Internal challenge for Ukraine was the need for a system, consistent, transparent and accountable to government policy aimed at ensuring energy security, which for 20 years of independence has not been formed. The main cause of threats of energy security for recent years, arising from conflicts and harsh conditions of supply of Russian natural gas, is precisely the absence of such policies, despite the fact that Ukraine has all the objective conditions (sufficient own resources base of fuel) for its implementation.

In Ukraine, the structure of consumption of primary energy resources was formed during the time of the former Soviet Union and is focused mainly on Russian energy resource. In the total volume of consumption of energy resources remains too large share of natural gas, which is mostly imported (41% against average – 24%), and insufficient weight of their own energy resources, especially coal (26% vs. 29%). This structure does not correspond to the modern international trends and the interests of our country, because it does not take into account the peculiarities of its own raw material fuel base and a huge potential of fuel and energy industries.

If the global structure of the geological reserves of fuel, there are significant advantages in favor of coal (65.2%) compared with other fuels – oil (16.8%), gas (13.5% methane with coal deposits), uranium (4.5%), Ukraine has the advantages of an even more significant in favor of coal (85.2%) and uranium (9.8%) with limited oil reserves (0.9%) and natural gas (4.1%).

Countries and regions of the world's energy security problem solve in different ways depending on the availability of raw materials and resource base for fuel and energy, geographic location, the currency of economic efficiency.

For countries and regions with low (less than 30%) level of self-sufficiency by the fuel and energy resources is an important safe and guaranty of energy import. Successfully solve the problem of the country with a developed export-oriented economy through the diversification of energy supply – Japan (the level of self-sufficiency is about 7%), Italy (18%), most of the European Union (EU).

For middle level of self-sufficiency (30-70%), the main objective of energy policy is to ensure energy security in terms of opportunities to manage their own resources at a loss or reduction of the external supply (Ukraine, Germany, Sweden).

For the countries of their own well-to-energy (self-sufficiency level of 70%), the main objective is the development of fuel-energy complex (FEC), which can not only meet the domestic demand for high quality and affordable energy, but also economically justified to export (U.S., Russia, Norway, Australia).

In the world are turned out the following ways to strengthen energy security:

– reduction of energy demand by increasing efficiency and energy consumption;

– improvement of energy self-sufficiency;

– diversification of energy with an increase in the use of competitive local and renewable fuels and energy;

– diversification of sources and supply routes of imported energy resources etc.

Unfortunately, Ukraine had not achieved tangible positive results in none of these directions. Lack of incentives and motivation in introducing new resource-saving technologies and innovation led to the dynamic development of very low energy, economic and environmental performance of all sectors of the economy, including fuel and energy industries.

At present days, Ukraine's gross domestic product (GDP) of energy intensity in 2.6 times exceeds the average energy intensity of GDP in developed countries. The reason for this is lack of incentive for efficiency, a sufficient number of market instruments and other incentives, including taxation. However, this efficiency is the most effective way to enhance energy security and competitiveness, minimizing negative impacts on the environment and reduce greenhouse gas emissions. In forming the policy in energy-saving sector must be taken into account that energy efficiency criteria should apply to all spheres of economic and social development, including the redistribution of public funds.

One of the main documents the legal framework of FEC development is the Energy Strategy of Ukraine in period till 2030, which was adopted in 2006. Positive is that it is relies on the experience of different countries in ensuring energy security and has specific quantitative standards and measures for its support. In particular, reductions the energy dependence of the country to be achieved mainly by bringing the maximum potential energy savings increase in output (production) and consumption of energy and diversification of its own internal and external sources of supply. Moreover, diversification of external and internal supplies of energy (gas, oil, nuclear fuel) has been carried out with the ensuring of the European norm - at least three sources for each type of energy with the provision of 25-30% of the total.

However, the vast majority of the events that had to be implemented in the first five years of the Strategy are not met due to objective reasons (the onset of the financial and economic crisis, to provide at the time of its development was not possible) and subjective (no link action with some basic goals, priorities, objectives and responsibilities and government control of them).

There is currently updating the Strategy, in which, according to the experts of the Institute of Economics and Forecasting of NAS of Ukraine, laid down certain threats to energy security of Ukraine.

1. Till 2030, the basic scenario of the gross domestic product (GDP) of Ukraine is expected to grow to 913 billion UAH., in 2009 to 2563 milliard, in 2030 (in 2009 prices), in 2.7 times the level in 2010. According to GDP per capita in purchasing power parity (PPP) will increase in 2.88 times (forecast estimates the population of Ukraine in 2030 may reach 42.8 million). That is, in Ukraine in 2030 GDP per capita should be the same as today in Estonia, Latvia, Poland, Hungary, and thus consolidate the low standard of living in Ukraine that is totally unacceptable.

For the record. At present, the GDP per capita in Slovakia higher than in Ukraine in 3.4 times, and in Slovenia – in 4.3; in Greece – in 4.6, in Germany – in 5.6; in Austria – in 6.0, in Switzerland – in 6.4; in Norway – in 8.3; in Luxembourg – in 12 times.

These are the main macroeconomic indicators of inputs data, which are based on forecasts of energy supply and development of fuel and energy industries. Accordingly, if the basis is unacceptable, we could not analyze the proposed offers, because they are unacceptable too. However, consider some of them, which ultimately will increase the threat to Ukraine's energy security, increasing dependence on energy imports.

2. Installed electricity demand substantially underestimated: in 2030 – 272 billion kWh (versus 395 billion kWh in the basic scenario of Strategy – 2006), including losses in the electrical, own consumption, exports (in fact this is volume of production). It is understated, not only because of the inherent low GDP growth, but also because of the neglect of the required replacement of fuel (primarily natural gas), electricity and other forms of energy (in the industry, domestic sector, etc.).

Consumption of electricity industry in 2030 will amount to 133 billion kWh, we note that in the pre-crisis in 2007 it amounted to 106 billion kWh, it is means increase by only 25%.

The population, which now consume 35 billion kWh at a very low level of equipment appliances will consume only 54 billion kWh (a 59% increase), which also indicates laid low standard of living in Ukraine 20 years later.

Almost does not provide for the development and use of export potential of electricity that would be beneficial for Ukraine, because the electricity tariff, for example, in European countries is 3-5 times higher than in Ukraine. In 2030, electricity exports will total 6 billion kWh (compared to 25 billion kWh in line with the Strategy-2006), whereas in the pre-crisis in 2007 it amounted to 13 billion kWh.

3. Except low power demand additional risks in a stable and sufficient to ensure the needs of the economy is suggested by the structure of power production- only 33% of electricity will be produced from coal-fired thermal power plants, which can be safely secured their own fuel, 49% – on nuclear power with all the risks relative to potential natural and anthropogenic disasters, many unresolved issues for the production of fuel for nuclear power plants and waste treatment, 7% – on the gas thermal power plants and cogeneration plants, 5% – on non-conventional and renewable sources of energy. This electricity is now very expensive and not competitive, increase its share will lead to a significant increase in electricity tariffs, which could lead to its economic inaccessibility for many categories of consumers, a loss of competitiveness.

For the record. In the main direction of the world consumption of coal is the production of electricity and thermal energy, and about 65% of which is used of its production. Over the past 18 years have seen a steady increase the share of electricity production with using coal from 37% to 40%, and the leading coal-producing countries, it has reached: in Poland – 97%, South Africa – 93%, Australia – 82%, China – 80% India – 75%, USA – 57%. In Ukraine it is 27%.

Now, many new coal power plants in Germany (which has its own coal reserves) – with the total capacity about 12 GW, in Belgium – 0.8, in India – 1.25, in Vietnam – 7.0, in Taiwan – 0.8, in Indonesia – 2.63, in Australia – 0.75 GW. The largest number of projects in the U.S. – 77 power plants.

According to estimates during the period till 2025 production of electricity from coal will increase in the U.S. – an average in 1.5% per year, China – in 4.5%, South Korea – in 2.5%, India – in 2.3%. After the accident at a Japanese nuclear power plant "Fukushima-1", arguments in favor of more secure electricity production from coal are increased, for the European coal-producing countries too.

4. Gas – it is the only energy source, which has demand the same as in strategy – 2006 (49.5 billion m^3 in 2030), although, now quite clear that it is necessary to reduce or replace by other forms of energy. The volume of its own gas production is expected to increase in 2030 is 1.5-2 times from the current level – up to 40 billion cubic meters of gas per year, which will consist of about 24-26 billion cubic meters of gas to existing and new conventional deposits, including deposits in the shallow shelf of the Black Sea, and 6-14 billion cubic meters, to be extracted from the deep shelf of the Black Sea and the non-traditional fields (methane from coal deposits, shale gas, tight gas central pool type). It should be noted that the volumes of 6-14 billion cubic meters - risky, it is very likely do not receive them, because currently there are not even needed more or less reliable geological data on non-traditional fields and deep shelf, is not carried out stock assessments, research the feasibility of its production.

Regarding to shale gas, should be noted that the technology of its production of environmentally harmful, especially to water resources. Therefore necessary to conduct a preliminary deep comprehensive mining impact research to on water and land resources and economic feasibility of shale gas in comparison with other alternatives as power supply and economic specialization of regions.

5. Coal – unfortunately, is not recognized energy resource that can not only guaranteed to maintain a certain level of energy security, but in the modern face of rising world prices for all types of energy to contain some price pressure on the Ukrainian consumers due to moderate level of prices for domestic coal and electricity, which produced with its use. Demand for commodities steam coal is reduced to 54 million tons (you can get them with 75 million tons of production) against 89 million tonnes under the 2006 Strategy. This reduction in demand due to reduced demand for electricity and the proposed structure of the electricity production, shifted in favor of nuclear energy and alternative and renewable energy sources, as well as the neglect of substitution of natural gas steam coal in

metallurgy and other sectors of the economy.

Thus, it can be argued that the proposals to update the Energy Strategy of Ukraine till 2030 in the electricity sector will not lead to increased energy security, but rather will reduce the reliability of energy supplies, divert huge resources to investment in Ukraine secondary energy sources, increase the risks increasing dependence on energy imports and thus create the prerequisites for the formation of new, very tough call for the economy of Ukraine in matters of energy supply.

To search for effective responses to modern challenges and threats to energy security, it is advisable to take into account the world experience in solving this problem and to determine the characteristics of energy supply of Ukraine in view of global trends.

Energy Policies of the European Union is formed in the absence of a powerful source of raw materials and fuels is aimed at solving complex problems – providing competitive environmentally clean energy on a background of climate change, growing global demand and energy prices, and uncertainty about their future supplies. To solve this problem with the European Union in 2006 began implementing a new energy policy, which aims to create an effective regulatory framework in the areas of supply and energy consumption, enhance competitiveness and consolidation of the EU energy sector on the principles of market liberalization and minimize the impact on climate change.

In 2008, the European Commission proposed a "Plan for the EU's energy security and solidarity action", which includes five areas of energy security:

– increasing efficiency of energy;

– diversification of energy supplies;

– maximum use of its own resources;

– stockpiling of crude oil (petroleum) and natural gas;

– improving the external energy relations;

– the creation of mechanisms for the resolution of crises.

A strategic approach to solving problems of energy security includes in ambitious new strategy for the EU by 2020 (Strategy "20-20-20"), which provides up to 2020, in particular, the reduction in energy consumption by 20%, reduce greenhouse gas emissions by 20%, bringing the share of renewable energy sources in EU energy mix to 20%.

In general, for Ukraine are acceptable these, adopted in the European Union, solutions of solving problems of energy security. However, comparative analysis of fuel resource base made for Ukraine and EU shows that, in contrast to EU, Ukraine has strong raw material base of its own fuel (4.1% of world coal reserves with a population of 46 million people and GDP 196 billion dollars. U.S. PPP, the EU-27, respectively, 3.6%; 492 million people; 13.3 billion dollars USA; components of nuclear fuel – uranium and zirconium). For Ukraine, does not use such advantages, as other countries do, means do not solve problems of reliable energy supply and reduce dependence on imported energy.

Thus, considering written above, features of the fuel resource base, and monopolistic dependence on energy imports, lower currency efficiency of the economy (which excludes energy imports in sufficient quantities), the limited opportunities for diversification of energy imports due to lack of infrastructure and lack of investment resources for the creation of , you can argue that in the coming years, the most effective opposition to the modern internal and external challenges and threats to energy security are the following:

– activation energy-saving and effective energy-saving government policies in all areas;

– increasing production and consumption of domestic energy resources and the substitution of natural gas;

– acceleration of the modernization of coal mines and increase production of coal;

– modernization of energy-intensive industries with the introduction of energy saving technologies.

At present time the greatest potential for increasing the volume of domestic production of energy has the coal industry. The result of this potential at existing mines can expect almost complete elimination of unprofitability of mines and achieve financial imbalances in the industry – the main argument for non-recognition of coal as an energy carrier priority in ensuring energy security and management decisions on early closure of the so-called "unviable" mines.

In 2011, almost all the privatized mines (24 mines) was working cost-effective, and only one in ten state-owned mines. The average price of coal was 551 uah / t, the total cost – 843 uah / t. If in 2000 the cost price has been covered by income from the sale of coal production by 91.3% in 2005 – by 80.8%, in 2010 – by 65.3%. Losses of state-owned mines in 2011 amounted to 7.2 billion uah, Partially (80.7%) were compensated by means of government support.

However, analysis of unprofitable mines, it is formed under the simultaneous influence of many factors like the objective (the complexity of geological conditions, poor quality coal reserves, etc.), subjective (physical deterioration and moral obsolescence of fixed assets, low levels of production capacities, a low level of coal preparation, creating low relative to world prices for domestic steam coal for thermal power plants, price distortions, poor management). The degree of influence from the latter factors, which can be removed, is decisive.

Currently, 90 mines (77%) are working without the reconstruction more than 30 years, 49 mines (47%) of manufacturing facilities are using less than 50% of possible use , 15 mines sell coal in the form of a member without preparation at low prices (the proportion of raw coal in marketable production is more than 70%).

However, the vast majority of operating mines has a powerful resource, production and economic potential, which can be activated by increasing the level of facilities, modernization (reconstruction, technical re-equipment) to improve the quality of coal production, optimization of production costs, and gradual transition to world prices. The practical use of this potential could provide 85% of the work on the mines profitable at or close to it. Proof of this is the current cost-effective operation of the privatized mines that have the same complexity of geological conditions, but once was upgraded and implemented measures to improve the effective work.

The potential for increasing the efficiency of the mines is estimated as follows:

– increasing the coal production due to more intensive use of production capacities and modernization of production, as a result – productivity growth, a significant (25-30%) decrease the prime cost of coal production (this result is explained by a high degree of quasi-fixed costs in the prime cost structure – to 70-75%);

– optimization of staff number after modernization and installation of modern high-performance hardware and decrease the proportion of manual labor (cost can be reduced in 10-15%);

– improving the quality of coal production as a result – increasing the coal prices and income from its selling.

One reason of formation the unprofitable mines is associated with reduced prices on the domestic steam coal for thermal power plants (in 2007-2010 they were 55.0-85.0%, compared with world prices and the prices of coal imported to Ukraine the same quality). The result is less revenue from the sale of coal products in amounts similar to the amount of government support for partial reimbursement of costs. This indicates that under conditions of the free market prices in the openness of the domestic coal market (that is required by WTO rules), most of the mines would have no losses.

Regarding to the possible volume of coal production, that potential of existing mines, which are secured by reserves for 20 years or more, which can be activated after the upgrade is 120 million tonnes (in 2011 production is 82.2 million tons). Currently in progress is the construction of five mines with total capacity of 6.9 million tonnes, three of them has high (more than 70%) degree of readiness. In addition, as raw materials are no restrictions for the construction of new mines.

To increase the volume of coal production and increase efficiency of the industry's priority should be the following:

– activation of activities to attract investment from various sources (public, private investors, including through the privatization of the mines) for technical upgrading and rehabilitation of mines;

– increased coefficient of use the modern mining technology of the new technical standards, which operates by existing mines;

– the modernization of mines and thermal powerstation and solving the problems of minimization of negative impact of coal production and consumption on the environment in the same time: energy generating facilities equipped with dust filters to bring emissions of greenhouse gases and other pollutants to the European standards, the introduction of cleaner coal technologies, improve the quality of coal products;

– became to the comprehensive development of coal mines (coal mine methane utilization, use of heat mine water, rock dumps and mine processing plants), which will transfer a portion of the cost of mining coal by-products, earn additional income from its selling and positive environmental outcome;

– improving the control of effective use of public funds in the coal industry, the development of new order of distribution of state support between the mines in various areas of its use, taking into account the requirements of WTO (World Trade Organization)and EU, which would have increased the incentive to reduce production costs;

– improving the pricing of coal according to requirements of WTO and the EU.

Geomechanical Processes During Underground Mining – Pivnyak, Bondarenko, Kovalevs'ka & Illiashov (eds)
© 2012 Taylor & Francis Group, London, ISBN 978-0-415-66174-4

Features of carrying out experiment using finite-element method at multivariate calculation of "mine massif – combined support" system

V. Bondarenko, I. Kovalevs'ka & V. Fomychov
National Mining University, Dnipropetrovs'k, Ukraine

ABSTRACT: General requirements for induction heating of machine parts connections with a purpose of their dismantling are substantiated. The methodology for the specific surface power and other mode parameters determining that meet formulated requirements is developed. The influence of electromagnetic field parameters on the character of thermal process development is shown.

The growth of problem difficulty solving by applied mechanics has led to required mutual analyses of various factors affecting on the research object as its combination of time and space. Creating objects and analyses of carrying out experiment by analytical researches have been become unexplainably unprofitable as at time and as to implement its calculation. Laboratorial modeling does not allow conducting large series of experiments and in some cases requires super sophisticated expansive equipment, which can allow to carry out qualitative tests in specific scientific laboratories only. Appearing net numerical methods has allowed to put computing experiment into the new level. The possibility of dividing the research object into separated sites (elements) allows to form its difficult geometry and physical structure combined with various system of outer factors influencing on calculating area with and without an account of time factor. The realization example of such technique is the calculation of the bearing girder stability of oil mining platform. The girder is undergone by the platform weight, side wind pressure in an upper part, water fluctuation in a down part, ice corks and aggressive sea salt. The calculation in time with an account of all mentioned factors had led to the result when bearing girder capacity decreased in one third during five year exploitation. In this case, defined outer influence had been the sea salt, which destroyed cross elements of a girder; therefore, it had led to constructive disturbances. As you noticed, during solving certain problem different factors from various fields of knowledge were considered. The experiment had been carried out not only step by step, but also with crossing the construction mechanics and the chemistry of aggressive environments. Thereby, frequently used approach in calculating composite constructions being undergone by exterior factors is finite element method. Its relatively simple and quiet universal mathematical base with wide range of describing physical environments allows modeling complicated systems from geometry, material properties, features applying initial condition and exterior loadings. And algorithmic approach at realizing any experiments remains the single-type and requires extremely reconstructing methodology of calculating experiment.

However, the connection between laboratorial and computing experiments remains absolutely high. To carry out calculations it is necessary to get series of physical quantities which have to be set during calculation process as coefficients of usual and differential equations. In addition, the adequacy of given results can be evaluated by comparing results of computing experiments and observations, which can be a physical experiment also. Due to computing experiment many series of construction analyses can be carried out varying and geometry of calculating area and interaction conditions of this area with external environment. Therefore, the computing experiment allows choosing the range of varying technological factors, which within experiments can give the possibility to get optimum result with low costs.

Defining SSS of any not primitive system under action of difficult loading, with meaning that the quantity of that loading and its distribution can be changed in space and time, is trivial task requiring clear description about conditions to model the interaction of separated elements of calculating area. Implementing operations of forming such model is divided into three stages: the first is creating and developing of models of separated elements, which are "simple" in terms of geometry and loading application; the second is combining simple models and conditions of its connection within general calculating area; the third is tying between external loadings, boundary conditions of calculating area and its separated elements.

Taking into account the duration and the development of such models radical solution generally takes places. In this case, in geomechanics the approach allowing carrying out SSS calculation of drift support and rock massif is realized separately. The efficiency of this approach sharply reduces at the growth of difficulty of support elements interaction with the massif, surrounding a mine working. Similar difficulty is also the nonlinearity of support construction behavior, removable contact efforts and taking into account exorbitant state of mine massif rocks. In such conditions physically reasonable solution of given problem may only be at entering fully description of polyvalent system of stress distribution between support elements and rock layers.

Let consider the difficulties and the methodology of similar solution for volume calculating area, including a part of lateral working outside and inside the zone of its connection with a longwall face, it is maintained by the combination of bearing-bolt and frame-and-bolt supports (Kovalevs'ka 2011). The simplest elements of such calculating system, in terms of physical and geometrical meanings, are rock bolts, frames, connection points of bolts and frames, wire mesh, backing-up and a model of rock layer in a view of parallelepiped.

Since on a stage of a mine exploitation of Ukraine the resin rock bolt is the widest-spread, modeling of which causes interest. Given bolt type possesses the simplest geometry and mechanism of interaction with rocks around a hole. That's why, in computing experiments, especially carrying out in elastic state, it is necessary to simplify this object as much as possible, until to remove it from calculating area due to external loading replacement. This approach can not provide required accuracy of defining SSS system in conditions of creating little nonlinearity during carrying out computing experiments.

The investigation of resin bolt behavior has shown that the active phase of bolt resistance to ground pressure begins after partial softening of surrounded rocks. In result, redistribution of efforts as in bolt construction and as at all drift contour occurs. Modeling at different mine-geologic conditions has revealed the biggest adequacy of the following calculating schemes: the first is the bolt having a steel rod and closely installed in a hole with the same diameter; the second is a bearing plate with big diameter, rigidly contacting with the bolt, added to the previous scheme; the third is the same as the second scheme, but includes the increase of hole diameter up to initial dimensions and its filling by the model of a hardened resin.

At quite geometric modeling it is difficult to forecast the reaction of resin bolt at time and at selecting elastic-plastic scheme of material behavior. If at elastic calculation the interaction scheme of applying loading of initial conditions can not be changed, then in case of sufficient relative displacements and the vector of applying loadings and the quantity of the zone of contact interaction with rocks can be changed rarely at the linear law. It stimulates to apply non-static conditions of the contact between elements of rock bolt model and mine massif model at carrying out a calculating experiment.

Let us consider the physics of the process of rock bolt interaction with the surface of a hole. At the moment of resin bolt installation between steel, resin and rocks of the hole surface the contacts based on chemical-molecular influence are set. As a rule, the contact between steel and resin has even and predictable character that defined by high rate homogeneity of given materials. The contact between resin and rock massif is chaotic, in terms of geometry, chips and microcracks on the surface of a hole, and material mechanics, rocks are made from a complex of materials having strength properties and chemical characteristics. In results, for realizing adequate modeling of rock bolt exploitation conditions it is necessary, within single calculating scheme, to use wide complex of initial conditions and types of contacts (Fomychov 2012).

In the simplest geometric scheme of modeling resin bolt a single contact surface only exists, for which the contact is set with compliance of model continuity of "bolt – rock" system in conditions of high diametric loading or extremely rigidness of surrounded rocks. In case, when contact loss expects, or happens, between rocks and bolt body, combined scheme is applied, in which the part of hole model surface has rigid connection with a rock bolt, and the part of its surface forms the contact with a bolt in conditions of mutual slip. Geometric dimensions of such areas, its interrelation, are adjusted by the ratio of relative bolt extension and contained rocks, with an account of plastic characteristics of polymeric resin. If preliminary bolts tension is considered, then in the calculating scheme this condition, as a rule, will be realized by applying required effort along the normal on the end of a rock bolt. However, such approach is not fully adequate relatively to efforts and displacements distribution on a surface of mine working contour. And, in case of sufficient efforts of the preliminary bolt tension, it is necessary to move to the second geometric scheme of modeling a steel-polymeric rock bolt.

In the second calculating scheme of steel-polymeric bolt the modeling of preliminary tension occurs due to overlaying contact conditions for a rock bolt bearing plate. Two main approaches are applied: the first is a rock bolt installing in a borehole that internal surface of a bearing plate gets deep into the rock contour of a mine working (so called "hot" set), where the quantity of deepening defines the quantity of preliminary tension; the second is the bearing plate and the bolt rod contacting due to "bolt connection", efforts of which define efforts of bolt tension and the contact between a bearing plate and rock massif can not be rigid.

In variants of the calculation of bolts and rock massif interaction problems exist when cutting efforts influencing on the model of a bolt prevailing relative to longitudinal loading. In such cases strength and deformational characteristics of polymeric resin, using at bolt installation, significantly influence on bearing bolt capacity. In results, for providing the adequacy of the calculation the third scheme of modeling bolt is needed. Due to using such schemes it is possible to model displacements of bolt contour with quite big deformations in any chosen directions. As a rule, the contact between the bolt rod and polymeric resin is rigid, and the contact between resin and rocks is either rigid or "hot" set. In total with experimental selection of mechanical characteristics of polymeric resin model, such system of contacts allow "smoothly" taking into account of real rock massif state features (cracking, transversality, scale effect and watering) at modeling.

From above mentioned it becomes understood that the formation of finite element net, for common case of modeling steel-polymeric bolt, is unconventional. The first problem, common for any support elements of a mine working, is small size of the finite element in the mine working support relative to sizes of finite elements are used at describing a rock massif. The combination of the finite element size, the total number of finite elements, conditions of finite element connection on separated object boundaries of calculating area determines not only the rapidity of carrying out calculation, but and the quality of receiving displacement field. In general it is considered that the reduction of finite element linear sizes always leads to the quality increase of receiving result. It is fully conformed to reality of relatively simple calculations.

At smaller sizes of finite elements describing the bolt geometry get the form of tetrahedron. And the whole picture of finite element net has clearly defined irregular character. During calculating such net can lead to the zone formation of fluctuating stresses, that distorts its common picture. As experiment showed, for most solving problems it is optimal to select twenty-ties finite elements. All these elements have the same geometry. The cross-section of a rock bolt is divided by such finite elements into four similar sectors, and the axe of bolt symmetry coincides with common edge of all four finite elements. Therefore, the whole bolt becomes the complex of cylinders having certain heights and composing of four similar finite elements. Due to height variation of these cylinders the density of finite element net can be changed, selecting optimal correlation of finite elements and accuracy of describing bolt geometry at setting initial and boundary conditions.

As known, in most coal mines of Ukraine for supporting mine workings rock bolts, as single support, are used rarely. In general bolts are exploited in combination with a classical flexible frame support. Modeling a frame support is separately quite difficult task, which can be divided into few stages.

Let us start from the consideration of the main question: how accurately is it necessary to describe the geometry of cross frame support profile? Generally the geometry of (flexible) frame support is omitted or simplified. For doing it, as a rule, two approaches are used: when all constructive support elements are replaced by simple geometrical figure with fixed width in cross-section and average mechanical characteristics (the combination of these features is formed on the basis of predictable rebuff of the real support), and when the support is replaced by uniform pressure along the mine working contour. In both cases receiving pictures of SSS calculating area can be fully corresponded to elastic state of real rock massif only. As at developing non-linear processes in the behavior of any material the accurate localization of its centers within selecting geometrical objects is required.

Replacing real frame cross-section by simple model in a view of rectangle, from one side, and reducing calculating area dimension, from another side, increase the stability of carrying out calculations. However, at calculating with an account of limiting and beyond-limiting material states, as a test showed, such simplification perceptibly influences on the raising stress along a drift in certain frame cross-section. Therefore, using real geometry of frame cross-section, though it increases the difficulty of calculus, and in the same time fully allows to describe factors influencing on SSS of "massif – support" system elements.

Now let us pass to the consideration of influence of qualitative modeling flexible tie of a frame support on the stress and deformation distribution in a frame itself and nearest rock massif. Applying flexible tie in the frame construction allowed sufficiently increase its work characteristics. However, this tie, as a factor influencing on SSS system, complicated the process going in real conditions. Constructive feature of given support element at high rate modeling (Figure 1a) extremely increases computing costs and decreases the stability of results. At certain conditions the upper frame part can be rotated relatively to bars, that negatively effects on static balance of all calculating area. Thereby, a simple model of flexible tie is applied (Figure 1b), which provides for the continuity of support model with saving last geometric model characteristics of a frame. But considered technology of modeling can not guarantee the possibility of big frame element movements relatively to each other in all range of examining geomechanical problems. When considerable movements of mine working contour are foreseen, then the another flexible tie model is applied (Figure 1c). In this case, the geometric authenticity of external frame contour is disturbed insignificantly. For different support models linear deviations do not exceed 50 mm, but now the model accepts significant linear (up to 400 mm) and radial (up to 20°) movements (Fomychov 2012 & Bondarenko 2010).

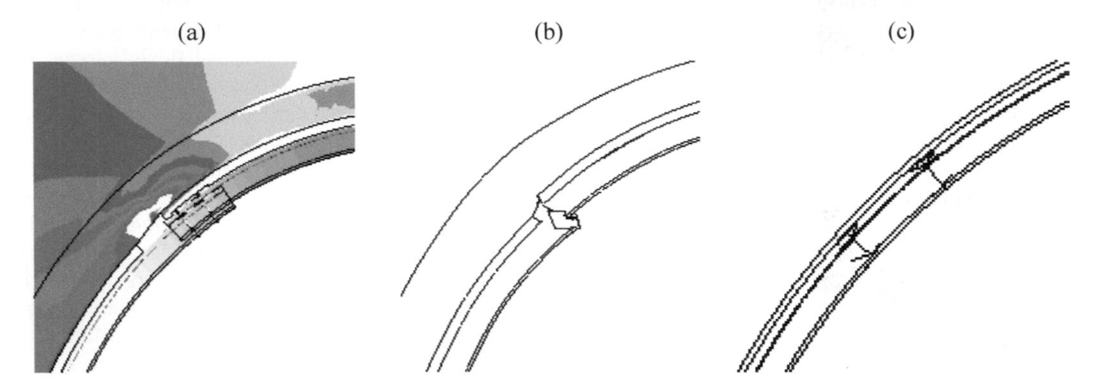

Figure 1. Modeling the flexible tie of a frame support: (a) geometric accurate model; (b) the tie model providing for its movement without taking into account of dynamics with saving origin frame contour; (c) the tie model providing for big movements without saving dynamics taken into account.

Therefore, modeling the frame support flexibility can take into account of big movements. It is achieved due to flexible tie modeling, which is fixed in and made from material possessing low resistance indicators to compressive efforts. On Figure 1, b the movements of upper frame part, which lowered down and simultaneously pressed the flexible tie model into a side bar, are well shown. In result of the arch movement, presented on Figure 1, c, reaching up to 300 mm (Bondarenko 2010), that causes to increase zones of limiting rock state forming the mine working arch.

Frame elements providing the contact of the frame and drift contour should be modeled with low detailing level. As such objects are entered in the construction for providing effort transmission and perceive external loading at a simple scheme (Kovalevs'ka 2011). Omitting these objects can sufficiently change SSS of a frame and around mine working rocks. These elements are steel concrete lagging, rock filling, wire mesh with wide cells, etc.

A steel concrete and a filling can be as a single averaged by mechanical characteristics object, the geometry of which conforms to geometric characteristics of real objects. In most cases such approach guarantees reasonable adequacy of given primitive and real object characteristics. For version, when separating a lagging and a filling in calculation is required, it is necessary to describe the geometry of concrete blocks with external frame curvature appropriately. It allows not to solve the task of the contact between bodies, touching along the surface of different curvature. For emulating the behavior of loose environment the lagging is described by mechanical parameters, corresponding with the behavior of solid cnvironment having high deformational measures and using the law of linear-piecewise connection of deformation and stress. Also Young modulus is chosen to tend to 0.5.

Describing the models of rock layers as the complex of separated geometric elements is not needed; geometrically, it is the complex of parallelepipeds, and mechanically, it is averaged characteristics, gained experimentally or substantiated by laboratorial and full-scale investigations. From another side, if to consider rock massif as a complex of rock layers, the description of its interaction within certain problem becomes applied problem. Depending on conditions of problem statement the same properties of rock massif heterogeneity can influence on an accuracy of results (Bondarenko 2010).

Foliation of rock massif extremely changes stress distribution, as around mine working contour, and in zones next to boundaries of rock layers. In this case, influence degree on stress field for different system components can fluctuate from 10 to 270% (Bondarenko 2010).

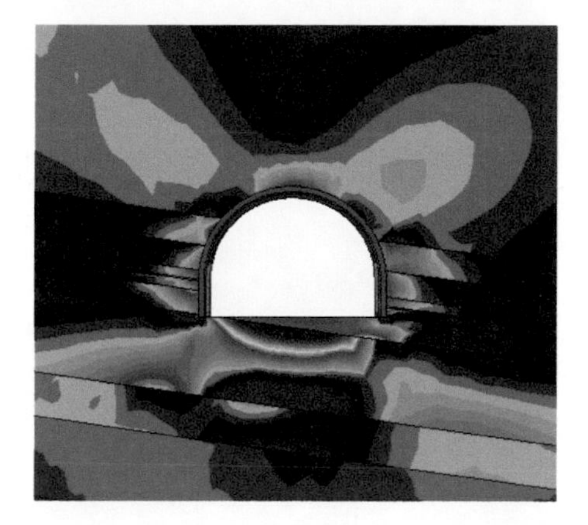

Figure 2. Stress distribution in thin-layered massif near in-seam workings.

Let us consider stress diagram presented on a Figure 2. The main reason, because of which the foliation influence is quite high in given calculating model, is the difference between rock layer strength. Rock layers forming foot and sides of a drift have the increased rigidness relative to rock layers, which forms upper and down areas of calculating model. It causes the stress concentration within geometric area of certain layer.

Taking into account of foliation the problem of reasonability of coal seam deep angle for adequacy of received results happens. In most analytical and numerical math solutions this factor is omitted, as it allows sufficiently decreasing the difficulty of finite math equation and/or simplify calculating model due to applying for symmetry equation.

To confirm all above mentioned the picture of stress distribution on a Figure 2 is analyzed. The major feature of this stress diagram is the absence of symmetry relative vertical axis of in-seam working, that is, the stress distribution in sides of a drift has different qualitative and quantitative character. At that, the quantity of similar disbalance depends on angle of dip and rock layer physical characteristics. For various calculating models such disbalance can consist 40% of quantitative and 180% of qualitative measures. Such measures for real in-seam working can be achieved when dip angle of seam in calculating model is taken into account only, and the quantity of this angle is over 3° (Bondarenko 2010).

In spite of rock layers can have different geometry and physical characteristics during modeling to increase the adequacy of received results, it has to change contact conditions on boundaries of these layers. In total three types of such contacts can be highlighted. There are rigid contact, contact with sliding and contact with frictional force. Using any type of contact can cause qualitative and quantitative changes of stress distribution, which presented on a Figure 3.

In case of rigid contact (Figure 3a), at horizontal occurrence and not wide range of physical rock layer characteristics, the stress distribution is almost similar to the stress distribution in the model without foliation. And in the case of Figure 3b the stress distribution is significantly different from the previous one. Such changes in qualitative and quantitative measures of stress distribution happened by entering mutual sliding of rock layers in the calculating model. Now, the stress level is higher in coal seam, than in surrounded rocks and its role of increasing drift stability extremely rises.

(a) (b)

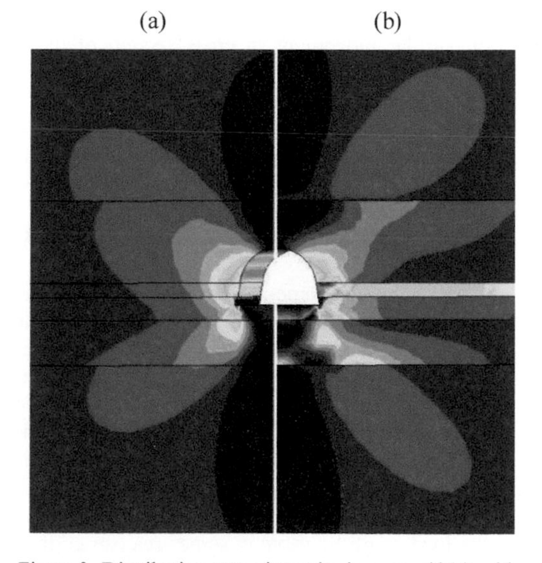

Figure 3. Distributing stress intensity in a massif (a) without taking into account of sliding on rock layer boundaries, (b) with taking into account of sliding on rock layer boundaries.

Difficulties of modeling the element interaction of "massif-bearing-rock bolt support" system are concentrated in zones of interaction of bolts and maintaining drift contour support elements. These elements are wire mesh with big cells, bearing metal bars having few bolt-holes. Metal bars allow combining few bolts, being installed in the drift roof between frames, into single rigid load-bearing structure. The simplest method of combining a bolt and bearing metal bar is rigid contact. At that, the height in cross-section of bar model has to be increased relative to real meaning. It is necessary to ignore bearing plates of each contacting bolts in the whole support model.

The contact between surfaces of bearing bar and rock massif has to be defined as free, without using friction force. It is linked with two moments. Firstly, during the calculation of individual bolt movements, relatively to each other, absolute quantities can have different directions, that will cause to appear additional stresses in the bar. At that, if the bar rigidly connects with a massif, it will cause to wrong correction of SSS on mine working contour. Secondly, when wire mesh takes place between the bar and rock massif surface, it will not limit movements of the bearing bar along the surface of rock massif.

In the considered support construction the wire mesh plays the key role of integration factor providing for effort distribution between installed bolts and frames in mine working roof. However, this mesh limits free near-the-contour rock movement into a mine working. And the quantity of resistance on similar movement in cross section of the construction is slightly small. It allows to consider given element of the construction as rigid membrane, set in place of its contact with bolts and frame supports. In results, the wire mesh is modeled as an object repeating the drift contour with width not exceeding the diameter of bar mesh, and mechanical characteristics corresponding to flexible material.

Modeling the frame-bolt support the case is different (Kovalevs'ka 2011). In this case, the additional problem is the provision of correctly modeling the systems of effort transmission between a frame bar and side bolts, which are combined by a flexible tie (cable). In real condition such cable freely rounds a frame and takes place between bearing and additional bolt plate. Thereby, the cable can slide easy along its length relatively to side bolts and frame bars. Then, the loading of the cable in some area can go through it on distance. This frame-bolt support feature, being unquestionable benefit from a point of engineer view, looks as a complex of problems from a point of computing modeling view.

Name these problems: absence of rigid cable connection – static uncertainty of the system; difficulty of adequately describing cable geometry – form selection of longitudinal axis of cable model; effort transmission in zones of bolt contact – a cable and a frame – a cable – solution of contact problem; cable deformation in conditions of zones of plastic flow occurrence in elements of calculating model.

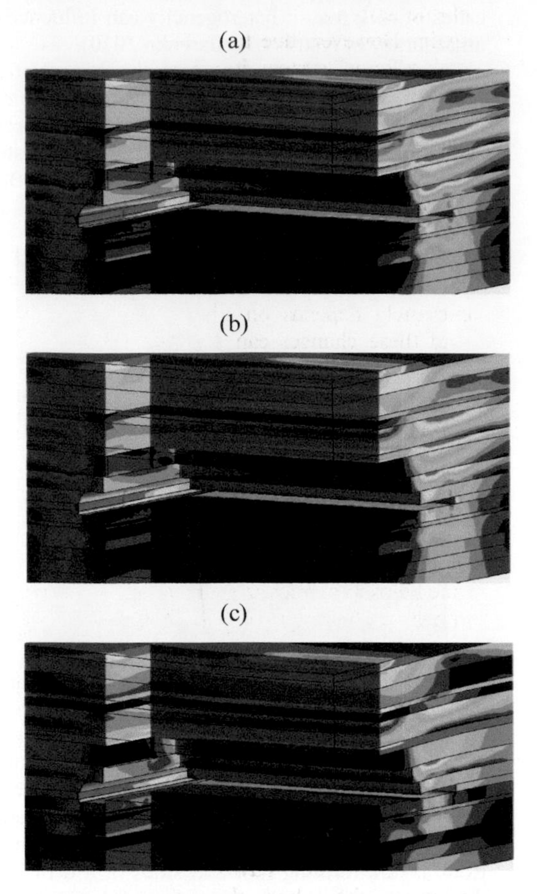

Figure 4. Modeling the connection of a drift and a longwall: (a) elastoplastic calculation; (b) calculation with account of rheology (10 hours); (c) calculation with account of rheology (90 hours).

Modeling the cable can be considered as the most adequate, implemented by the following: first – the cable combined with a bolt, tying the body of cable with bolt end; second – the form of longitudinal cable axis was chosen as sinusoid with straight sections on function maximum; third – the preliminary cable tension was modeled due to penetrating the cable into the frame.

However, in this variant of modeling insignificant problem is hidden. As the limitation of longitudinal movements and small elasticity of cable model do not allow to keep the continuity of calculating model at various measures of bolts and frame movements, the adequacy of the model is disturbed at initial stage of the calculation. At relatively small deformations the bolt end moves into the drift more intensive, than the frame. In results, the contact between the cable and the frame is broken. With going further elastoplastic calculation and the growth of plastic deformation the cable presses into the frame again. However, due to earlier lost contact in the "cable-frame" system, it becomes impossible to calculate effort transmission through formed-again contact section. Consequently, this problem has to be solved due to two calculating steps: the first is the SSS calculation until the moment of cable tension; the second is correcting parameters of contacts between the cable and the frame and the calculation.

All above mentioned relates to constructive support elements. However, functioning of the support extremely depends on SSS rock massif changes. And these changes can occur in result of driving mine workings, manifestation of rheological rock properties, etc. Fundamental factor taking into account of such changes influence in the model becomes the time factor.

The results of taken time factor during the calculation of stress and deformation in the rock massive become effort redistribution being perceived by support elements. Also, changing drift geometry on time helps to correct orientation of main stress areas in near-contour rock massif. In total, taking into account these factors can cause to fully redistribute efforts on the surface of support and rock massif contact, and change the conditions of these areas formation and as a result, convert the deformational picture of a mine working.

The brightest example of rheological rock properties influence on the changes of SSS rock massif is the series of calculations carried out for "Stepnaya" mine, PLC "DTEK Pavlogradugol". Results of calculations show how the stress distribution and movements are changed in the zone of drift and longwall crossing (Figure 4) on time. On a Figure 4, a the stress intensiveness is shown, which gained for the task, solved due to elastoplastic set. For this variant of solution characterizes the biggest stresses and the minimum absolute movements. The calculation implemented with account of rheology had been showing within real 10 hours that absolute movements in the model were raised insignificantly, and the stress distribution suffered as qualitative and as quantitative changes (Figure 4b). Finally, since 90 hours the growth of movements in roof and foot of longwall (Figure 4c) is markedly seen, with load redistribution of rock layers of the model. It can be said, that the next step of caving rocks is formed.

Manifestation of mine workings crossing influence also has localized character on time. Dependently on mutual mine working size and existence duration, such influence on SSS system can have as insignificantly (e.g. drift and passage crossing), and as notable influence (an example on Figure 4c).

From all above mentioned the SSS calculation in the zone of longwall influence for mine workings with bolt-frame support in the conditions of Donbass region's mines can be implemented only by taking into account next factors: constructive flexibility of the frame support; high movements of massif contour; stress relaxation, in result of rheological rock properties, and, finally, mutual mine working influence. In this case compensation of efforts and geometric parameters of calculating area will allow to get averaged, and consequently, probable picture of SSS in the zone of real object location.

REFERENCES

Kovalevs'ka, I., Vivcharenko, O. & Fomychov, V. 2011. Optimization of frame-bolt support in the development workings, using computer modeling method. Istanbul: XXII World mining congress & Expo (11-16 September). Volume I: 267-278.

Fomychov, V. 2012. Backgrounds of modeling frame-bolt support with an account of nonlinear characteristics of physical environment behavior. Dnipropetrovs'k: Naukovyi visnyk, 3: 13-18.

Bondarenko, V., Kovalevs'ka, I., Martovitskiy, A. & Fomychov, V. 2010. Developing scientific basics of increasing mine working stability of Western Donbass mines. Monograph. Dnipropetrovs'k: LizunovPress: 340.

Bondarenko, V., Kovalevs'ka, I., Symanovych, G., Martovitskiy, A. & Kopylov, A. 2010. Methods of calculating displacements and surrounded rocks reinforcement of Western Donbass mine workings. Monograph: Dnipropetrovs'k: Pidpriemstvo "Driant" Ltd.: 328.

Geomechanical Processes During Underground Mining – Pivnyak, Bondarenko, Kovalevs'ka & Illiashov (eds)
© 2012 Taylor & Francis Group, London, ISBN 978-0-415-66174-4

Parameters of shear zone and methods of their conditions control at underground mining of steep-dipping iron ore deposits in Kryvyi Rig basin

N. Stupnik & V. Kalinichenko
Kryvyi Rig National University, Kryvyi Rig, Ukraine

ABSTRACT: Representative analysis of collected data according to surveying measurements of the earth surface deformations within the mine baffles of Kryvyi Rig iron ore basin is carried out. The relationships between the lowering of the stoping level and change of dimension of basin subsidence and surface collapse across the strike are found. Practical aspects of implementing the methods of diagnosis and monitoring of geoengineering state of rock mass in the lines of shear and collapse zones are considered.

1 INTRODUCTION

In Kryvyi Rig iron-ore basin throughout decades a unique situation with preservation of daylight area within the boundaries of operating and closed mines.

Due to long and intensive exploitation of Krivbass deposits by underground operations considerable areas damaged by mining operations were founded. Surface damages with craters, caves and caving zones were caused by application of different methods of development systems with ore caving and adjacent rocks at underground mining as well as the single-stall system development, when mining the overlying horizons.

In the first case, with the unstable soft ores rather smooth day surface subsidence with the formation of the projected caving zones were observed.

When using the single-stall system development the formation of caving zones occurred unevenly, depending on the volume of room and pillar mining, ore ceiling and host rocks hardness. In this case, the prediction of caving zones was and still is more difficult task, because of taking into account physical and mechanical properties of rocks with higher strength characteristics. In this regard, incomplete ceiling settling is possible at the room roof caving, which in turn can lead to mini-rooms formation, which accounting and control is practically impossible. Settling of such mini-rooms located near the day surface (for example, the conditions of deposit development of the former MA n. a. Ilyicha) can lead to unplanned day surface caves many years later after deposit declining and complete mine closure.

In addition, a rather complex situation in the prediction dead blind deposits located in the overlying horizons is. These deposits, as a rule, are small in size in relation to the main ore bodies, were mined by special projects. Taking into account their small size and isolation from the main deposit we can't state with confidence that, with respect to them all necessary arrangements for full depreciation (or stowing) of dead rooms were made. Therefore, such potential cavities may also be a potential hazard to the day surface.

In addition, magnetite quartzite underground mining on overlying horizons of mines by "room-pillar" technology has led to the formation in the depths of a huge number of voids calculated in millions of cubic meters. These voids are potentially dangerous in the event of their caving. Calculating characteristics theoretically guarantee their stability, however, as it was shown by an example of the day surface caving on the mine n. a. "Ordzhonikidze", a practice sometimes refutes the theory.

Currently, underground mining has gone to deeper horizons, at mining of which the effect of stoping on the day surface is considerably reduced. In this case, it sudden and uneven caves on the surface are practically impossible. At the same time, there is a gradual subsidence of a considerably large area due to the increasing size of trough movement of overlying rocks with a decrease in underground mining.

The purpose of this work is to identify the relationship between the lowering of stoping level and change the size of trough movement and the day surface caving, as well as the definition of guidelines and methods for diagnosis and monitoring of geoengeneering state of rock massif in the lines of displacement and caving in the areas of active and dead mines.

2 THE WAYS OF PROBLEM SOLVING

The area of undermined territory estimated by the State Design Institute "Krivbassproject" is 3600 ha, including the area of pitcraters within the rock trough movement in the areas of active and dead mines is about 1030 hectares (Modern technologies...2012). These areas have a tendency to expand due to continuous iron ore underground mining.

So, the monitoring of the existing voids state and undermined territories is a priority for future development of the Kryvyi Rig iron ore basin.

The control of geomechanical rock massif state in the lines of displacement and caving zones in the areas of active and dead mines is possible by creating a geoinformation system for prevention and monitoring the day surface settling. At geoinformation system development for prevention and monitoring of over large areas subsidence, which is the Kryvyi Rig iron ore basin, it is necessary to solve the following tasks:

1. To create an electronic database of existing underground voids on the basis of the preserved mining and engineering documentation.
2. To implement a system for monitoring deformations of the day surface with the construction of dynamic digital models of terrain changes in real time.
3. To identify the main patterns and the magnitude of the day surface deformation depending on physical and mechanical properties of rocks and the size of the existing underground voids.
4. Using various methods (for example, the method of analogies) to identify possible locations of the unknown voids and their projected volumes.
5. To create an electronic database of existent and projected underground voids on the basis of the preserved mining and engineering documentation and completed investigations.
6. Based on the results of monitoring of the day surface deformations and the electronic database of existent and projected underground voids to create a geoinformation system for early warning of possible surface subsidence in a particular area of the Kryvyi Rig basin.
7. To clarify the boundaries of the potentially dangerous zones of the day surface displacement on the basis of geoinformation system for prevention and monitoring of over large areas subsidence.

Currently the basic task is to determine the optimal system for monitoring deformations of the day surface with the construction of dynamic digital models of terrain changes.

Such well-known corporations of company Hexagon, as Leica Geosystems, Intergraph and GeoMos suggest using GPS for accurate geo-referencing with the help of InSAR and LIDAR technologies.

InSAR is a radar-location method using interferometric radar. This geodetic method uses two or more radar images to generate maps of surface deformation or digital terrain models, using the differences in the phase of the waves, which return to the satellite or aircraft. The method allows to measure potentially the terrain change in the centimeter deformation scale over a long period of time.

LIDAR is a laser radar infrared – range with remote sensing optical technology, which allows you to create detailed maps of the terrain day surface.

On the basis of remote sensing the day surface a spatial analysis of the obtained results and data in real time processing is performed, which allows to obtain a detailed picture of the deformation and to predict the possible risk zones.

In our view, the development of geoinformation system for prevention and monitoring of Kryvyi Rig iron ore basin areas subsidence must solve two major problems: the prevention of possible deformations and the determination of the causes and patterns of these deformations.

In our opinion, the first step is to perform a preliminary simulation and launch a pilot project in a small area, such as the eastern edge of Gleevatskiy quarry of PJSC CGOK in the zone of underground mining influence, or the property of the closed and liquidated MA n.a. Lenin.

This pilot project would allow interested parties to determine the capabilities and desires of partners, and would be a part of the Memorandum of Understanding between the City, Kryvyi Rig National University and GeoMos AG. The latter serves as a consultant and represents the interests of the Hexagon Group, and partners.

3 CONCLUSIONS

The practice and experience of mining enterprises show that the current costs of implementing measures to prevent possible emergencies is much lower than the cost of their liquidation in the future.

P.S. It is our deep conviction that the only way to guarantee the impossibility of the day surface caving and subsidence within the boundaries of the Krivbass mine areas is the introduction on the mines of the basin the development systems with hardening stowage. When the momentary apparent increase in production costs, these systems development will provide savings in the future, ensuring the preservation of the day surface and the safety of the residents of Kryvyi Rig basin.

REFERENCES

Modern technologies of ore deposits development. 2012. Scientific Papers on the work of the Second International Scientific-Technical Conference: Kryvy Rig: Publishing: 140.

Geomechanical Processes During Underground Mining – Pivnyak, Bondarenko, Kovalevs'ka & Illiashov (eds)
© 2012 Taylor & Francis Group, London, ISBN 978-0-415-66174-4

Concept and assumptions for developing underground brown coal gasification plant for supplying synthesized gas to heat and power plant

J. Nowak
KGHM CUPRUM Ltd. – Research & Development Centre, Wroclaw, Poland

ABSTRACT: Concept of developing underground brown coal gasification plant for supplying synthesized gas to heat and power plant operating as a cogeneration unit is described in this paper. Calculations of energy balance connected with the demand for gas fuel are shown. Dimensions and number of underground reactors are given as well. Basic assumptions for energy plant are discussed. Some elements of project economic assessment are also presented.

1 INTRODUCTION

Using the synthesized gas from underground hard coal or lignite gasification system has not been employed in Poland yet. In case of lignite it is especially important due the large reserves of this fuel (Kozlowski et al., 2008; Kudelko, Nowak, 2010). Using lignite for synthesized gas production would result in applying it in power and chemical industry. In many cases it would be possible to get the power from solid fuel without its extraction on the surface. It is extremely important in regions, where local community protests against open pit mines. On the areas where surface infrastructure is developed, there would be an opportunity to obtain the chemical energy contained in brown coal with minimal interference in local plans of land development. Therefore there is a great interest in this mining method as a clean coal technology (Kudelko & Nowak 2010; Kudelko & Nowak 2007; Nowak 2007).

Heat and power plant in Głogów Smelter and Refinery uses the blast furnace gas as a fuel. It is a by-product of smelting process (Nowak et al 2011). Since the technology change is planned i.e. replacing the shaft furnaces by the flush furnaces, it will not be possible to apply the former system of supplying the heat and power plant. The further operation of the power plant will be continued basing on GZ-41.5 network gas, which is currently delivered to the supply system in order to enrich the blast furnace gas which is still used.

After abandoning supplying the plant with blast furnace gas it is planned to use only the network gas. The alternative is, however, the concept of introducing the synthesized gas from the planned installation for underground gasification of lignite from the deposit located in the vicinity of Glogów. That type of gas is a better power carrier than blast furnace gas and its production is relatively cheap.

The preliminary concept of using the synthesized gas instead of previously used blast furnace gas and the requirements concerning the reserves of solid fuel necessary for gasification as well as all related with economic and technical consequences are presented in this paper. The proposed solution is highly advantageous for the investor. It allows for using the existing power plant to verify the installation for underground lignite gasification (ULG) as a source of fuel for energy production. It is also a safe way of carrying out the technological tests due to the opportunity of using the network gas in case of irregular deliveries from ULG installation or too low of synthesized gas. Such technological pattern will operate on the full technical scale. The success of operation of both plants will confirm the possibility of using the synthesized gas for energy production purpose as well as the possibility of using the underground gasification systems in the industrial practice.

2 CHARACTERISTS OF ENERGY PRODUCTION PLANT AND QUANTIFYING ITS FUEL DEMAND

Initially the heat and power plant used only the hard coal as a fuel (Nowak et al 2011). During the next years, the blast furnace gas from the smelter was used to heat the boilers together with coal. The network gas is also delivered to the installation and it is used as a kindling gas. Future changes assume the successive abandonment of the hard coal.

Currently Głogów Heat and Power Plant is owned by Energetyka Lubin, which is a part of KGHM Polska Miedź SA Capital Group. It is supplied by three types of fuel:

- **hard coal;**
- **blast furnace gas;**
- ***Lw* (GZ-41.5) natural gas.**

Main receiver of energy from the plant is Głogów Smelter&Refinery and town of Głogów. The plant is successively modernized. The intent of those changes is to increase the power energy production with providing simultaneously the heat safety for Głogów Smelter and Refinery and town of Głogów. It is assumed that in future the hard coal fuel will be totally replaced by the natural, network gas. It is also planned to install new boilers adjusted for co-combustion of blast furnace gas from Głogów smelter and (in later future) new steam turbine.

In future it is planned also to construct the gas-steam power unit with 40 MW_e of power. The aim of this project will be:

1. Production of power and heat energy in high efficient gas cogeneration.
2. Providing the energy in emergency situations – power safety of smelter and mines.
3. Economic effectiveness.

Power and Heat plant of Głogów Smelter and Refinery produces the power energy in three turbine sets with the following parameters of electric power:

- TG-1 10.0 MW_e;
- TG-2 13.0 MW_e;
- TG-3 25.0 MW_e.

Total power is 48.0 MW_e, and annual production amounts about 220 000 MWh. Average load is equal 25 MW_e.

Heat produced in Głogów Heat and Power Plant amounts 176.6 MW_t. It is equipped with 7 steam boilers:

- 5 boilers of ORg-32 type, combusting blast furnace gas with hard coal;
- 2 modernized boilers of K-7 type (2009) and K-6 type (2011) combusting blast furnace gas with natural gas.

They supply three turbine sets, one produces the heat and is adjusted to condensation work, while two are of bleeding – condensation type.

Characteristic of previous fuel

Power and heat plant in Głogów smelter after modernization for co-combustion uses the natural nitrogen-saturated gas of *Lw* (previously GZ-41.5).

Blast furnace gas is produced in shaft furnaces and is delivered to the power and heat plant to be utilized through the dust-cleaning and transferring installation. Flow rate is about 180 000 m^3 / h, and the pipeline diameter is. The gas components are:

- CO_2 – carbon dioxide
- CS_2 – carbon bisulfide
- CO – carbon monoxide
- H_2 – hydrogen
- N_2 – nitrogen
- H_2S – hydrogen sulfide
- CH_4 – methane
- O_2 – oxygen

The percentage of each component depends of the blast furnace charge. Although this gas is a byproduct, it is a base of heating system of many smelters, like in Głogów Smelter and Refinery. Standard blast furnace gas has a density about 1.169 kg / m^3, and one ton of iron may give even 4000 m^3 of it. Its heat value ranges from 3.4 to 3.7 MJ / m^3.

It should be noticed that blast furnace gas produced in Głogów Smelter and Refinery has inferior quality that the one produced for example in ironworks. Its heat value is 2.1-2.9 MJ / m^3. Additionally it needed to be cleaned before delivery to the power and heat plant.

Since the blast furnace gas parameters are not good enough for its effective combustion, it is conditioned by natural, network gas, mainly to adjust and stabilize it. The regulatory and standard requirements for *Lw* type gas are as follows:

*heat value – at least 27.0 MJ / m^3,
*example composition:
– methane (CH_4) – about 79%;
– ethane, propane, butane – about 1%;
– nitrogen (N_2) – about 19.5%;
– carbon dioxide (CO_2) and other components – 0.5%.

The hard coal is hauled in small railway cars to the stock-yard, from where it is delivered to the heat and power plant and combusted together with blast furnace gas. Exhaust fumes, due the high level of contamination are desulfurized and dedusted.

It is proposed to use in the heat and power plant the synthesized gas having low to medium parameters. The selection will be possible after detailed analyze of capabilities and requirements of boilers, in which the gas will be used. Probably it will be the gas having the heat value from 3.4 MJ / m^3 to maximum 7.4 MJ / m^3 with majority of lower values. It results from applying the supply flow on the near level and the higher profitability of using the less purified gases. The issue will be analyzed in details in future, because now there are not any findings concerning the lignite, allowing for predicting the gas quality after its conversion. Those tests are under preparation now.

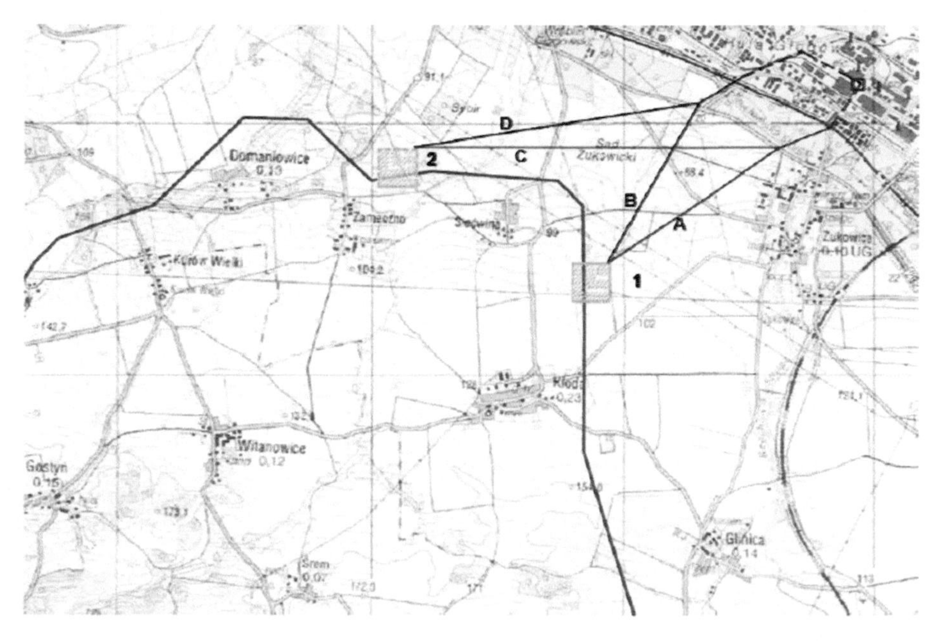

Figure 1. Planned locations of underground gasification sites and gas pipeline routes delivering gas to heat and power plant: 1 – first version of location, 2 – alternative version; A, B, C, D – alternative routes of gas pipelines.

Current balance of gas energy

In order to ensure the proper operation of steam boilers in heat and power plant in Głogów smelter, the parameters of syngas, including composition, heat value and flow rate, should be the same as energy parameters of currently used process gas. Parameters of the gas supplying the steam, pulverized fuel boilers, where the coal-dust is combusted, are adjusted and stabilized due to mixing the blast furnace gas with natural network gas. Total demand for gas is 1 612 million Nm^3 / year i.e. 201 500 Nm^3 / h, including:

– blast furnace gas – 1 600 million Nm^3 / year i.e. 200 000 Nm^3 / h;

– natural gas – 12 million Nm^3 / year i.e. 1 500 Nm^3 / h.

Current balance based on blast furnace gas and natural gas consumption is:

– blast furnace gas, 1600 million Nm^3 / year × 2.29 MJ / m^3 = 3664 million MJ;

– natural gas, 12 million Nm^3 / year × 27.183 MJ / m^3 = 326.20 million MJ;

– annual demand for energy from gas is 3990.2 million MJ.

Defining the demand for synthesized gas

At the air injection, the syngas heat value amounting 3.4 ÷ 4.0 MJ / m^3 can be expected. Due to the CO_2 aid, low humidity of gas (almost zero) the higher value may be expected. Then, the demand for gas will be 997.6÷1174 million Nm^3 / year.

Defining the volume of solid fuel

Assuming that 1 kg of lignite gives 2.06 m^3 of gas (1 ton of lignite ≈ 2 000 m^3 syngas) the value of 499 ÷ 587 thousand tons of lignite per year is obtained.

Correction of fuel volume due to its quality

Because the part of deposit may be the layer of carbonaceous rocks but with relatively high energy value and that the part of fuel will be used by the facility for its own needs, the correction parameters, concerning the solid fuel are introduced.

The following assumptions were made:

– 30% part of carbonaceous deposit having the heat value lower than pure brown coal, at least 4 MJ / kg;

– syngas consumption for own needs of gasification facility, about 10%.

Additionally the gas loses caused by its migration into the rock mass must be taken into consideration. They are difficult to be defined with regard to the Polish geological conditions. Several factors, both geological and operational may have the impact on this phenomenon. Using the data from already operating gasification systems, the correction factor of 15% was used there. Then the demand for solid fuel is 738.3÷868.3 thousand tons / year.

It is assumed that during the further studies, especially after the detailed geological survey and laboratory tests, these estimates may need the revision.

Total demand for solid fuel, assuming the life time of power production facility $T = 30$ years, is on the level of 26.05 million tons of lignite. Assuming that the lignite density is 1.1 t / m^3, 23.7 million m^3 of solid fuel "in situ" is necessary.

Supposing the beds discontinuity, different types of tectonic disturbances, possible change of lignite density value after the detailed analyzes, 30 million m^3 ("in situ") or 35 million tons of solid fuel must be found.

For the needs of combusting the syngas it will be possible to use two modernized boilers (K-6 and K-7), having the following parameters:

– volume of combusted blast furnace gas, 130 000 Nm^3 / h;

– annual volume of combusted blast furnace gas, 1040 000 000 Nm^3;

– annual value of chemical energy of blast furnace gas, 2080000 GJ;

– annual value of chemical energy of natural gas, 208000 GJ;

– total value of chemical energy, 2288000 GJ.

Calculation of syngas possible to be combusted:

– heat value of syngas 4.0 MJ / Nm^3 (assumption for low-purified gas),

– volume of syngas

2 288 000 GJ : 4.0 GJ / 1000 Nm^3 = 572 million Nm^3 / year;

572 million Nm^3 : 8 000 h = 71 500 Nm^3 / h.

For the needs of UCG project, the minimum load of 1 boiler of 50% (17 800 Nm^3 / h) should be additionally assumed.

Minimum efficiency of lignite gasification system should amount approximately 20 000 Nm^3 / h. The construction of gasification installation may proceed in stages, each stage ca. 20 000 Nm^3 / h.

The syngas delivered to the heat and power plant will be used to produce:

– in summer: power energy in condensation,

– during heating season: power energy together with heat energy using existing technological systems (boilers, turbines, exchangers).

Planning the annual production value:

– power energy:

15 MW_e × 8 000 h / year = 120 000 MWh / year;

– heat energy: 200 000 GJ / year.

Possible stops in syngas delivery, will not cause any negative effects during the initial period of Głogów Heat and Power Plant operation.

Since the ULG installation will be commissioned successively, the deliveries of gas will rise. It is assumed that at first, geo-reactors producing about 20 000 Nm^3 / h of syngas will be started. It is suggested to increase the volume of gas by 10 000 Nm^3 / h in the next stages until the required, target level is reached.

Now it is difficult to specify the stages in synthesized gas delivery, since the data necessary to calculate the operating parameters of reactor are not available. First, the results of lignite tests as well as defining the quality and composition of blow gases are necessary.

3 GEOLOGICAL CONDITIONS OF LIGNITE OCCURRENCE IN HEAT AND POWER PLANT REGION

On the area of planned surveys any special exploration works concerning the lignite were carried out before and the knowledge about lignite occurrence is based on the results of copper exploration drillings. In copper exploration boreholes in Głogów and Ścinawa vicinity, drilled in the middle of twentieth century, the brown coal bearing formation with lignite layers were drilled through in the cover. Test of lignite were made only occasionally and in the placed located far away from the fixed site (Figure 1).

In order to draw up the Plan of geological exploration and prospecting surveys, the profiles of copper boreholes from the areas located south, south-west and west from Głogów, were collected and analyzed. Location and boundaries of the area designed to be studied are shown on Figure 2. The surface of the whole survey area, covered by exploration boreholes is about 26 km^2.

The area of planned geological surveys is now under populated. The space development is formed by several villages with compact settlement. The significant part of buildings is in bad technical condition, especially historical houses. New buildings, mainly single family houses supplement the existing space development or are constructed on new appointed investment areas adjacent to the already built-up area.

The beds of lignite can be correlated on big area in Głogów region, but is also happens that in places they are replaced by highly carbonaceous rocks or they do not occur at all. The biggest continuity has the "Łużyce" bed, found in all boreholes drilled in this region. In some cases the carbonaceous rocks replaced "Ścinawa" and "Głogów" bed. While "Henryk" bed in many old boreholes was not found at all or was substituted by carbonaceous rocks.

Beds (strata) of brown coal are separated by sand, clay or silt layers. Basing on geological report for Głogów I copper deposit, the following can be stated (Preidl & Mikula 1997):

- depth of individual beds occurrence is:
  - Henryk from about 75.0 m to 148.4 m;
  - Łużyce from 130.0 m to 270.8 m;
  - Ścinawa from 153.6 m to 308.6 m;
  - Głogów from 231.6 m to 412.0 m.
- each lignite bed has from one to four layers, and their total thickness ranges from 0.4 to 14.0 m. The distance between the individual lignite layers in the bed reaches 30 meters in places.
- total thickness of lignite layers in beds is:
  - Henryk from 0.0 m to 8.5 m;
  - Łużyce from 1.0 m to 12.3 m;
  - Ścinawa from 0.0 m to 14.7 m;
  - Głogów from 0.0 m to 11.0 m.

Such differences in lignite occurrence will cause many problems in its utilization. The possibility of employing the open pit mining seems to be especially unfavorable. From the other side it is convenient due to the possibility testing the different technical solutions within several methods of underground lignite gasification. It offers also the wide range of research opportunities for studies on using the bio-gasification of brown coal (Szubert, Nowak & Grotowski 2011).

4 SELECTION OF DEPOSIT AREA FOR THE NEEDS OF UNDERGROUND GASIFICATION INSTALLATION

The area of lignite, designed for future gasification, will be defined precisely after completing the exploration works. It will be placed within the region presented on Figure 2.

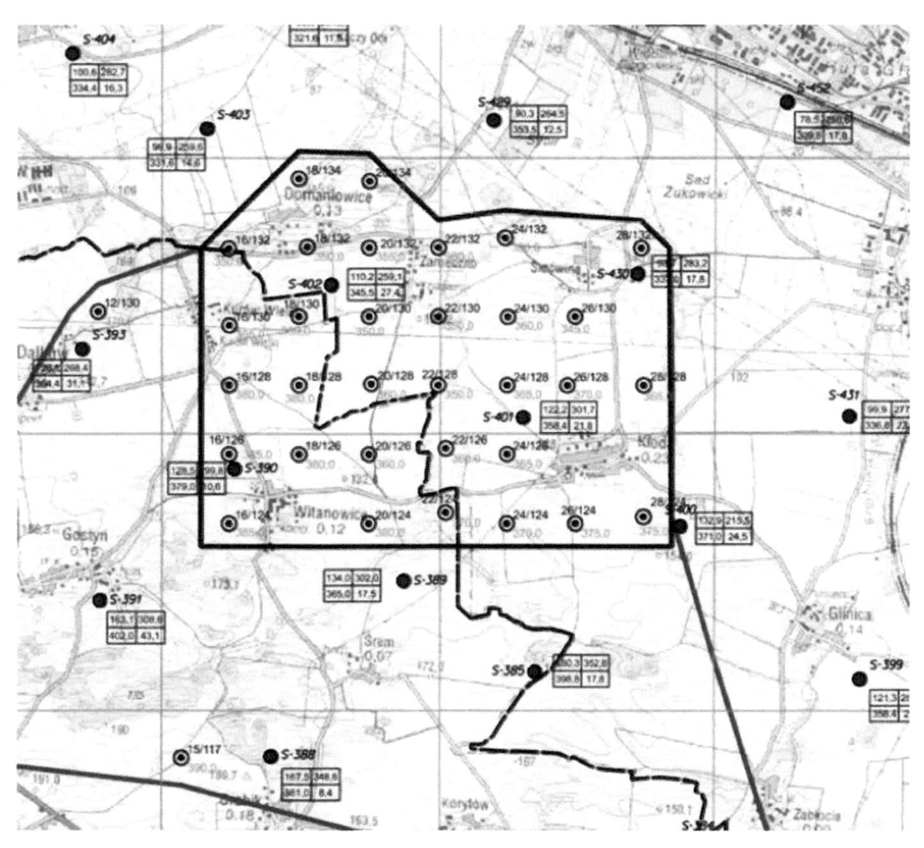

Figure 2. Boundary of estimated lignite reserves within the selected survey area.

The area was selected after the analyze of population and land development within the possible future location close to Głogów Smelter and Refinery. Additionally the data about lignite occurrence in copper exploration boreholes on this area and outside it, were also taken into consideration (Derkacz & Sztromwasser 1984; Kozula & Golczak 1988; Piwocki 1996).

5 DEFINING THE PARAMETERS OF UNDERGROUND GASIFICATION INSTALLATION

Dimensioning the installation for underground lignite gasification depends on the heat and power plant demand with regard of fuel. Installation must provide the volume of synthesized gas necessary to produce, by the power plant, 120 GWh / year of the electric power and about 200 000 GJ / year of heat power. As it was calculated in chapter 2, about 35 million tons of lignite, having the volume of about 30 million m^3, are necessary for that purpose.

Basing on preliminary defined geological conditions and a concept of gasification lines development, it was assumed that underground reactors will be constructed successively. They will be used during many stages of the engineering process. That planning of their usage should limit the number of boreholes and thus reduce the costs.

Part of construction works of underground structures of installation will be carried out simultaneously with construction of surface facilities. Drilling works and construction of extraction columns are very expensive and time consuming.

During the first year it is planned to make:

– 14 extraction boreholes;

– 4 dewatering boreholes;

– 4 backfill injection boreholes;

– additionally – low diameters monitoring boreholes.

During next years the following holes will be drilled:

– 54 extraction boreholes per year;

– 8 dewatering boreholes;

– 10 backfill boreholes.

Those numbers are only estimates. The precise determining of their number will be possible after completing the geological surveys and defining the gasification technology.

Reaching the assumed, minimum gas parameters, equal the energy parameters of gas currently used to supply the steam boilers in Głogów heat and power plant, will not be a vital problem after condensing the water vapor, even in case of applying the air injection. In order to maintain, constant in time, flow rate, pressure and gas composition, the gasification process should be carried out simultaneously in many reactors with the time displacement of operation cycle phase together with applying the storage tanks with capacity equal the minimum daily gas production.

6 ECONOMIC EVALUATION OF THE PROJECT

Economic analyze of the underground lignite gasification project is very difficult, especially when the information about the deposit and the location and reserves in individual beds is incomplete. Hence it was necessary, in some cases, to use the data from publications, most often included in KGHM CUPRUM – RDC studies (Nowak et al 2010; Nowak et al 2011 & Baranska-Buslik et al 2011).

Additionally the similar pattern and principles of defining the data as in the study (Nowak et al, 2010), what gives the opportunity to compare both co conceptual solutions. Since there are not any practical experiences in applying this method, different interpretations of financial regulation are possible, however it is not very important from the point of financial evaluation at that stage of work.

Economic evaluation was made basing on:

- technical and cost data contained in the study (Nowak et al 2010);
- cost data obtained from KGHM CUPRUM and foreign specialists dealing with this problem;
- information about material's prices – Sekocenbud;
- cost estimates normal and integrated for earth, construction, installation and assembly works;
- sets of unit price rates of industrial construction for investment operations – Bistyp Consulting.

In the analyze the following factors were estimated:

– expected incomes from selling the gas produced by the lignite gasification facility;

– investment outlays on planned project;

– depreciation of fixed assets;

– operation cost related with proposed project;

– residual value in last year of account.

The analyze was made according to the same rules as in the second (II) part of gasification, i.e. in the study "Pre-feasibility study of pilot installation using the underground gasification of lignite deposit and project commercialization" (Nowak et al 2010).

Construction of gasification facility is and investment project, which will operate basing on certain equipment and technical installations. The life time of basic equipment and capital assets ranges from 10 to 20 years. At this stage of work it is difficult to state if with regard of the period, for which the estimate is made, other factors, for example business or economic, may have the substantial impact. This type of project are evaluated within the perspective of about 15÷20 years. It reflects the approximate estimate of their economic life time. Attempts of forecast for longer periods of time will be useless.

With regard of the given project the time horizon of 20 years (1 year for construction and 19 years for operation) was taken.

The level investment expenditure, operating costs related with gas production and incomes from selling the gas were defined basing on technical concept of the facility operation and, assumed there, engineering parameters of lignite gasification system.

The investment expenditures are planned for:

– completing the geological report and design;

– investment supervision;

– purchasing the parcel of land for industrial activity (construction of lignite gasification facility and its development);

– preparation and organizing the site;

– construction of storage tanks;

– construction of engineering facilities within the station;

– installation of power, water, sewage and other networks;

– station for preparing backfilling material.

The investment outlays were estimated on about 136.06 million PLN (45.35 million USD) during the first year. This cost with further development are 181.5 million PLN (60.5 million USD). The construction costs are gathered in the following groups:

– expenditures for construction of facilities within the station;

– expenditures for installation of equipment and lignite gasification as well as water purification system;

– expenditures for constructing the gas reservoir, reducing station and gas pipeline;

– expenditures for backfill preparation facility with pump-pipeline system for transporting the tailings and storage tank for the backfill.

During the next years of station development the expenses for preparing the land, drilling the boreholes and up building the installations are planned. However the main outlays will be costs of drilling the boreholes (see chapter 4).

While calculating the operation costs in order to define the financial internal rate of return, the item which are not related with real cash expense, except depreciation, were ignored. These are:

- reserves for future restoration of assets because are not equal the real depreciation of goods or services;
- reserves for potential occurrences.

The following cost items were considered:

– taxes;

– environmental fees;

– royalties;

– consumables;

– technical utilities;

– labor costs;

– repairs and maintenance;

– liquidation fund;

– depreciation.

The following was taken into consideration and evaluated:

– costs of house-tax and land tax;

– fee for excluding the land from farming i.e. 10% / year of the dues – the so called single payment due to permanent excluding the land from production (Dz.U.04.121.1266 from 3 February 1995 r. about cropland and forests protection with changers for 2010 Dz.U.2009.115.967);

– environmental duties related with exhaust gases emission to the atmosphere (Dz.U.08.25.150 from 27 April 2001 r. Environmental Law, with rates according to the appendix no. 1 to the Environment Ministry Announcement dated 18 August 2009 about the amount of fees for using the environment in 2010 – M.P.09.57.780);

– royalty depending on the volume of produced gas (4.58 PLN / thousand m^3);

– payment for current reclamation depending on the volume of void formed after the lignite gasification (0.14 PLN / m^3).

The fee for excluding the land from production was taken on the level of 29 145 PLN / ha.

The house tax is assumed according to the regulations i.e. 2% of their initial value. Rates of house tax of 20.51 PLN / m^2, and land tax of 0.77 PLN / m^2 are taken according to the Ministry of Finance (2010).

The total operation costs were estimated as 34 764 thousand PLN / year i.e. 11 588 thousand USD / year.

The incomes were established considering the volume of gas delivered to the heat and power plant. At the same time the changed energy value of gas in comparison with the network gas was taken into consideration. The price of gas calculated this way was 0.115374 PLN / m^3, i.e. 38.458 USD / 1000 Nm^3.

Using the above data the quantity of income was estimated on about 44.6 million USD. Other receipts, such as sale or reuse of reclaimed land, sale of the own heat and electric power, were not taken into account.

The analyze effectiveness was made using the UNIDO method at the following assumptions:

– constant prices from I[st] quarter of 2010 (i.e. not taking into account the inflation factor);

– commodity and service prices are net prices, i.e. without VAT;

– calculation period is 20 years, including one year for constructing the station;

– investment will be financed by the ownership capital;

– exchange rate of EURO for delivery of imported machines – 3.8 PLN;

– income tax – 19%;

– discount rate of 11.24% is equal the capital costs taken in KGHM Polska Miedź SA.

Results of financial analyze:

- IRR is about 20.17%;
- NPV for 20 years calculation period is about 181 515 000 PLN.; NPV > 0;
- simple period of return is 8.2 years.

Among the income items demonstrated in the final year of the project is the residual value of investment i.e. not depreciated part of fixed assets such as buildings, outside installation etc. Residual value was taken into account in financial analyze, since it reflects the real receipt of resources. It was also comprised while calculating IRR and NPV indexes. Estimated residual value is about 62.5 million PLN what accounts for about 12.5% of total investment outlays. The way of determining the residual value assumed in the analyze is well-founded because the lignite gasification station will operate during the next years – the period of at least 30 years is envisaged. This economic analyze of the project indicated the high profitability of proposed mining venture.

7 CONCLUSIONS

Głogów Smelter and Refinery is located close to the site of ULG installation. It was stated that the most economy founded is delivering the gas to Głogów Heat and Power Plant through the pipeline in the volatile state. Connecting the installation with the heat and power plant should not be a problem. Four routes for the pipeline were proposed. Thc ground through which the pipeline is planned to pass includes the farmland and waste land without any important surface elevations.

Now the operation of Heat and Power Plant in Głogów Smelter and Refinery is based mainly on hard coal and blast furnace gas. It still strains to improve the operation and rise the effectiveness. Numerous modernizations and gradual reduction of hard coal share in the power production process, confirm that trend. In future the plant will be forced to abandon one of the main fuels. The reason is replacing the shaft furnace by the flash furnace in the Głogów smelter and change of smelting technology. Since the plant does to want to base only an the hard coal, it must find the alternative fuel, which may be the gas, especially the syngas.

The components and caloricity analyze showed that it is a gas which can fully replace, currently used blast furnace gas, and may even be much more efficient and cleaner. Assuming the lowest predicted parameters of syngas, it has about 50% higher heat value than the gas currently supplying the boilers.

New gas fuel may have the impact on the set up parameters in combustion chambers. It is a typical procedure in case of changing the feeding gas. The adjustment will result from the multiple options of utilizing the syngas.

Financial analyze of the project under examination, showed that the investment is profitable, and the return of investment outlays should appear after about 8.2 years. Net present value (NPV) is about 151.5 million PLN, and internal rate of return (IRR) about 20.17%.

The presented economic evaluation of the project is of estimative nature. It results from the absence of geological data, detailed extraction parameters, problems with defining the process effectiveness at this stage of the project. Unquestionably the results of economic evaluation indicate that the further development works are reasonable taking into consideration the venture profitability aspect.

That result is not a surprise. Almost all known ULG projects, where economic data are presented, has very good estimates and high profitability of the investment.

Additionally it will be to use the part of CO_2 emitted by Głogów smelter and force it to the gas generator, increasing consequently the quality of syngas. It requires, however the considerable financial expenditures, necessary for construction the installation for CO_2 capturing and delivery to the site of underground lignite gasification.

The presented proposal of applying the underground lignite gasification to produce the synthetized gas and its delivery to the heat and power plant in the smelter is a cost-effective project and the works on the project are continued. They will be probably finished in 2015 where the first delivery of gas to heat and power plant is planned.

REFERENCES

Derkacz, J. & Sztromwasser, E. 1984. *Plan of geological and exploration works for lignite in Ścinawa – Bytom Odrzański region.* CAG PIG-PIB Warszawa, Poland.

Kozula R. & Golczak I. 1988. *Geological report of "By-*

tom Odrzański" copper deposit in C-1 + C-2. Category, CAG PIG-PIB OD Wrocław, Poland.

Kudelko, J. & Nowak, J. 2010. *Conditions for safe underground gasification of lignite in Poland.* School of Underground Mining. Dnipropetrovskk/Yalta, Ukraine 2010, New Techniques and Technologies in Mining, Taylor & Francis Group: 97-101.

Kudełko, J. & Nowak, J. 2009. *Multiple evaluation of lignite deposits management in Lower Silesia through their underground gasification.* XIX Conf. Actualities and perspectives of mineral resources economy, Rytro, 4-6 November 2009, PAN IGSMiE, Kraków, Poland.

Kudelko, J. & Nowak, J. 2007. *Geo-sozological conditions for strategy and selection of Legnica region lignite deposits management.* CUPRUM no. 1/2007, Ore Mining Scientific and Technical Magazine, Wrocław, Poland.

Nowak, J. 2007. *Legnica lignite deposit management strategy including underground coal gasification.* Publ. Of National Mining University, Dnipropetrovs'k, Ukraine: 225-231.

Piwocki, M. 1996. Evaluation of Legnica-Ścinawa lignite deposit management possibilities. Task 5: Reserves calculation according to the current operative economic criteria, PIG, Poland.

Szubert, A., Nowak, J. & Grotowski, A. 2011. *Possibilities of methane recovery from non-industrial lignite seams using bio-gasification method.* 15th Conference on Environment and Mineral Processing 8-10.06.2011, Ostrava, Czech Republic: 235-242.

Kozłowski, Z., Nowak, J., Kudelko, J., Uberman, R., Kasiński, J. & Sobociński, J. 2008. *Technical-economical ranking of lignite deposits management from the perspective of Polish power energy policy.* Book. Technical University of Wrocław Publ., Poland.

Team work under leadership Nowak, J. 2010. *Preliminary concept of lignite deposits management considering the construction of pilot underground gasification installation –* Stage II. *Pre-feasibility study of pilot installation using the underground gasification of lignite deposit and project commercialization,* KGHM CUPRUM Ltd. – RDC, Wrocław, Poland, not published.

Nowak, J., Szafran, R., Kobak, P., Strzelecki, M., Barańska-Buslik, A. & Zaremba L. 2011. *Analyze of possibilities of using syngas from ULG installation in KGHM PM SA production cycle within the future energy policy of KGHM CUPRUM Ltd – RDC,* Wrocław, Poland, not published.

Barańska-Buslik, A., Nowak, J., Nowik, T., Strzelecki, M. & Kobak, P. 2011. *Detailed location analyze and preliminary calculation of reserves data concerning the solid fuels for ULG installation and proving the delivery of synthesized gas to the production line.* KGHM CUPRUM Ltd – RDC, Wrocław, Poland, not published.

Preidl, M. & Mikuła, S. 1976. *Geological report for "Głogów I" copper deposit C-2 cathegory.* CAG PIG-PIB OD Wrocław, Poland, not published.

Geomechanical Processes During Underground Mining – Pivnyak, Bondarenko, Kovalevs'ka & Illiashov (eds)
 ISBN 978-0-415-66174-4

Study of rock displacement with the help of equivalent materials using room-and-pillar mining method

V. Buzylo, T. Savelieva & V. Saveliev
National Mining University, Dnipropetrovs'k, Ukraine

ABSTRACT: Field study made with the help of equivalent materials to determine minimum dimension of interchamber and barrier pillars and limiting chamber span was carried out. Modeling was made for gypsum quarry.

1 INTRODUCTION

The aim of modeling rock displacement made with the help of equivalent materials is to determine general regularities of interchamber and barrier pillars operation as elements of underground construction.

Modeling results were used to design calculation diagrams to determine loads acting on barrier pillar. Furthermore, modeling enabled to determine limiting chamber span and minimum width of barrier pillars.

Geometrical and force similarity were taking into account in modeling (Kuznetsov 1959 & Nasonov 1978).

Force similarity were determined by the following equation

$$N_M = C_e \cdot C_\gamma \cdot N_H ,$$

where C_e – geometrical model scale; C_γ – density scale; N_M, N_H – mechanical characteristic of model material and nature.

Breaking compressive stress was taken as a mechanical material characteristic.

Modeling was made in conditions of Olekminsk quarry. Density of gypsum $2.3 \cdot 10^{-3}$ kg / m^3, dolomite $2.5 \cdot 10^{-3}$ kg / m^3, silt stone $2.55 \cdot 10^{-3}$ kg / m^3, mudstone $2.5 \cdot 10^{-3}$ kg / m^3 was taken into account. Quartz sand with small amount of alabaster was used as an equivalent material. While studying set of tests materials which characteristic is given in Tables 1, 2 were selected.

Table 1. Breaking compressive stress for rock and model material at geometrical scale 1:100.

Working model within rocks	Component name	Componen *t* weight , g	Model, 10^5 MPa		Nature, 10^5 MPa
			calculation	factual	
Gypsum	Sand	1000			
	Alabaster	4	0.66	0.54	110
Mudstone	Sand	1000			
	Alabaster	2.5	0.22	0.22	37
Silt stone	Sand	1000			
	Alabaster	6	0.78	0.75	130
Dolomite	Sand	1000			
	Alabaster	10	1.3	1.31	219

Density of model material was slightly changed at various component correlation and was $1.4\text{-}1.5 \cdot 10^{-3}$ kg / m^3. Strength of equivalent material at geometrical scale 1:200 was twice as little as that one given in the Table 1.

The following stand dimensions to patternmaking were accepted: length is 2 m, height is 1 m, width is 0.25 m. The front wall of the pattern was made of glass.

Models had 8 layers imitating corresponding rock stratification within the nature. Layer characteristic at scale 1:100 is given in the Table 3.

The process of patternmaking is the following: material was arranged by layers with 2-3-cm width and compressed by roller (10 cycles).

Table 2. Breaking compressive stress for rocks and model material at geometrical scale 1:200.

Working model within rocks	Component name	Component weight, g	Model, 10^5 MPa		Nature, 10^5 MPa
			calculation	factual	
Gypsum	Sand	1000			
	Alabaster	2	0.33	0.30	110
Mudstone	Sand	1000			
	Alabaster	1.3	0.11	0.10	37
Silt	Sand	1000			
	Alabaster	3	0.39	0.38	130
Dolomite	Sand	1000			
	Alabaster	5	0.65	0.60	219

Table 3. Layers characteristics.

Rock name	Layer height within working, cm
Dolomite	10
Gypsum	8
Mudstone	3
Silt	2
Dolomite	2
Gypsum	4
Dolomite	4
Pumps	56

There are 2 panels with barrier pillars between them within the working. Chamber span in all models in nature is 8 m, pillar width is 4 m. Ceiling within chamber of the roof is made of gypsum with 1-m thickness. The width of barrier pillars was 20 and 30 m (all dimensions here and then are given in terms of nature).

2 MODELING ROOM AND PILLAR MINING METHOD WITH 20-METER WIDTH OF BARRIER PILLAR

Model imitated the area of deposit. Barrier pillar is in the center of it. Chambers are worked-out to the left and to the right within 2 panels. This is initial position. Model scale is 1:100.

The width of barrier pillar is 20 m (in nature). Six pillars were initially worked-out on the both sides of barrier pillar. Construction was in the stable state, there was no caving. Gradual interchamber pillar caving was imitated then. As a result, load acting on barrier pillar was increased. First, the width of 2 interchamber pillars was reduced up to 2 m (pillars 7-8 and 8-9) (Figure 1a). There were no disturbances. It is natural, because pillar size was accepted with safety margin 2-3. Then these pillars were completely removed and chamber span was reduced up to 32 m. Such span was stable (Figure 1b).

Experiment showed that 8-m chamber span was accepted with rather high level of safety margin.

Such workout of interchamber pillar was carried out to the left of barrier pillar. Destruction of interchamber pillars has started. Pillar 4-5 was destructed first, then the rest interchamber pillars and at last the barrier pillar (Figure 1c).

There was not arch formation within chamber roof. Entire rock mass above gypsum layer has completely collapsed.

Model 1 showed that 4-m width pillars and 8-m chamber span have rather high level of safety margin but barrier pillar hasn't performed its function. This pillar has destructed and couldn't prevent roof rock displacement.

Furthermore, this modeling showed that arch formation is not an obligatory element of roof collapse. Rock mass displacement up to the surface took place due to pillar destruction without arch formation in roof chamber and in the panel in a whole. On the basis of study described above width of barrier pillar in subsequent models was increased up to 30 m.

3 MODELING ROOM-AND-PILLAR MINING METHOD WITH 30-METER WIDTH OF BARRIER PILLAR

General construction of the model is the same as the previous one. Barrier pillar is in the center of the model, but its width was increased up to 30 m (all dimensions are given in terms of nature). Chambers are worked-out both to the right and to the left of it. 10 chambers are worked-out within the right panel and 7 chambers are worked-out within the left one. Chamber scale is 1:200.

The interchamber pillars were removed in the left panel and chamber span was increased up to 44 m (Figure 2a). After that pillar destruction between chambers 6, 5 and 4 took place. Rocks within the panel collapsed up to the model surface (Figure 2b). Barrier pillar remained the same and caving didn't spread to the neighboring model. So, width of barrier pillar was adequate to the level of strength. There was not arch formation in this model as well as in the previous one. Caving spread to the model surface at once.

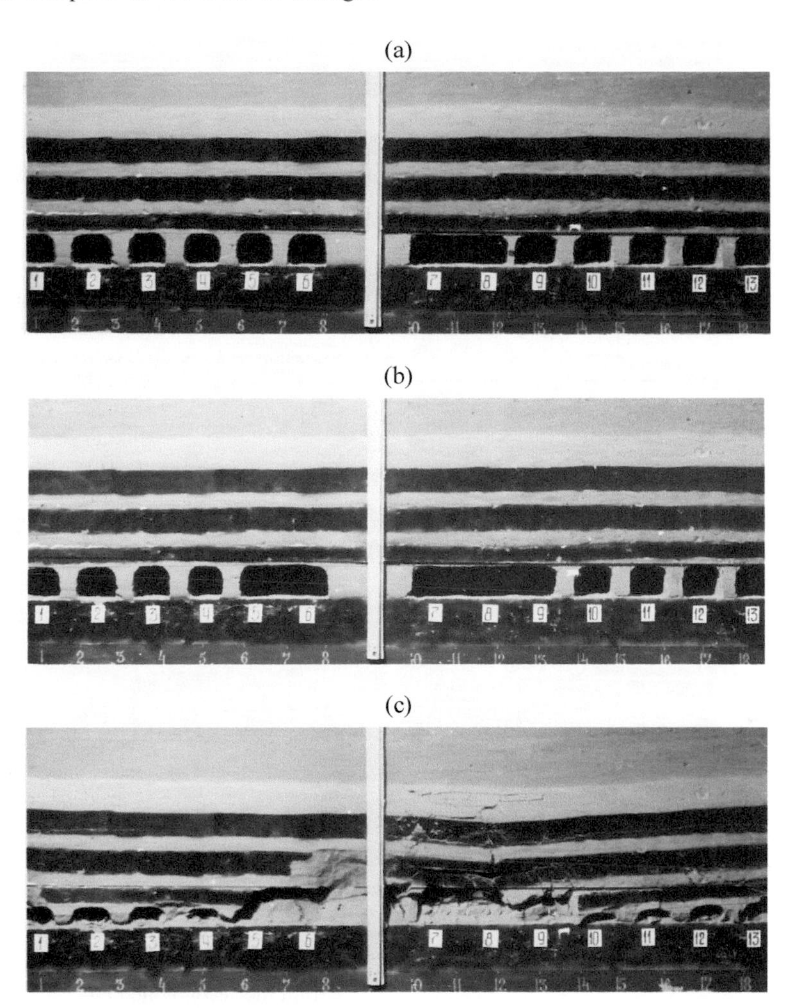

Figure 1. Model 1. Modeling concerning adequacy of pillars and chamber span to the level of strength: (a) interchamber pillar was worked-out to the right of barrier one; (b) interchamber pillar was worked-out to the left of barrier one; (c) model after roof rock collapse.

After that pillar 15-16 was removed within the first panel. It was not enough to collapse the rest pillars and shifting rock mass to the surface (Figure 2c). There was not arch formation. Barrier pillar left the same. It confirmed an adequacy of its dimensions to the level of strength in case of emergency.

4 MODELING LIMITING CHAMBER SPAN

In the first two cases stable chamber span was 32 and 44 m. To check this result one more time model No 3 was worked-out (Figure 3a). Span of a single chamber was being gradually increased within this model. Roof collapse took place at 44-m span that confirmed results obtained in the models 1 and 2. It should be noted that under roof collapse within such single chamber, arch was formed but its contour is indistinct (Figure 3b).

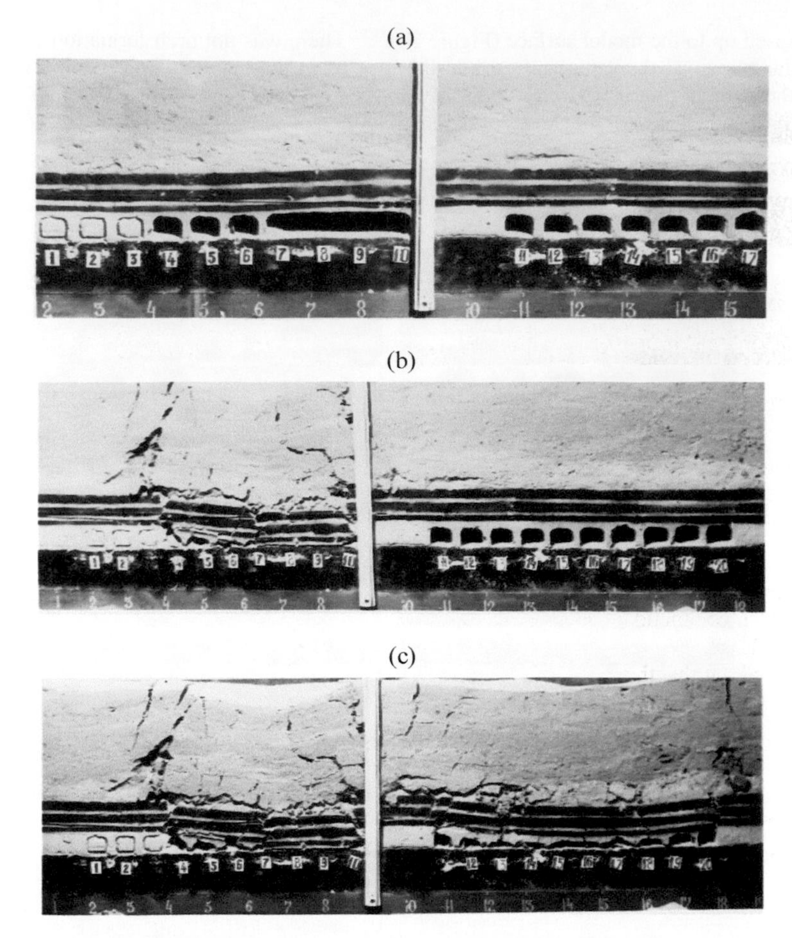

Figure 2. Model 2. Modeling the process of roof collapse in case of interchamber pillar destruction: (a) model before roof collapse within the left panel; (b) roof collapse within the left panel; (c) roof collapse within the right panel after working-out pillars 15–16.

5 DETERMINING LOADS ACTING ON BARRIER PILLAR

It was supposed to determine the character of enclosing rock displacement within the panel confined by barrier pillars with the help of modeling in case of destruction of interchamber pillars. It is required to design the diagram determining loads acting on barrier pillar.

It was determined that after destruction load of three or four interchamber pillars acting on neighboring pillars increases and they are destructed as well. Entire rock mass displacement to the surface takes place then. Moreover, rocks are cut along the boundary of barrier pillar. Arch is not formed. Entire rock mass is shifted to the surface at once. If the width of barrier pillar is insufficient, it can be collapsed.

20-m width barrier pillar collapsed within the first model. 30-m width pillar appeared to be stable within the second one.

How to make barrier pillar stable after interchamber pillar destruction? Obviously, it should take load from entire rock mass within panel, i.e. the load which interchambers pillars took earlier. According to the study described above calculation diagram to determine load R on running meter of barrier pillar is offered (Figure 4a). This diagram shows that load calculation acting barrier pillar is

$$P = (L + B)H\gamma ,$$

where L – panel width, m; B – width of barrier pillar, m; H – depth from the surface to the roof and the layer, m; γ – density of roof rocks, kg / m^3.

(a) (b)

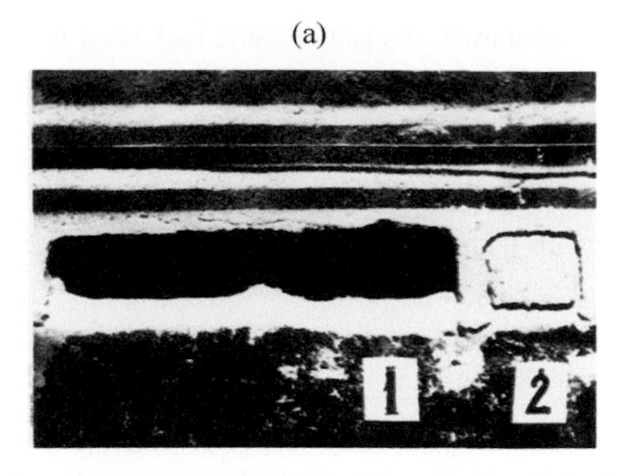

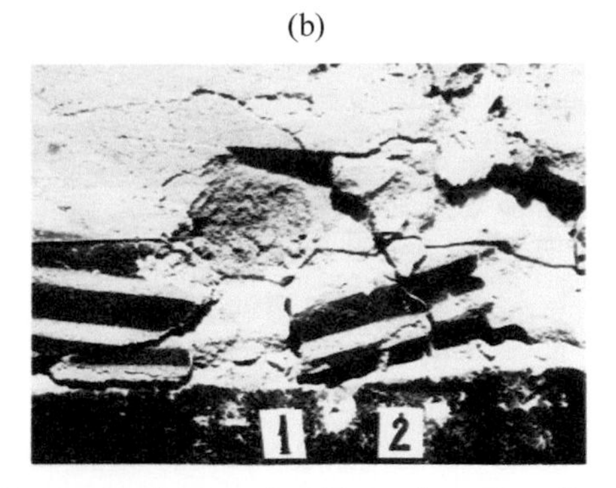

Figure 3. Model 3. Modeling limiting span of a single chamber: (a) span chamber extension; (b) chambers after collapse.

Compressed stresses within barrier pillar will be

$$\sigma = \frac{(L+B)H\gamma}{B}.$$

These stresses should be less than assumed ones. It enables to recommend small level of safety margin equal to 1.5-2.

It is pointed out that this conclusion doesn't correspond to that one offered by V.V. Kulikov (Figure 4b) in his paper (Kulikov 1978).

It is supposed that V.V. Kulikov's hypothesis should not be considered as universal one and acceptable to all mining conditions. Probably, it can be true in definite conditions. Modeling carried out for gypsum quarries proved that the work at these quarries differs from the diagram offered by V.V. Kulikov.

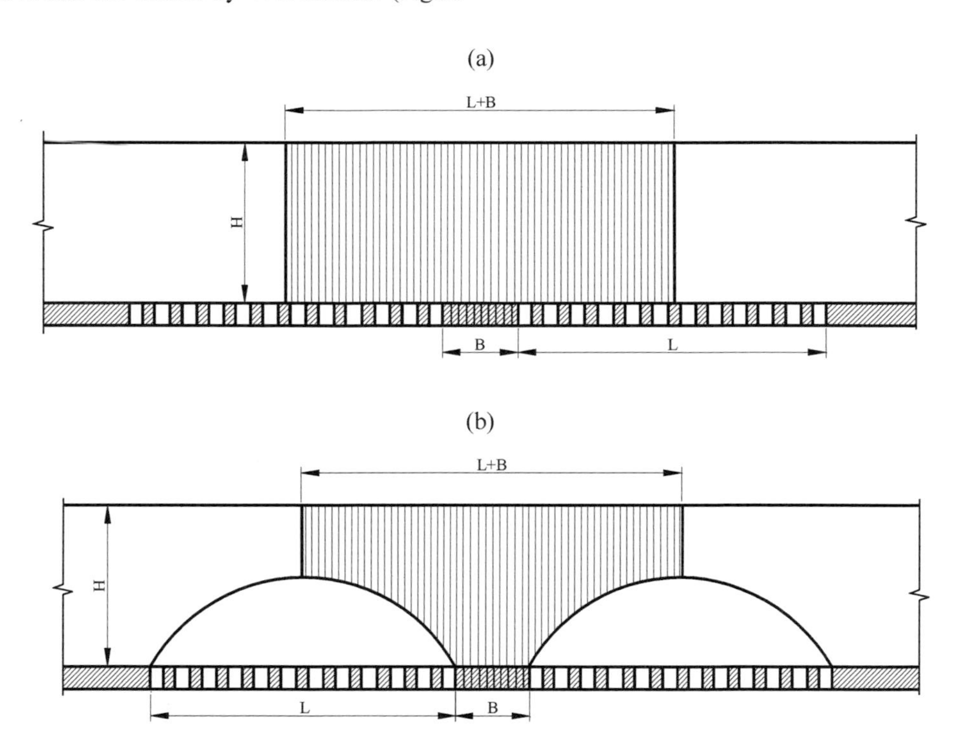

Figure 4. Calculation diagrams to determine load acting on barrier pillar: (a) according to modeling results; (b) according to V.V. Kulikov's data.

6 CONCLUSIONS

Calculation diagram to determine load on running meter of barrier pillar was obtained by modeling made with the help of equivalent material. This calculation diagram of barrier pillar is true to conditions where modeling was carried out, that is, rock mass above roof rock is 60-70 m, panel width is 100-200 m with particular rock stratification.

It was determined that arch formation under interchamber pillar caving at gypsum quarries doesn't take place.

Some studies showed that load acting on pillars fewer than rock mass from layer to surface. It is determined by rock weight from the roof layer to pressure curve. However, diagram of determining load acting on pillar is confirmed by this study. According to this diagram load acting on pillar is determined by the weight of entire rock mass from layer to surface.

Roof of the chambers seemed to be stable while increasing span up to 44 m. It is unexpected conclusion. In future it will be required to carry out the study of tension stresses within the roof of chambers with the same span using the method of elasticity theory to explain validity of such large spans.

REFERENCES

Kuznetsov, M.N., Budko, M.N., Philipova A.A. & Shklyarskiy, M Ph. 1959. *Study of rock manifestation on the models*. Moscow: Ugletechizdat: 223.

Nasonov, I.D. 1978. *Modeling mining processes*. Moscow: Nedra: 256.

Kulikov, V.V. 1978. *Textbook on production processes and production technology*. Moscow: Nedra: 25.

 ISBN 978-0-415-66174-4

Ecological problems of post-industrial mining areas

A. Gorova, A. Pavlychenko & S. Kulyna
National Mining University, Dnipropetrovs'k, Ukraine

O. Shkremetko
St. Petersburg Energy Institute, St. Petersburg, Russia

ABSTRACT: The ecological consequences of liquidation of unprofitable mining enterprises have been considered. The peculiarities of the management of environmental risks to humans and the environment in post-industrial areas have been analyzed. The necessity of using methods of bioindication to control the state of environmental objects and health of the population on post-industrial mining areas has been substantiated.

Restructuring of the coal industry of Ukraine in 1995-2000 is the most ambitious attempt to recover and reconstruct a huge economic sector of the state in a relatively short time. This was a positive aspect of the mass liquidation of unpromising and unprofitable mines. As a result of reconstruction, the coal industry was supposed to become a more compact and market-oriented branch. However, in practice the positive results of restructuring of the coal industry faded into insignificance. The drastic negative socio-economic and environmental impacts that led to the emergence of regions with persistent symptoms of depression have come to the fore.

One of the important arguments that seemed to be the key argument for liquidation of the mines was reduction of maintenance costs and recovery of mining enterprises and national economy as a whole. But, as practice shows, such liquidation of mines has neither improved the ecological situation, nor solved any other problems that emerged during their exploitation and liquidation. All of this is related to the fact that the mine closures were not preceded by the comprehensive scientific assessments and forecasts for the state of environment in these regions as well as no consequences of further influence of the liquidated mines on the components of the environment were studied. In particular, such unreasoned and unreasonable closure of mines may lead to catastrophic consequences both ecological and social as the coal industry is a very complicated multi-industrial production and economic complex, which is a heavy industry not only in content but also in terms of a high level of environmental risk.

It should be noted that there are some issues related to the liquidation process that are still not resolved, particularly:

– there is no clear notion of the volume of mining of resources at which the operation of mining enterprises makes no economic sense;

– the issue of liability of economic entities for the consequences of mine closure has not been cleared up.

Currently there are no clear mechanisms for mines closure, which would take into account all environmental impacts from the moment of shutdown of production equipment to the development of strategies for sustainable operation of post-industrial areas. Besides, in most cases the ecological problems arising at different stages of liquidation of unprofitable mines have considerable impact on the future development and functions of adjacent territories.

The process of liquidation of mining enterprises is ecologically dangerous, that is why a continuous monitoring of the environmental state should be provided. With the advent of new directions of influence on the environment, there is a need to organize:

1) management of the hydrogeological regime of the high density mining areas;

2) continuous treatment of highly mineralized mining water discharged into the hydrographic network;

3) neutralization of hazardous and radioactive waste and land rehabilitation;

4) accounting of mining holes and works for their stabilization;

5) bringing of waste heaps to an environmentally safe condition.

The problem is also in the adaptation of existing environmental standards to the realities of coal mining areas where the situation close to ecological disaster has occurred. It is necessary to develop the methods of environmental risk assessment at mine closure and the system of standards that would take into account the process of gradual improvement of ecological situation due to measures that are taken (Sliadnev 2001).

Till now the liquidation of the coal mines has been funded residually and with the violations of environmental laws. Neglecting of environmental safety regulations during the closure of mines leads to significant changes and violation of hydrological regime of areas, pollution of surface and groundwater, land subsidence, etc.

The purpose of this study was to analyze the state of the environment in the post-industrial mining regions of Ukraine, where the mines are closed.

The coal industry is the foundation for sustainable operation of the national economy and its energy safety. Major coal reserves are concentrated in Donetsk Coal Basin, Lviv-Volyn Coal Basin and Dnieper Brown Coal Basin.

Coal mining has been carried out on the territory of the state for more than 200 years. Thus, in particular, at the times of the Soviet Union Ukraine has formed almost 22-24% of GDP of the USSR. As a result, the significant area of mining operations has been formed – 20 thousand km^2 or 3.3% of the total area of the state. Currently, the mining and processing of mineral resources in Ukraine covers almost 1/3 part of production assets, 20% of employment and 25% of GDP of the state. Mining industry also dominates in the GDP structure of independent Ukraine, although for comparison in the U.S. it is 2.6% of GDP, in Germany – 1.1%, in France – 0.8%, and in Japan – 0.6%.

Since 70 years of the twentieth century there has been a decline in the coal industry of Ukraine. This has been primarily due to the overall difficult economic situation in the country, instability of the industry, exhaustion of balance reserves, unprofitability of most mines, difficult mining and geological conditions and the lack of a new mine construction. This has led to mass closure of mines. The long-term use of mineral resources in the mining regions resulted in significant changes in the environment and occurrence of emergency situations.

Closure of coal mines leads to negative environmental consequences, such as hydrogeological, hydrochemical, gas chemical, engineering and geological.

Considerable part of the liquidated mines has been closed by "wet" conservation method, which has resulted in significant environmental and geological problems, since the flooding of mines increases a man-caused impact on the lithosphere and hydrosphere. Large areas of coal-measure rocks intercepted by mine workings led to deformation of the earth's surface and destruction of residential and industrial objects. Besides, most of liquidated mines are associated with the operating ones and the changes occurring in them affect the existing mines (Hoshovski 2000).

Coal mining is accompanied by significant inflows of water into mine workings, since the production of 1 tonn of Ukrainian coal involves about 3 m^3 of groundwater. It should be noted that for decades no proper attention was paid to the issues of treatment of mine waters on the mining enterprises. Also there remain unresolved issues of treatment of mine waters collected in gathering ponds of closed mines as well as no standards for their carrying back to the hydrographic network have been developed. In 2010 about 36.5 million m^3 of highly mineralized mine waters discharged to the surface from the closed mines of Donbas. As for other regions, on the earth's surface of Chervonohrad mining region, which is one of the largest region in the Lviv-Volyn coal basin, there are 200 million m^3 of mine water with a total mineralization of 6-8 g / l. Besides, mine waters are contaminated with mineral salts, suspended solids, sulphates that leads to pollution of water tables and violation of their hydrological regime (Phillip Pack 2009).

In case of "dry" conservation of mines the adverse effect on the environment lies in a discharge of mine waters into the surface water bodies. This method of mine closure leads to exhaustion and pollution of water resources, both surface and underground.

In general, every year about 400 million m^3 of rock are extracted in the process of underground coal mining in Ukraine. According to the data of Scientific Production Association "Mechanic", for over 200 years of the development of coal deposits in Ukraine 1100 heaps covering 6.300 hectares of fertile land were formed. According to the experts each million tons of Ukrainian coal requires about four hectares of land for storage of rock. Lack of vegetation on the waste heaps causes their erosion both by water and air, which further leads to a negative impact on the environment.

The issue of further handling of waste heaps after closure of mines remains unresolved. 121 coal mines are now in the process of closure, and there are 341 dumps on their territories, and about 105 of them are the burning heaps (see Table 1).

During open burning of waste heaps the carbon oxides and dioxides, nitrogen and sulfur oxides and polyaromatic hydrocarbons are released to the atmosphere. One actively burning heap is a source for release of about 25 to 250 t / year of pollutants to the atmospheric air, the concentration of which 10 times exceeds the maximum permissible limits (MPL). During 24 hours 10 t of carbon monoxide, 1.5 t of sulphuric anhydride and considerable quantity of other gases and heave metals are released to the atmospheric air from one heap.

Table 1. Distribution of burning heaps, by coal mining regions (Phillip Pack 2009).

Coal mining region	Number of closed mines	Waste heaps	
		Total number	Burning heaps of total number
Donec'k	52	177	69
Luhans'k	36	244	34
Lviv-Volyn'	8	7	2

Among other ecological problems caused by the waste heaps, the following should be mentioned:

– runoffs from the waste heaps containing salts of acids and heavy metals;

– withdrawal of large areas of fertile land;

– pollution of atmospheric air, surface and underground waters.

Every year more than 400 t of solid parts are washed and eroded and about 6 t of salts are leached from every waste heap. This is due to the oxidation of pyrites contained in the heaps, which further leads to transformation of some metals, particularly Fe, Al, Mn, Zn, into mobile forms.

The deformation processes occurring on the waste heaps of mines are characterized by formation of gullies, the width and depth of which are ranging from 2-4 m and from 1-3.5 m, respectively, and in which the rock is moving for a distance reaching sometimes 6 m. In particular, on the territory of the Chervonohrad mining region there is one of the largest heaps in Europe, where the rock refuse of the Chervonohrad Central Concentrating Mill (hereinafter – ChCCM) are stored. The area of this heap is 89 hectares, and its height is 68 m.

The researches carried out by the Institute of Geology and Geochemistry of Combustible Minerals of National Academy of Sciences of Ukraine in Lviv shows that the rock of the waste heaps contains a high content of chromium, manganese, molybdenum, scandium, exceeding background values for soils by 2-7 times. Taking in consideration the above, a special attention should be paid to the chemical composition of the waste heap of the ChCCM, as it occupies the largest area among all the heaps – 89 ha, and its height is 68 m. Table 2 and 3 show the content of micro- and ultramicroelements in the rock of the the ChCCM waste heap (Informational... 2011).

Table 2. Content of microelements in the rock of the ChCCM waste heap, g / t (Informational...2011).

Content of elements	*Cu*	*Zn*	*Mn*	*Pb*	*Mo*	*Ni*	*Ba*	*Cr*	*Ti*
Minimum	12.59	0	86.78	7.638	0	16.53	106.8	31.188	2717
Average	89.04	35.7	2353.6	35.66	1.64	37.56	369.3	235.36	1234.2
Maximum	244.16	62.1	4484.4	273.24	3.97	79.48	583.5	2159.0	4595.2

Table 3. Content of rare earth elements in the rock of the ChCCM waste heap, g / t (Informational...2011).

Content of elements	*Be*	*Bi*	*Yb*	*Y*	*Sc*	*Ga*	*V*	*Sn*
Minimum	0.543	0	2.278	2.278	3,728	7.797	76.38	0
Average	2.645	9.435	4.65	44.11	15.02	25.43	151,88	4.48
Maximum	4.566	24.3	10.87	108.7	53.98	48.88	387.95	36.93

As Tables 2 and 3 show, the content of heavy metals far exceeds MPL, for example, by 45,5 times for plumbum (MPL – 6 mg / kg), by 81.3 times for copper (MPL – 3 mg / kg), by 19.8 times for nickel (MPL – 4 mg / kg), and by 31 times for zinc (MPL – 2 mg / kg).

The problems of biological recultivation of the waste heaps are not solved at closing of mines. Most of them are slowly overgrowing, though 30 to 50 years are required for re-vegetation.

The land subsidence is another problem not taken into consideration at closing of mines. The area of subsidence in coal mining regions is 8,000 km^2, where the subsidence and destruction of surface above the underground workings have been recorded in the area of over 2.4 km^2. The depth of subsidence averages 0.2-1.2 m and in some places reaches 5.0 m. All of this results in instability of a soil body followed by the fracturing, waterlogging and flooding processes. As a result of land subsidence a destruction of buildings and structures, failure of utilities, waterlogging of agricultural land, underflooding of settlements, etc. take place.

The largest negative impact of the land subsidence is observed within the industrial-urban agglomerations, because the mine workings often pass under the built-up areas. The undermining areas cover such cities as Donets'k, Makiivka, Horlivka, Yenakieve, Bilozers'k, etc. In Donets'k, Makiivka, Horlivka, Yenakieve, Torez, Marianivka as a result of mines restructuring by partial or complete inundation, the flooding process is developed in conjunction with the land subsidence under the mining workings. The reason for the large scale of the development of negative ecological and geological processes, including flooding, is that the mines of Donbas have common aerodynamic and hydraulic networks, and, in fact, it is one mining-geological system. For example, in the Dnipropetrovs'k region 0.74 km² of mining works area are flooded; in the Donetsk region flooding is fixed at the area of 1.66 thousand km². 30 cities with total area of 230.0 km² are underflooded. Among the most flooded cities are Slaviansk (72% of the total area), Bilozers'k (72.2%), Telmanove (100%), Velyka Novoselivka (35%), Sivers'k (29.4%). The total area of 347 flooded villages is 68.46 km² (Koskov 1999). The permanent increase of flooded areas is mainly caused by man-made factors. In Chervonohrad mining region the total area that in some or other way has experienced the negative consequences of subsidence and local landscape changes is almost 62 km² and covers Chervonohrad and Sokal Cities and Bendiuha, Volsvyn, Hirnyk, Hlukhiv, Dobriachyn, Mezhyrichia, Silets, and Sosnivka villages. The depth of subsidence and local landscape changes is 3.5-4.0 m. The largest changes are observed on the areas of closed mines – "Bendiuzka", No. 5 "Velykomostivs'ka", No. 1 "Chervonohrads'ka" and "Visean". The land subsidence also leads to the formation of man-made lakes, such as the lake of over 2.5 m depth located in the region of Silets and Sosnivka Villages. In general the flooding has mostly occurred in the central part of Chervonohrad mining region where more industrial objects and linear utilities are located.

Besides, as a result of destruction of sewage and mine water drainage systems and washing by the groundwater of toxic components from the rocks of waste heaps, which are widely used in the region for filling up of impounded territories, the groundwater became contaminated and unsuitable for household water use.

Also in 2010 the karst activation in the mining regions of Ukraine is in progress, which is due to the effect of mine drainage on the coal mining areas (Informational ... 2011).

Another important factor requiring attention at closing of mines is methane emission to the atmospheric air as the methane is a powerful greenhouse gas which causes climate changes. Methane emission in Ukraine makes 16% of all anthropogenic greenhouse gases emissions. The main amount of methane is released to the atmosphere during the coal mining process. Currently Ukraine takes fourth place in the world by volume of methane emitted to the atmosphere from the coal deposits. In general methane emissions in mining regions make up about 5.6 billion m³. Thus, 172.5 million m³ of methane in Dnipropetrovs'k basin, 3.7 billion m³ in Donetsk basin, 1.8 billion m³ in Luhans'k basin and about 60 million m³ in Lviv-Volyn basin get into the atmosphere, and only 8% of its total volume is used as fuel, the rest is discharged to the atmosphere. Besides, the danger of methane discharge at the closing of mines is that even after the liquidation of coal mines, it continues to escape from coal veins to the surface. It is migrated though the pores, cracks and faults in rocks. Duration and intensity of gas emission from abandoned places of liquidated mines depends on many geological factors. For example, after the closure of the "Central-Pervomais'k" mine in Pervomais'k (Luhans'k region) methane has been getting to the surface for more than 20 years. During this period, the explosions and ignition of methane in residential and industrial premises have been observed within this mining region. The emission of gas from the holes drilled for the protection of surface structures was about 60 million m³, and its concentration was 24-25%. Besides, the emission of radon and taron is observed at some mines (Kasymov 2001). Therefore, an important component of mine closure should be control over mine gases getting to the earth surface in hazardous concentrations of $CH_4 > 1\,\%$ and low oxygen content in the basements of houses, water wells and other underground structures.

All the processes specified above have negative impact on the state of the environmental objects and human health. The termination of economic activity of mining enterprises does not guarantee the elimination of their impact to the environmental components. That is why the closure of mines should be carried out in a way allowing for elimination of negative impact on the environmental objects and health of population living in these areas.

For over ten years in Ukraine the identification of high-risk facilities, including coal mining enterprises, has been provided. However, till now there are no methodological principles for analysis and evaluation of ecological risks to the environment and public health during the liquidation of mines (Gorova 2011).

Ecological risk is defined as the probability of deliberate or accidental, gradual or catastrophic anthropogenic changes of the existing environmental objects, factors and ecological resources. Thus, environmental risk can be represented by two components: risk to the environmental objects and the risk to humans. Each of the components is quite important for assessing the state of the environment, because the scale of risk occurrence and the level of its danger depend largely on the perception of the risk by the subject of assessment (micro-organisms, flora and fauna, population).

For assessment of ecological state of post-industrial areas it is reasonable to use the bioindication methods that allow conducting of complex assessment for all factors by taking into consideration their modifications and interaction.

The risk management should be based on economic and technical feasibility analysis as well as on legal and normative acts. It should also be noted that the important factor of risk management is provision of information on ecological situation of post-industrial areas to the public through mass media.

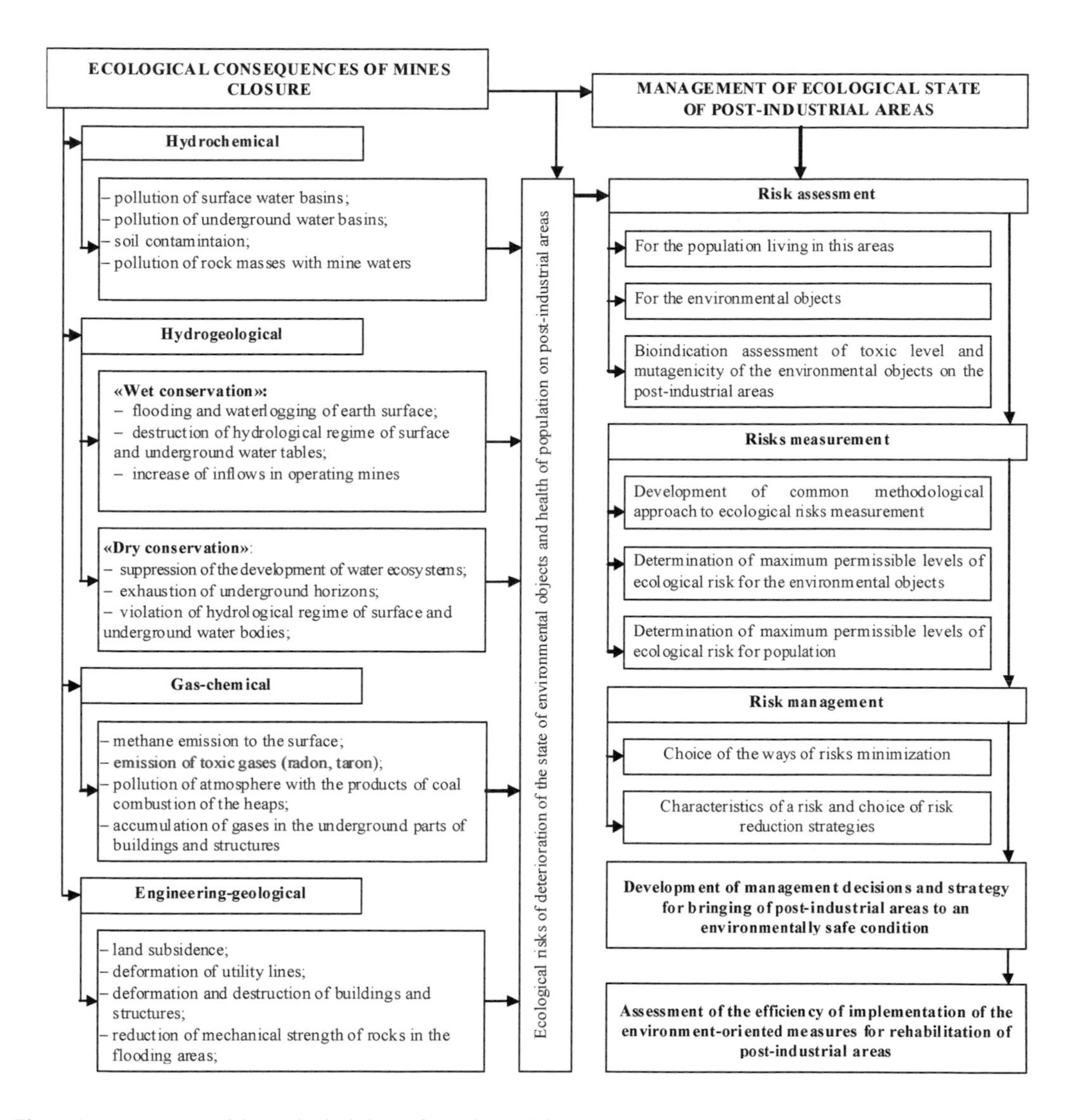

Figure 1. Management of the ecological risks of post-industrial mining areas.

Figure 1 shows the structural flowchart of management of ecological risks emerging on the post-industrial areas at liquidation of mining enterprises.

According to the proposed methodology we carry out long-term investigations of environmental risk levels in several coal-mining regions of Ukraine. The bioindication researches are conducted at the cellular, organ, organism and population level of organization of living matter (Guidelines ... 2007). The conducted studies have revealed significant differences between the levels of damage of bioindicators located in the area of influence of the objects of liquidated mines and control territories. Bioindicators that have been grown in post-industrial areas have shown the high levels of damage that were 2-15 higher than in clean areas. The use of biotesting methods allows for the defining of directions of further use of post-industrial areas and quick and accurate assessment of efficiency of implementation of environmental protection measures.

The conducted analysis of the impact of closure of mining enterprises on the environmental objects allows for the following conclusions:

– mass closing of mines during the previous 15 years was conducted without consideration of further consequences for the environmental objects and population health;

– the closing of mines involves different ecological risks for the environmental objects and population living in post-industrial areas;

– the closure of mines should be planned and economically and ecologically justified;

– the development of feasibility projects for mines closure should be accompanied with the development of projects for liquidation of environmentally hazardous facilities – dumps, septic tanks etc.

The permanent ecological monitoring centres with obligatory application of bioindication research methods should be created in post-industrial areas. For timely detection and prevention of negative effects on the environmental objects and human health the comprehensive studies of ecological risk levels should be conducted on the territories of liquidated mining enterprises. Continuous environmental monitoring of post-industrial areas will allow citizens to effectively exercise their rights to environmental security, require improvement of environmental quality, and force the management of mining enterprises to take active actions for prevention of environmental threats and neutralisation of various ecological risks.

In summary, the solving of ecological problems of post-industrial territories will facilitate stable operation and development of mining regions of Ukraine.

REFERENCES

Sliadnev, V.A. 2001. *Factors of Impact of Mass Closure of the Mines on Ecological and Geological State of Donbas.* The Coal of Ukraine, 7: 18-20.

Hoshovski, S.V. 2000. *Hydrogeological and Geochemical Problems at Liquidation of the Coal Mines.* The Coal of Ukraine, 7: 37-38.

Phillip, Pack. 2009. *Risk Assessment in Donieck Basin: Closure of Mines and Waste Heaps.* Prepared for UNEP, GRID Arendal: 21-22.

Informational Yearbook regarding Activization of Hazardous Exogenous Geological Processes in Ukraine, according to the monitoring of EGP. 2011. Kyiv: Public Service of Geology and Mineral Resources of Ukraine, State Research and Production Enterprise "State Information Geological Fund of Ukraine" (ill.): 88.

Koskov, I.G., Dokukin, O.S. & Kononenko, N.A. 1999. *Conceptual Foundations of Ecological Security in the Regions of Mines Closure.* The Coal of Ukraine, 2: 15-18.

Kasymov, O.N., Kasianov, V.V. & Radchenko, V.V. 2001. *The Experience and Perspectives of Use of Methane Released from Abandoned Mines.* The Coal of Ukraine, 4: 38-40.

Gorova, A.I., Kulyna, S.L. & Shkremetko, O.L. 2011. *The Analysis of Ecological Situation in the Mining Regions of Ukraine.* Ecological Bulletin, 3: 10-12.

Gorova, A.I., Ryzhenko, S.A. & Skvortsova, T.V. 2007. *Guidelines 2.2.12-141-2007. Survey and Zoning by the Degree of Influence of Anthropogenic Factors on the State of Environmental Objects with the Application of Integrated Cytogenetic Assessment Methods.* Kyiv: Polimed: 35.

Guidelines for the Assessment of the Environment Quality by the State of Living Beings (Estimation of Stability of Living Organisms in terms of the Asymmetry of Morphological Structures). 2003. Moscow: Approved by the Resolution of RosEcology, 460: 25.

Geomechanical Processes During Underground Mining – Pivnyak, Bondarenko, Kovalevs'ka & Illiashov (eds)
© 2012 Taylor & Francis Group, London, ISBN 978-0-415-66174-4

The mechanism of over-coal thin-layered massif deformation of weak rocks in a longwall

V. Bondarenko & G. Symanovych
National Mining University, Dnipropetrovs'k, Ukraine

O. Koval
LLC «DTEK Sverdlovanthracite», Sverdlovs'k, Ukraine

ABSTRACT: The formation process of zone of articulated-block movement of weak rock layered massif of over-coal strata during the inclined coal longwall mining is considered.

The features of formation mechanism of zone I made of over-coal strata, broken by vertical fractures in the area of tensile stresses, which forms so called the area of articulated-block rock displacements, are analyzed. (Zborshik & Nazimko 1991; Savost'yanov & Klochkov 1992; Akimov 1907; Zborshik & Nazimko 1986). On a Figure 1, the qualitative picture of rock block force interaction of the first layer of the zone I, the thickness of which is m_1^I with rocks of zone II of disorderly caving, from one side, and rock block of the second layer with thickness m_2^I, from another, is shown. On the basis of classical concepts, deformation behavior of rock layer does not allow it to keep integrity while the continuous movement goes down till the beginning of contact with soft and expanded caved rocks of zone II. Therefore, in a rock layer, on sections of maximum bending moments $(C_1 - C_1')$ and $(D_1 - D_1')$ normal-to-bedding fractures of disruption made by the activity of stretch horizontal stresses as a result of low rock mass resistance to such type of loading occur. In the rest (according to thickness of rock layer m_1^I) of section part compression stresses σ_x^I continue to be, because of which the rock block keeps some resistance to over-coal strata movement. During the disturbance of rock volume near considered section due to tight condition of deformation the section keeps residual resistance to ground pressure, and formed quasi-plastic joint allows rock block to displace along axe Y on required quantity. Thus, expansion-articulated system possessing certain bearing capacity, the quantity of which extremely depends on a local field of stresses around quasi-plastic joint, is formed from rock blocks.

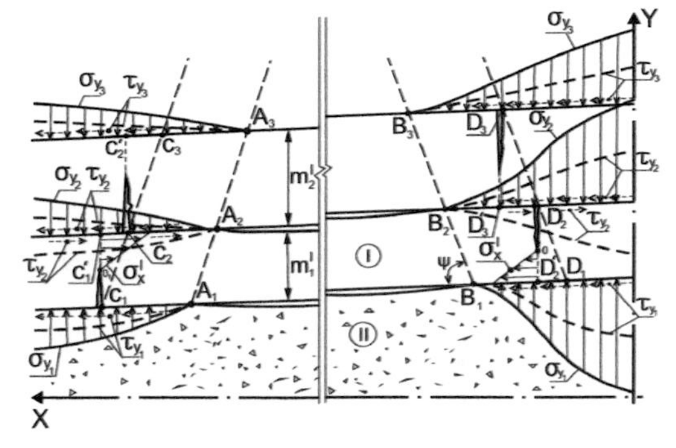

Figure 1. The mechanism of deformation of articulated-block expansion system of zone I rocks.

The mechanism of quasi-plastic joint formation and operation, on our opinion, is presented by following (Figure 2). During the bending of rock layer of the zone I with thickness m_1^I its neutral axe X_N moves towards to the activity of compression stresses σ_x^I by the reason of objectively excess of rock deformation modulus on compression and modulus of deformation on tension. On a site $0 - Y_2$ stresses σ_x^I are absent as a result of fracture formation; on a site $0 - Y_1$ compression stresses σ_x^I, distributed according to the law, closing to linear (Pysarenko 1979) along coordinate Y, with maximum σ_{x1}^I on the surface of rock stratum Y_1 (curve 1), operate. On the contact of rock layer in vertical direction stresses σ_{y1}^I operate from the side of caved rocks of the zone II. Using well-known for rocks, the Coulomb-Moor theory of strength (Ruppeneit 1954), maximum quantity of stresses σ_{x1}^I will be

$$\sigma_{x1}^{I} = \sigma_{com}^{I} + \frac{1+\sin\varphi^{I}}{1-\sin\varphi^{I}}\sigma_{y1}^{I}, \qquad (1)$$

where σ_{com}^{I} – limitation of rock compressive strength of the first rock layer of the zone I; φ^{I} – internal friction angle of rocks.

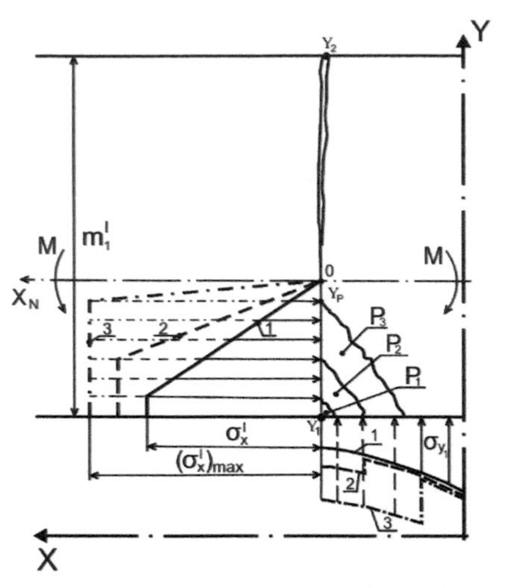

Figure 2. Mechanism of quasi-plastic joint formation on boundaries of the rock block of the zone I.

During further deformation of rock layer under the influence of ground pressure its disturbance occurs in the area of coordinate Y_1. In the first approaching the process of destruction can be presented as the formation of prism's chip P_1 (Ruppeneit 1954), which under the influence of stresses σ_{x1}^{I} is going to move to the side of caved rocks. This displacement is opposed by rocks of the zone II and stresses on the contact with prism's chip σ_{y1}^{I} are increased (curve 2) that causes the increase of rock resistance to stresses σ_{x1}^{I}, according to condition (1). At the same time rock block is turned relatively to point 0 on the angle, exceeding its possibilities at elastic-plastic deformations. In this case the process of chipping rock prisms will develop until the moment, when another end in the worked-out area (Figure 1) of rock block will settle on caved rocks of the zone II. Due to displacement of rock prisms stresses σ_{y1}^{I} are raised and stresses σ_{x1}^{I} are increased accordingly, and the form of curve σ_{x}^{I} approximates to rectangular (curve 3), that is given in the name of "quasi-plastic joint". Coordinate Y_P, characterizing the moment of joint resistance, is determined by the rotation of rock block relatively to point 0 till the moment of its contact with rocks of the zone II, disorderly caving.

The second parameter characterizing the moment M of quasi-plastic joint resistance is the maximum of horizontal stresses (σ_{x1}^{I}), the quantity of which in the first approaching can be determined by the method (Ruppeneit 1954), subject to the compatibility principle of prism's chip movements P_i and contacting with it rocks of the zone II.

The second feature of the zone I of articulated-block rock movement is a local contact of rock layers with each other and the formation of cavities between adjacent layers in the central part of space between blocks occurs (Figure 1). Taking into account not only vertical, but also horizontal rock shifts of over-coal mass analyzing feature sufficiently influences on the parameters of force interaction of rock blocks composing the zone I.

In term of the local contact we mean the first rock block interaction with caved rocks of the zone II and the interaction of the rest (at thickness of the zone I) of rock blocks between each other in the place of quasi-plastic joints basically, which are the boundaries of these blocks.

Generally acknowledged fact is the formation and the opening of normal to stratification fractures, in the area of tensile stresses activity, occur angularly of complete displacements ψ, the quantity of which in Donbass region for flat-lying seams is estimated 70-75° ($C_1 - C_2 - C_3$ and $D_1 - D_2 - D_3$ are shown by dotted line on Figure 1). Deviation of the normal to stratification ψ is obviously linked with the parameters of jamming rock layers on the boundaries of the worked-out area.

Rock layers in the zone I form articulated-block expansion system consequently from a rock layer near the zone II to long-distance on the boundary of a zone of slow rock lowering. Also in each following rock layer the cross-section with a fracture, which is appropriate to the maximum of bending moment, moves to the worked-out area. It depends, on our opinion, on soft jamming rock layer, caused by intensive displacement and partial softening of rocks in the zone of ground pressure activity on the boundary of the worked-out area.

To examine the process in detail, the formation of non-rigid jamming for the first articulated-block rock layer with thickness m_1^{I} has been analyzed (Figure 1). Rock layer, which is deformed by ground pressure interaction starting from point B_1, undergoes the reaction σ_{y_1} from a caved rock side of the zone II. This part of the curve of the reaction σ_{y_1} deviates to worked-out area due to increased deformability of caved rocks of the zone II. Summary activity of stresses σ_{y_1} creates the bending moment pointed at the increase of rock layer stability and moves the maximum of the bending moment to the side of worked-out area.

The second rock layer with thickness m_2^I, which is deformed together with the first layer, undergoes similar (to a certain extent) geomechanical processes. To the right of point D_2 (Figure 1) due to combined vertical movement of the first and the second layers along the contact between them normal stresses σ_{y_2} operate. To the left of point D_2 due to the console turn of the first layer the contact of two layers has to disappear and the quasi-plastic joint will be formed in the second layer along a line $D_2^I - D_2$. However, mine investigations (Akimov 1970; Zarya & Muzapharov 1966) and simulation of equivalent materials (Zborshik & Nazimko 1991; Savost'yanov & Klochkov 1992) show that the quasi-plastic joint occurs along a line $D_3^I - D_3$ shifted to the side of worked-out area relatively to the line $D_2^I - D_2$ in the second rock layer. We believe it is conditional by the following reasons. From one side, the rotation of rock block of the first layer is realized not around the point D_2, but around the point 0 on the neutral axe of the layer (Figure 1 and 2). From another side, during the formation of articulated-block joint expansion system in the first rock layer its partial unloading happens due to ground pressure. It also provides moving rock block surface of the first layer towards to the second deforming rock layer.

Mentioned factors provide for the contact of layers on the line $B_2 - D_2$ and the normal stress activity σ_{y_2} which converts the diagram of bending moments in the second layer as the maximum of bending moment moves to the worked-out area, where the formation of the quasi-plastic joint of the second rock layer of the zone II occurs.

Physically analogous geomechanical processes occur on other ends of rock blocks of the zone I layers located in the worked-out area (Figure 1, lines $C_1 - C_2 - C_3$, $C_1 - C_1'$ and $C_2 - C_2'$). In results, in the zone I the articulated-block expansion system is formed. It has certain bearing capacity and is characterized by local contact of adjacent layers in the place of quasi-plastic joints, situated along the height of the zone I with an angle of complete displacements ψ. In the first approaching the boundaries of cavities between layers (Figure 1, lines $A_1 - A_2 - A_3$ and $B_1 - B_2 - B_3$) which are also situated angularly ψ, as the mechanism of its formation is similar to the mechanism of the local contact formation in the place of quasi-plastic joints.

The third feature of articulated-block rock movements of the zone I is the development of horizontal rock layer movements relative to each other. In works (Zborshik & Nazimko 1991; Zborshik & Nazimko 1986) during modeling of equivalent elements and finite element method the laws of horizontal rock layer displacements of over-coal massif have been determined: rocks of the immediate and the main roof displace towards to coal seam. Therefore, the quantity of displacements is decreased during the movement in direction to day surface and at height $(6-15)\,m_y$ of coal seam thickness the direction of horizontal rock layer displacements is changed opposite, i.e. it goes towards to the worked-out area. The mechanism of influence of horizontal rock layer movements on its stability has been considered by using the scheme of Figure 1. It is set that the horizontal movements of the first layer (thickness m_1^I) of the zone I, moving towards to coal seam, are bigger than the second one and have $(0.05-0.30)\,m_y$ (Zborshik & Nazimko 1991; Savost'yanov & Klochkov 1992; Zborshik & Nazimko 1986). While such quantity of neighboring rock layer displacements relative to each other occur its connection is broken and layers interact between each other at stratification plane only by means of friction forces $\tau_{y_i}(x)$ originating due to the activity of normal stresses $\sigma_{y_i}(x)$ at plane i. As the first rock layer of the zone I displaces to coal seam side, it exposes by reactive tangential stresses $\tau_{y_i}(x)$ from the side of caved rocks of the zone II, towards to the worked-out area. Summary action of tangential stresses $\tau_{y_i}(x)$ on a site $B_1 - D_2'$ creates the moment relative to the point 0 of the rock block turn, directed at holding of further displacements to caved rocks of the zone II. Thereby, due to tangential stresses τ_{y_1} so called "recovery" moment is formed raising the stability of articulated-block expansion system of the first rock layer and increasing the resistance to the process of over-coal mass movement.

On a contact of the first and the second rock layers of the zone I tangential stresses of friction τ_{y2} also operate. Its direction of influence on the first rock layer is determined by primary quantity of horizontal displacement of the first layer with respect to the movement of the second one and it is shown on a Figure 1. Tangential stresses τ_{y2} operate towards to the worked-out area and its sum along the contact $B_1 - D_2$ creates the moment relative to the point 0 (decreasing the stability of the first layer rock block), which can be conditionally called "tipping". Relation of "tipping" and "recovery" moments depends on coordinates of the point 0 and total quantities of tangential stresses operating along lines $B_1 - D_2'$ and $B_1 - D_2$ (Figure 1).

For the second rock layer (thickness m_2^I), and as for following layers of the zone I, the moment of acting tangential stresses τ_{y2} and τ_{y3} is only "tipping" in place of quasi-plastic joint from coal seam side (Figure 1). It is conditional by decreasing and changing the direction of horizontal movements of rock layers at height of the zone I.

On the boundaries of rock blocks of the zone I, located on a worked-out area side, tangential stresses along stratification also create "tipping" and "recovery" moments relative to the point 0_1 in the first rock layer. In the second and following rock layers the moment is only "recovery" (Figure 1).

On our mind, above mentioned feature of tangential stress activity is the factor increasing the length of contacts (e.g. $C_2 - A_2$ and $B_2 - D_2$) of rock blocks of neighboring layers: the length of $C_2 - A_2$ is increased due to rock block deformation of the second layer by "tipping" moment; the length of $B_2 - D_2$ is increased due to bending rock block by "recovery" moment. In result, quasi-plastic joints (at height of the zone I) take place angularly ψ to stratification and an arch of the zone I at some distance from a coal seam closes forming an unloading zone.

Summing up the features of mechanism of rocks deformation of the zone I three main conclusions can be made, which were taken into account during the research of over-coal mass strain-stress state.

First – due to the quasi-plastic joint formation rock layers in a view of articulated-block expansion system keep the part of bearing capacity and resist to ground pressure.

Second – processes of forming quasi-plastic joints, local contacts and cavities between rock layers have to be considered interconnectedly, with the influence of abundant pressure and unloading zones that requires entering additional conditions and criteria during the research of strain-stress state of considered system at all.

Third – the limitation of height of the zone I is conditioned by forming an arch having different forms, geometrical parameters of which are gotten from results of system's strain-stress state research. All features of deforming the zone I influence on the process of forming the arch: the moment of quasi-plastic joint, the force of support due to rotation and unloading lower rock block, the moment of tangential stress friction activity in zones of local contacts at stratification of rock layers.

REFERENCES

Zborshik, M.P. & Nazimko, V.V. 1991. *Mine working protection of deep mine in zones of unloading.* Kyiv: Tekhnika: 248.

Savost'yanov, A.V. & Klochkov, V.G. 1992. *Rock mass control.* Kyiv: UMK VO: 276.

Akimov, A.G., Zemisov, V.N., Kantsel'son, N.N. & others. 1970. *Rock movement during underground coal and shale mining deposits.* Moscow: 224.

Zborshik, M.P. & Nazimko, V.V. 1986. *Mechanism of rock movement and stress redistribution around mine workings maintained in caved and compacted mass.* Scientific-technical collection "Mining mineral deposits": Vol.73: 48-52.

Pysarenko, G.S. 1979. *Material resistance.* Kyiv: Vysha shkola: 696.

Ruppeneit, K.V. 1954. *Some questions about rock mechanics.* Moscow: Ugletechizdat: 364.

Zarya, N.M. & Muzapharov, F.I. 1966. *Scheme of mechanism of rock mass movement during flat coal layer extraction by one longwall.* Coal of Ukraine: Vol.12: 9-12.

Zborshik, M.P. & Nazimko, V.V. 1986. *Laws of horizontal displacements of rock mass during flat layer extraction*: Vol.5: 18-22.

Geomechanical Processes During Underground Mining – Pivnyak, Bondarenko, Kovalevs'ka & Illiashov (eds)
© 2012 Taylor & Francis Group, London, ISBN 978-0-415-66174-4

Methodological principles of negative opencast mining influence increasing due to steady development

I. Gumenik, A. Lozhnikov & A. Maevskiy
National Mining University, Dnipropetrovs'k, Ukraine

ABSTRACT: The article deals with the liquidation of negative opencast mining works influence taking into account steady development of mining region. The data is given about annual areas that involved in opencasts operation and spoil dumping. It is shown that basis of strategy is the establishment of top and durative aims. The principles of information support creation are shown. Attention is given about evaluation of the changes in status of ecosystem is recommended to provide in definite sequence.

The process of getting by the way of opencast mining is one of the main courses for violation and pollution of the earth surface, open and ground water and atmosphere. As a result, there in spoil dumps of mining factories in Ukraine it is accumulated more then 8 milliard ton of waste. About 360 thousand hectares of land are disturber by mining operations. Yearly from 8 to 10 thousand hectares of land are taken for opencasts and spoil dumps. The dimensions of territories, which suffer the negative influence of opencast mining, exceed the very opencasts and spoil dumps' dimensions in 10-15 times. It yields to inefficient ecologic system's formation in these mining regions.

Specially ought to be noted the negative influence of mining operations on the environment during the sheet ground development in Ukraine. It was already mentioned that during the getting of 1 ton of mineral deposits, it is necessary to produce from 3 to 18 cubic meters of overburdens in these occurrences. It yields not just to large amount of waste formation but also to intensive violation of lands. For example, during the getting of 1 million ton of manganese ore it is disturbed from 16 to 30 hectares of land, for brown coal – from 6 to 12 hectares, for sulphur ore – from 24 to 35 hectares.

In respect that most of occurrences are placed in high-leveled agriculture zones, it is important to note that the spread of damage for agricultural complex achieves the significant ranges, which exceed the costs for lands' amortization and recultivation processes in several times. Over a relatively short period (last 25 years), the agricultural holdings' per person square in Ukraine decreased from 1.08 to 0.78 hectares.

Yearly during the opencasts' exploitation in Ukraine about 17 milliard cubic meters of opencasts' water is exhausted on day surface. Nearly half of this water is brackish water, moreover, 10% of sewage are considered to be of acid type (hydrogen ion concentration pH is below 6.5). Depending on mining and geological conditions, such water includes chlorides, sulfates, calcium and magnesium salines, ferrum and microelements. In common, every year about 80 thousand ton of suspended materials and salines are discharged by opencasts in natural reservoirs, what yields to breakdown of reservoirs' sanitary-hygiene state. Besides, during the opencasts' mining promotion, are significantly damaged separate components of hydrosphere.

It is enough to note that when the is height 150-200 meters, opencast's cones of influence spread over a distance more than 15 kilometers, thus at achieving of planned depth 450-500 meters of opencasts in Krivbass, the radius of cones of influence will work out 40-50 kilometers (or even more) when its square will work out 2000 square kilometers, what will yield to regional violation of open and ground water's regulations. Consequently, the regulations of minor rivers, lakes and other small reservoirs are disturbed; water-reducing and water-removing works cause the land's deformation; formation of cauldrons and flashes on the land surface; dissolution of solid of good solubility. Because of underside or outside saline-water encroachment, changes in ground waters' nutritional conditions can yield to durable changes of their quality.

The third component of natural environment, which suffers the significant influence of opencast mining, is atmosphere. It is known that atmospheric air is one of the essential elements for natural environment. For that reason, such quick paces of industrial production development and increasing scales of human's influence on the natural environment demand the close attention to the protection of atmospheric air. Yearly the atmospheric air gets dozens and hundreds million ton of hydrocarbons, nitrogen oxides, sulphur dioxides, industrial dust etc. The analyses of toxic material's content in the atmosphere near by mining plants' location area showed that the most polluted air is near by coke-chemical and primary metals establishments, which throw airwards not just industrial dust, but also SO_2 and CO, such toxins as phenol, ammonia, hydrogen sulphide, phenyl hydride etc. Atmosphere pollution form the negative influence not just on flora and fauna, but also promotes the premature breakdown of mining transferring equipments. The rate of pensioned off underreinforced equipment in these regions makes 40%, which includes 7% because of corrosion. The cut of periods between repairs and costs' increase fix the influence of atmosphere pollution on the equipments' wearing.

That is why, the constant development of mining complex's territory is possible just on the grounds of well-balanced development of human and environment. At that, all the principles and methods of management strategy's development in economical and social spheres, also in region's environment system, were subjected by new requests. The basis of strategy is the establishment of top and durative aims, the accounting of the environment's state and the definition of its usage order.

All that demands the changes in mining complex of Ukraine.

First of all:

1) creation of high-efficiency competitive enterprises, which will provide the supply of demand for first-grade products;

2) production infrastructure's renewal;

3) determination of investment priorities to involve native and foreign investors;

4) development and introduction of ecologically-oriented resource-recovery technologies in opencast mining;

5) solution of problems concerning closure of uneconomical and conservancy of unpromising mining institutions.

The integrated use of occurrences and mineral raw materials, and cut of opencasts' mining influence on the natural environment by the way of development and elaboration of ecologically-oriented resource-recovery technologies of occurrences' exploitation is the core direction of mining complex's development in Ukraine. To solve such problems it is necessary to identify the degree of negative influence on the environmental components of separate parts of technical and industrial opencast's objects. The role of information support increases in such conditions.

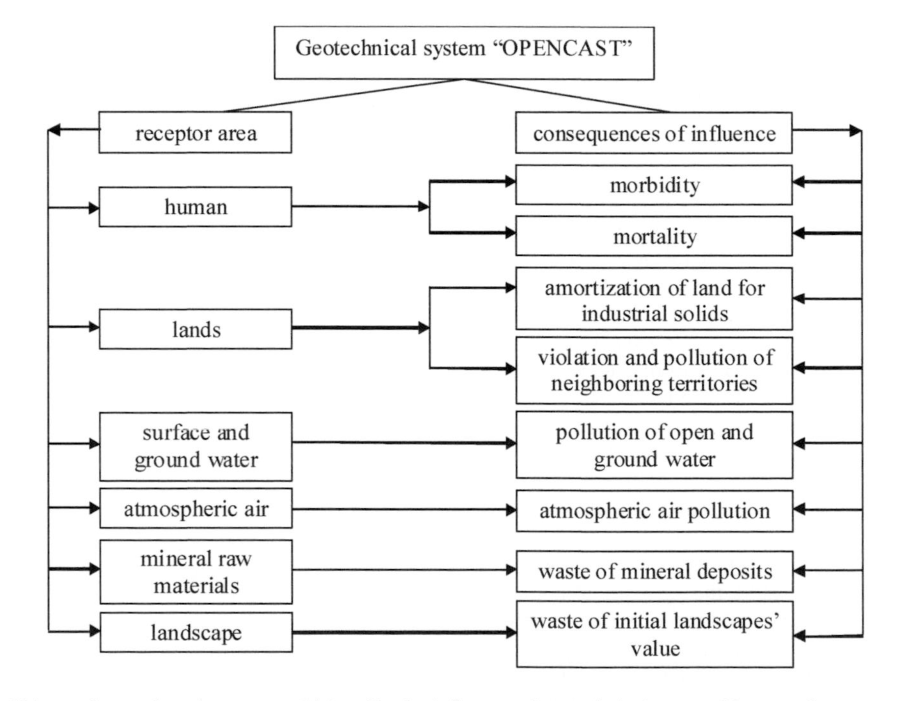

Figure 1. Objects of natural environment, which suffer the influence of geotechnical system "Opencast".

It is necessary to follow such main principles during the elaboration of information support for stout ecosystem's creation:

a) information support should be created as a human-computer system, wherein a personal computer functions as counter, and human – as a person to make an expert decision;

b) providing of system and information's accord, in other words, the usage of organizational and methodological statutes' complex, which provide the entirety of a system and subsystem during all the periods of creation, function and development;

c) openness and development of information support. Every step, which provides creation and function of the system, must provide the possibilities for addition, improving and renewal of subsystem; for replacement of every calculation module by the new one; for inclusion of additional accounts into the subsystem providing that entrance and exit information will remain the same. The principle of information support's development may be considered as a possibility to involve the subsystem or system into the more complex and high-leveled system;

d) compatibility with other traditional methods of information systems' designing.

Such objects of environment, which suffer the influence of geotechnical system "Opencast", are presented at Figure 1.

In the bottom of ecological situations' evaluation on regional and micro-regional level is placed the notion of territorial geoecologic norm and the optimal union of natural geosystems and imitative geotechnical systems, which all provide the regulation of ecology, along with economical and social achieving of values. Ecologic situation includes the consideration of relations between three blocks of system, they are natural ecosystems, human (humanity) and technosystems. Coincidently, every of them is as an object and as a subject of evaluation meantime. Thus, to find out the ecologic situation it is necessary to analyze the territorial totality of natural, imitative and industrial geosystems' statuses from all the sides. Figure 2 represents the principles of creation of information support.

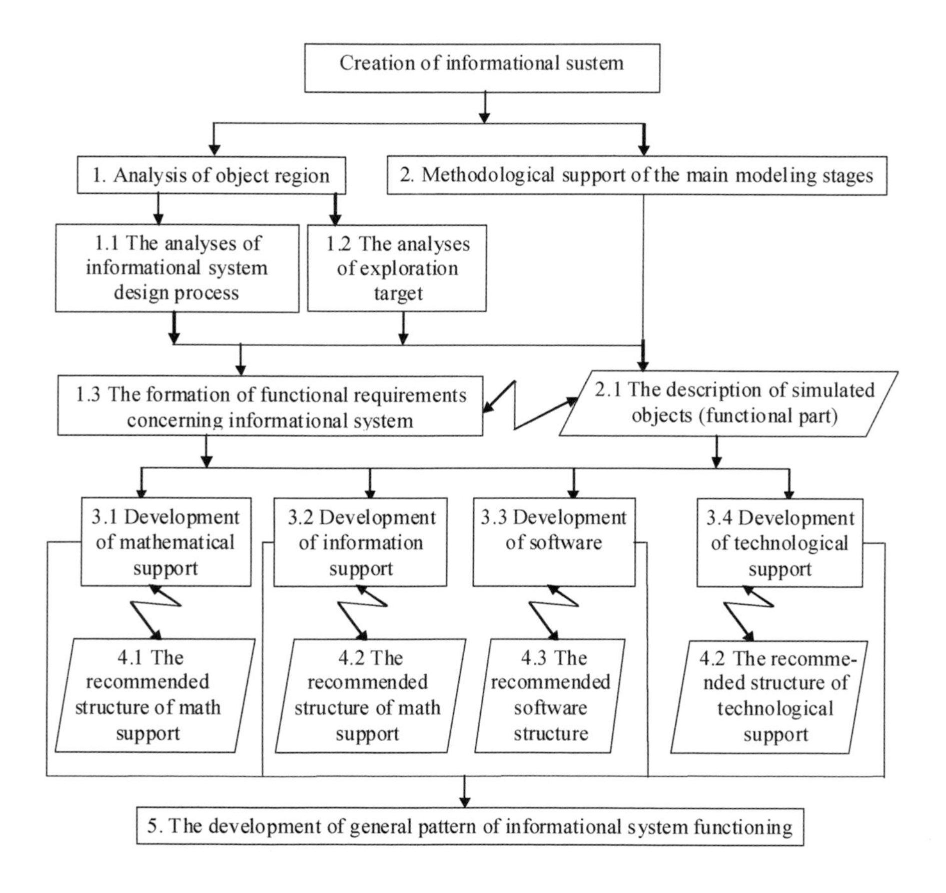

Figure 2. The framework of creation of information support.

Subject-object ecological terms reflect the environmental status concerning subjects, in other words, features of objects are evaluated from the point of view of their real or possible influence on the subjects. These terms characterize the subjects' statement. As a rule, during the evaluation of situation are used present restricted data concerning these environment and subjects' status but without efforts to evaluate the sufficiency of these data.

Dynamic aspects of informative support also ought to be noted. It is important to know not just the state of situation in definite time, but also anterior and expected situation and all possible tendencies of their development. Processes and phenomena, by which the ecological situations are identified, can be of constant, periodic, no repeat and episodic type. At that, systems' corresponding reactions are not as a mirror imaging of the influence, so as systems show spring power, speed of response and possibility to provide chain reactions etc. It should not be overlooked that the reaction of system on the influence is used to be late due to sluggishness. As a result, there in ecosystems are identified different tendencies of ecological status, for example balanced, progressive, chronic, pulsed, damping etc.

Evaluation of the changes in status of ecosystem (EC) is recommended to provide in definite sequence (Figure 3).

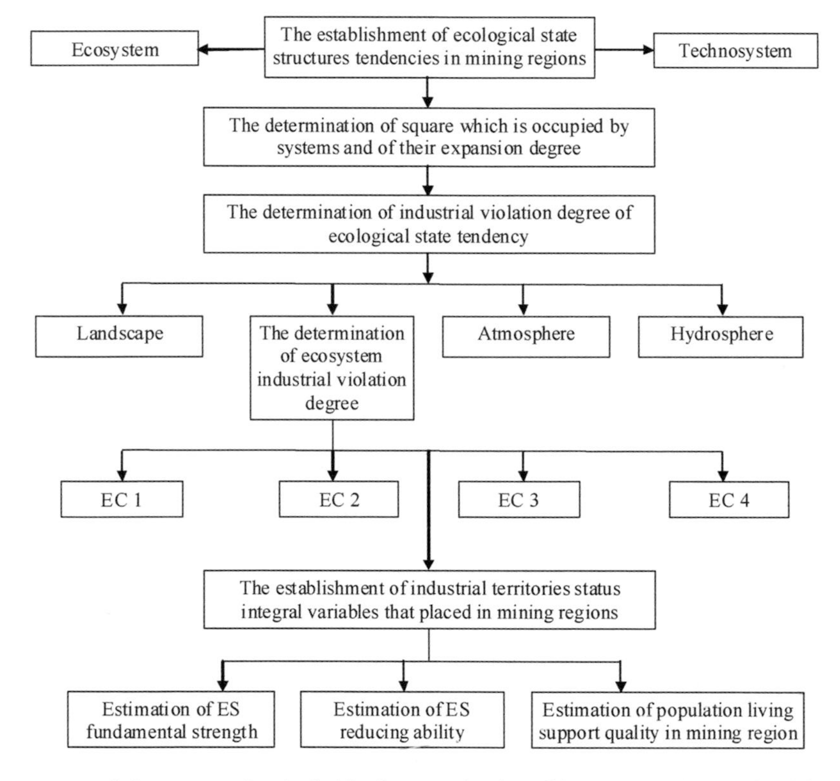

Figure 3. The recommended sequence of ecological-landscape estimation of the ecosystems' status in mining regions.

Ecosystems, which are common for our country and represent the main objects in ecologically-landscape estimation of the territories in mining regions, are introduced by agricultural croplands (EC_1); forest lands (EC_2); agricultural lands – meadows and green lands (EC_3); water resources (EC_4).

Possible ecological-landscape directions of rehabilitation (renewal) of damaged by opencast mining ecosystems in mining regions are: 1) creation of new ecosystems (EC_1 and EC_3); 2) building of technical constructions (car parks, garages, storages etc.); 3) creation of industrial occurrences; 4) building of reservoirs for industrial growth of fish; 5) construction of break areas; 6) creation of natural reserves etc.

Proceeding from fixed dynamics of agricultural holdings' squares and of ecosystems' characteristics, which are typical for Ukraine, it is possible to identify that the main prime direction of territories' renewal, which were damaged by development operations, is reconstruction of agricultural holdings, in other words, of ecosystems EC_1 and EC_3. Noted ecosystems can be placed either within the bounds of mining and land allotment or out of their bounds. In the first case, they represent new ecosystems, which are characterized by significantly less part of secondary industrial productiveness in comparison with preindustrial one. In the second case, EC_1 and EC_3 did not suffer the industrial physical influence. Their mutations are just of geochemical character, which demands the special preventive actions to avoid the negative development of influence of industrial processes.

Usage of suggested methodological statutes, principles and recommendations concerning the development of informative support of stout ecosystems' creation, ecological-landscape estimation of the ecosystems' status in mining regions will allow to achieve continuous and systematical forecasting of ecologic situation in the definite mining region. In the basis of current and long-term forecast of stability of ecosystems of mining regions are put the principles of determinations of rates' quality and quantity, which take into account the industrial changes of environment at all the stages of life cycle of occurrences' exploitation.

Geomechanical Processes During Underground Mining – Pivnyak, Bondarenko, Kovalevs'ka & Illiashov (eds)

Hydrodynamic cavitation in energy-saving technological processes of mining sector

Y. Zhulay
Institute of transport systems and technologies of National Academy of Sciences of Ukraine, Dnipropetrovs'k, Ukraine

V. Zberovskiy
Institute of Geotechnical Mechanics of the National Academy of Sciences of Ukraine, Dnipropetrovs'k, Ukraine

A. Angelovskiy & I. Chugunkov
POJSC "Krasnodonugol", Krasnodon, Ukraine

ABSTRACT: The estimation of efficiency of application of the ways and methods that realize the periodically stalling cavitation while drilling a well, expansion of casing pipes diameter increase, declogging of the water wells and hydroimpulsive loosening of coal seams has been carried out. It is noticed that intensification of technological processes and decrease of energy consumption are achieved by the transformation of a stationary fluid flow in a discrete-impulsive flow and by transfer of the high-frequency hydroimpulsive vibrations directly to the tool or rock massif at the surface of a well.

Implementation of energy-saving technologies especially in mining sector at present and in the nearest future has a definitive meaning in agricultural activity of the enterprises. Perspective technological process in this direction is the gain and realization of discrete-impulsive energy of high power in the fluid flow. The device converting stationary fluid flow into the periodically stalling cavitation mode is cavitational generator. This device allows to change character of the fluid flow only at the expense of geometrical and mode parameters without using any moving parts and additional sources of energy. Venturi tube-like generators create impulses of fluid pressure that excess output pressure of the pump unit in several times.

Examples of practical implementation of cavitational generators for intensification and energy-saving of various technological processes are shown in the work (Pilipenko, Man'ko & Zadontsev 1998). In metallurgy – for removal of scale during hot rolling of metal. In machinery building – for cleaning the surface from sharp edges, rust and other. In these cases energy-saving decreases down to 30%. In the chemical industry – during emulsification of fluids and production of fine-grained systems, energy-saving decreases by more than 50%.

The aim of the given work is to describe conceptual approaches for usage of hydrodynamic cavitation modes and evaluation of effectiveness of hydroimpulsive influence technical means in mining sector.

One of these trends is implementation of the periodically stalling cavitation in the technological fluid flow with conversion of pressure pulses into the working tool vibration load.

The second trend is the hydroimpulsive influence directly on the rock in the filtration part of the blast-hole or a well.

Technical means realizing these approaches have gone through a full complex of investigations from laboratory to experimental-industrial tests.

1 DRILLING OF WELLS

The most effective drilling method in the hard rocks is percussion-rotary method. The biggest disadvantage of this method is the presence of moving parts, springs and rubber couplings that get worn fast during exploitation. At this, the drilling effectiveness depends on the adjustment precision of the moving parts, and the period of equipment control and revision does not excess 25 hours.

Perspective trend that excludes all these disadvantages is the creation of hydrodynamic drilling tool strings (TS), pictures of which are shown on Figure 1: a) without core drilling; b) core drilling.

In the technological process the generator 2 converts stationary flow into discrete-impulsive, and the pulses energy of drilling fluid – into high-frequency longitudinal vibration acceleration of the breaking tool 3. Mentioned DS's have been studied under experimental conditions (Dzoz, Zhulay & Zapolskiy 2005) and under construction of hydrogeological and prospecting boreholes (Dzoz 2008).

Figure 2 presents copy of the oscillogram area with record of hydrodynamic parameters of DS experimental sample during its tests at the hydraulic bench unit. The parameters are: P_1 – input pressure in the cavitational generator; P_2 – pressure in a pipeline behind cavitational generator; a_1, a_2 – vibration acceleration in the sections before and after cavitational generator; a_3 – vibration acceleration on the breaking tool simulator body.

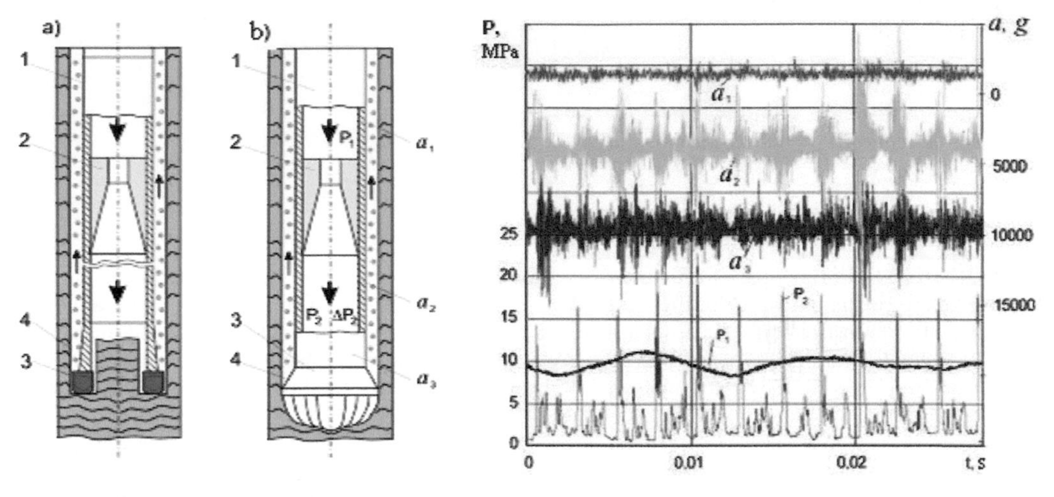

Figure 1. Principal scheme of hydrodynamic drill string (a) drilling with core extraction; (b) without core drilling: 1 – drill string; 2 – cavitational generator; 3 – breaking instrument; 4 – well.

Figure 2. Copy of oscillogram of parameters used during experiments of a test sample of hydraulic drill string on a hydraulic bench unit: $P_1 \approx 10$ MPa, $P_2 / P_1 \approx 0.123$.

The results of these tests have shown that in the DS pipeline behind the generator the mode of periodically stalling cavitation realizes with frequencies within the range from 70 to 3000 Hz and with swing of fluid pressure fluctuations ΔP_2 to $2.7 P_1$. The swing of fluid pressure fluctuations is the difference of maximal and minimal value of pressure in impulse $\Delta P = P_{max} - P_{min}$. Maximal values of vibrational accelerations on the breaking tool simulator body made up 1600 g (at $P_1 \approx 5$ MPa) and 2700 g (at $P_1 \approx 10$ MPa). It is established that drilling fluid delivery pressure increase leads to rise of vibrational accelerations and their frequency on the breaking tool.

Drilling efficiency with vibrational load on the breaking tool was confirmed during construction of hydrogeological boreholes of big diameter. The analysis of 190mm-diameter boreholes drilling has shown that while DS is working, drilling speed increases by 71.5% compared to rotary method. At this, the breaking tool wearing and energy consumption decrease down to 30%.

Based on the received results, the hydrodynamic DS's were created for drilling surveying wells with diameter of 76 mm. Correlation of these tests has shown that compared to hydraulic hammer – Г76ВО, hydrodynamic DS with drill bit -02И3 has provided 15.8% – increase of drilling speed and 13.2% – increase of the bit life. When used diamond bits of А4ДП type, the drilling speed increase made up 26.7% and drill bit life – 11.8%. The analysis of power loss for breaking tool rotation has shown their decrease during operation of hydrodynamic DS down to 20% at all drilling modes. Comparative analysis of working reliability of the hydrodynamic DS and hydraulic hammer Г76ВО has shown the following: during tests period 4 failures of hydraulic hammer was registered and 10 assembling-disassembling actions were done to adjust and remove tools. At the same time, there were no failures during tests of hydrodynamic DS. And by checking basic tools there was no sign of their wearing. Operational life of hydrodynamic DS has significantly excessed life of hydraulic hammer. Besides, DS has provided stabilization and stability of the drill pipes during drilling down to 522.5 m.

2 CASING DIAMETER WIDENING

"Weatherfopd" firm traditionally uses hydraulic method of casing diameter widening by injection of high-pressure fluid flow under the widening cone (cone is a tool used for pipes widening). It is known from the experience that around 50% of fluid energy is used on the pipes widening process itself. Rest of the energy is used to fight friction resistance. In order to reduce friction forces during cone movement, expensive MSDS lubrication is used with molybdenum additives that leads to significant economic cost. One of the ways to decrease friction resistance is an application of hydrodynamic vibrator (HDV) to impose vibration load on the widening cone.

The studies (Zhulay & Voroshylov 2010) have shown that percussion fluctuations of fluid pressure P_2 (Figure 3) occur at the exit of the generator. Such type of fluctuations is substantiated by the occurrence of periodically stalling cavitation at the flowing part of the generator. Vibrational accelerations realize at the widening cone (Figure 4).

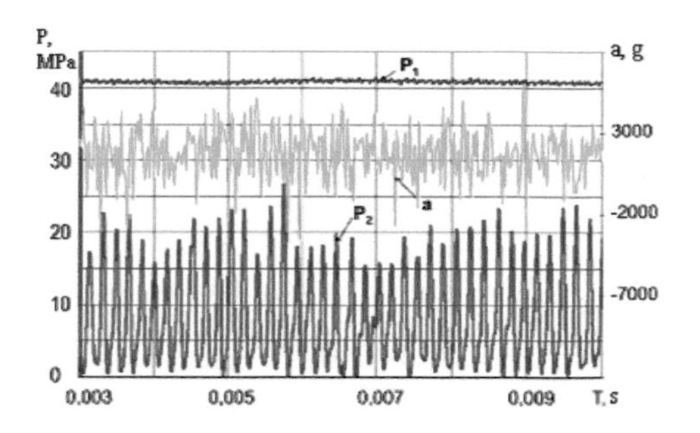

Figure 3. Copy of oscillogram of the HDV working parameters record $P_1 = 40$ MPa, $P_2 = 8.2$ MPa.

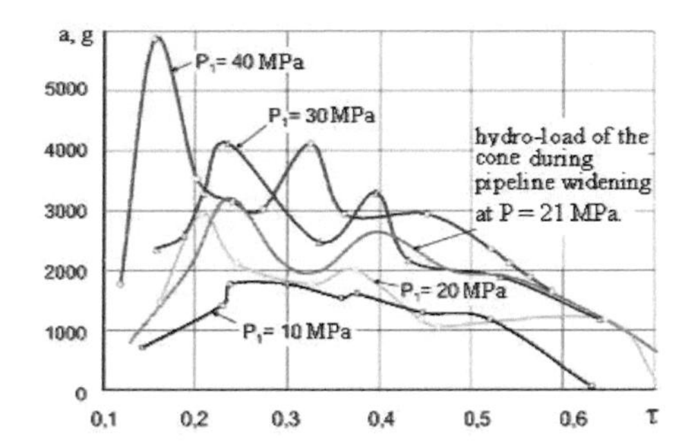

Figure 4. Dependence of vibrational load on the cone body on the HDV working mode.

Vibrational loads both in an axial and in radial directions realize with the range of the cavitation parameter change – $\tau = 0.15...0.6$. Dependence of the cone vibrational load during pipeline widening under pressure $P_1 = 21$ MPa is shown at the same Figure.

Evaluation of HDV application efficiency in the technological process was carried out at the samples of certified steel pipes with external diameter of 133 mm and wall thickness of 6 mm under the following conditions:

– test 1 – static influence without lubrication;
– experiment 2 – static influence with MSDS lubricant;
– test 3 – hydroimpulsive method during which fluid pressure pulses realize at the generator exit;
– test 4 – hydroimpulsive method during which vibrational loads realize at the cone (Dzenzerskiy).

Figure 5 shows aligned copies of oscillogram areas of pressure mean values before the cone at various conditions of the conducted experiment. External diameter of the pipelines after the widening made up 150.18-150.62 mm with wall thickness 5.7 mm with ellipsoid shape of pipes within the range of 0.09-0.44.

The results of studies have shown that hydroimpulsive widening of pipes realizing vibrational load on the cone decreases friction resistance by up to 93% compared to widening by static influence. Pressure value, at which the cone starts moving with following pipe widening, decreases from 26.9 MPa to 14.4 MPa. Unlike static influence, hydroimpulsive method prevents cone jamming, occurrence of sharp edges and cold hardening, and decreases energy consumption down to 40%.

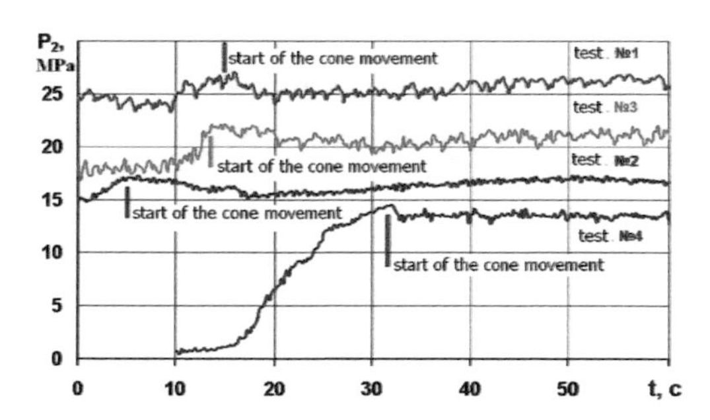

Figure 5. Aligned copies of oscillogram areas.

3 RECOVERY OF THE WATER WELLS PRODUCTIVITY

During the well functioning the clogging of its productive part occurs. To increase its debit or recover its functioning, it is necessary to conduct complex of measures directed to unclogging of near-the-well zone. There are various methods and means of influence on the well or host rocks are used for this: cleaning of well walls by roller bit, air-lift pump, mechanical influence by vibrators on the column with filter, creation of hydraulic thrusts during use of pneumatic (low-frequency) devices, acid influence on the filter, blast in the filter area or area of water-bearing horizon. In recent years, hydraulic whistles and magnet-strictional generators of ultrasound fluctuations have come into use.

It is worth mentioning that disadvantages of means and methods of wells unclogging define advantages of generator of fluid pressure fluctuations. There are no movable parts in the generator, it does not require additional energy source and it has longer operational life.

Figure 6 and 7 show schemes of surface equipment and cavitational generator unit (CU) location in the wells (Dzoz & Zhulay 2008). To control seams permeability of water-bearing horizon, such technical parameter as well debit is used.

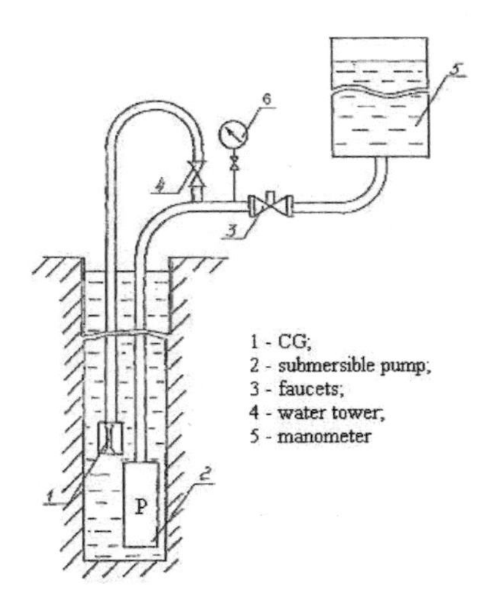

Figure 6. Scheme of a well unclogging with drive of cavitational generator powered from submersible pump.

Two wells operability with diameter of 406 mm and depth of 25 m was restored according to the technological scheme powering from submersible pump (Figure 6). Wells unclogging efficiency was being compared during the operation. Debit change was fixed at each measure before and after processing. At the beginning the wells were worked-out in the mode of water inflow through generator with closed faucet 3 and open faucet 4. Debit was raised by 40%. Then cleaning of the well from mud was provided by submersible water pump. Furthermore, at simultaneously open faucets 3 and 4, the pump was working for the generator and water outtake during 68 hours. Total debit increase has made up 200%

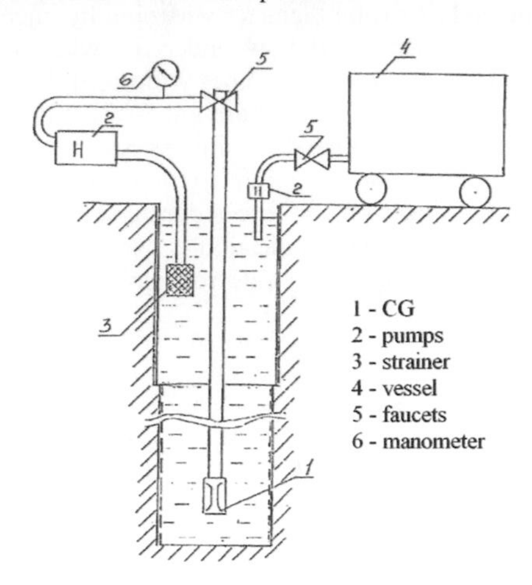

Figure 7. Scheme of a well unclogging with drive of cavitational generator powered free-running pump.

The boreholes with diameter of 160 mm and depth of 100 m (Figure 7) the water-bearing horizon was blocked by filter columns made in a form of perforated pipe. Unclogging was provided by interval-like movement of cavitational generator for full height of water-bearing horizon. It is connected with the fact that amplitude of pressure fluctuations decreases as it moves away from the generator exit. After influencing the boreholes by hydrodynamic method, their debit has increased up to 110%.

4 INTENSIFICATION OF COAL SEAMS HYDRO-LOOSENING

Up to now the basic trend directed to lowering dust formation and prevention of sudden outbursts of coal and gas is an injection of fluid into coal seams. The experience in the mining operations conduction allows to point out that application of methods based on static injection of fluid under conditions of big depths is low efficient. Widely used method of coal seams hydro-loosening is becoming not producible and does not provide coal humidification along full seam thickness.

Basic disadvantage of the given method is an uncontrollable process of fracture formation and fluid filtration along full thickness of layers and interlayers forming the seam. When filtration chamber is located within the unloaded zone the fluid freely filtrates along the cracks into the worked-out area. When the chamber is in the zone or behind the zone of an abutment pressure the hydro-fracture and water breakout occur along one of the layers of coal pack or also hydro sloughing of the seam side part.

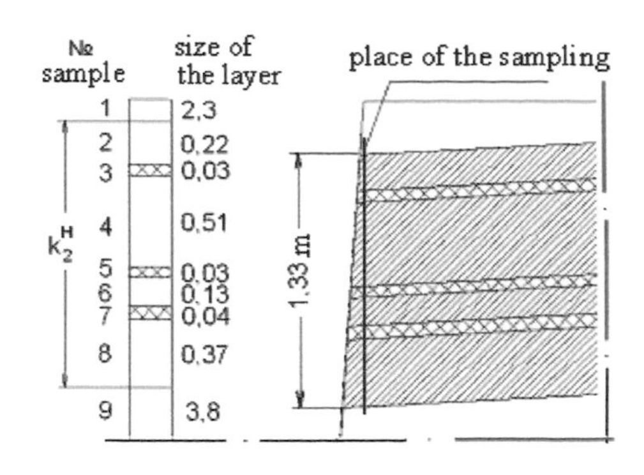

Figure 8. Structure of k_2^H seam at 617 m horizon of SE "Shakhtoupravleniye "Molodgvardeyskoye" POJSC "Krasnodonugol": 1 – seam roof; 2 – upper pack of seam; 3 – interlayer of coal slate; 4 – lower pack of seam; 5 – interlayer of coal slate; 6 – interlayer of coal seam; 7 – interlayer of coal slate; 8 – interlayer of coal seam; 9 – seam bottom.

One of the perspective trends allowing to exclude these disadvantages is an impulsive injection of fluid in mode of periodically stalling cavitation. At present, Institute of geotechnical mechanics of NAS of Ukraine together with POJSC "Krasnodonugol" leads mining-experimental works directed to research of hydroimpulsive loosening parameters of coal seams prone to outbursts – k_2^H, $k_2 + k_2^B$, i_3^1. These seams have complex multi-pack structure (Figure 8) and are considered to be prone to outbursts of coal and gas.

In addition, coal interlayers have various characteristics, construction and structure (Table 1). Hydro-loosening of these seams did not provide any expected results, thus during drivage of galleries by continuous miners, blasting works are used in mode of percussion explosion in dangerous zones.

Analysis of complex structure and properties of coal seams, for example $k_2 + k_2^B$, confirms that effective processing of a seam along its full thickness is practically impossible using normative scheme of hydro-loosening. In order to increase efficiency of hydro-loosening the unit of hydro-impulsive influence (further named as the device) has been developed (Patent 87038).

Hydrodynamic parameters of the unit have been investigated on a borehole model with fluid pressure at the exit of cavitational generator from 5.0 MPa to 30.0 MPa with fluid consumption being 40-60 l / min (Zhulay & Zberovskiy 2010). It is established that the periodically stalling cavitation provides working range of device according to cavitation parameter τ from 0.02 to 0.8. Right behind the generator, range of auto fluctuations makes up 1.8-2.4 of fluid pressure at the unit exit, and frequency – up to 12.0 kHz. Range of auto fluctuations in filtration part of the well decreases as it goes away from the generator diffusion cell and makes up 1.1-1.6 of fluid pressure at a distance of 1.5-2.0 m.

Table 1. Brief characteristics of coal in layers and interlayers forming the seam $k_2 + k_2^B$.

Number of test	Scheme of the seam	Layer thickness, m	Description of coal quality
1		0.30	Uniform, shiny, strong
2		0.1-0.15	Shiny, layered with interlayers of coal slate
3		0.20-0.22	Shiny, layered, soft
4		0.70	Uniform, shiny, strong
5		0.12	Uniform, shiny, mat in the layer roof, soft
6		0.18	Uniform, shiny with interlayers of mat, soft
7		0.05	Coal with inclusions of slate
8		0.15	Uniform, shiny
9		0.20	Uniform, shiny, strong

Mining-experimental works have allowed to evaluate efficiency of device application during hydro-loosening of complex structure coal seams prone to outburst. It is established that at rate of fluid injection equal to 40-60 l / min, the rational pressure of hydro-impulsive influence makes up 10-20 MPa. Signs of hydro-fracture and gas-dynamic phenomena do not occur when using technological scheme of hydro-impulsive loosening through boreholes with diameter of 43 mm and length from 6.0 to 7.0 m with sealing depth from 4.0 to 5.0 m. Unloading zone of the seam boundary part reaches 10 meters.

Efficiency comparison of static and impulsive modes of a fluid injection has shown that duration of a seam hydro-processing reduces to 50% during hydro-impulsive influence; fluid consumption reduces to 60%, safe zone of coal extraction increases from 6 to 10 m, energy-consumption decreases to 70%.

5 CONCLUSIONS

Cavitation phenomenon, studied in hydraulic systems as a negative and uncontrolled process, is efficiently used during intensification of technological processes in other branches of science and technique. Parameters of discrete-impulsive energy realization of big power within the fluid flow, considered in this work, allows to point out the following:

Efficiency of technological processes is reached by means of cavitation generators application that allow to transfer impulse energy from the source to the object of destruction with minimal cost. At this, there is no need for any additional costs. Stationary flow of the fluid converts into discrete-impulsive flow right in the unit and gets transferred to the tool or rock massif in form of high-frequency hydro-impulsive vibration.

Cavitational generator of the fluid pressure fluctuations has a line of advantages before other technical means of wave influence:

– simplicity of manufacture, absence of moving parts, additional sources of energy, duration of the resource, exclusion of the fluid fluctuations to the pump;

– construction of cavitation generator organically fits into various technologies and allows to intensify them at lower specific energy consumption and does not require labor-consuming redesign of the equipment.

REFERENCES

Pilipenko, V.V., Man'ko, I.K. & Zadontsev, V.A. 1998. *Cavitation self-oscillations intensify technological processes*. 1998. Proceedings of a Fluid Dynamics Panel Workshop. Kiev. Report 827: 32-34.

Dzoz, N.A., Zhulay, Y.A. & Zapolskiy, L.G. 2005. *Experimental evaluation of drilling string construction parameters influence with cavitation generator of a fluid fluctuations on level of vibro-load at a rock-breaking tool.* Materials of an international conference "Forum of miners-2005". Dnipropetrovs'k: NMU Vol. 2: 93-102.

Dzoz, N.A. & Zhulay, Y.A. 2008. *Intensification of drilling processes using hydro-dynamic cavitation.* Mining information-analytical bulletin. Moscow: MGGU, 4: 290-296.

Zhulay, Y.A. & Voroshylov, A.S. 2010. *Experimental definition of vibro-load on the instrument for pipes widening*. Geotechnical mechanics: collection of scientific works. Dnipropetrovs'k: IGTM of Ukraine's NAS, 89: 34-40.

Dzenzerskiy, V.A., Zhulay, Y.A., Khachapuridze, N.M., Redchits, D.A. & Voroshylov, A.S. *Efficiency determination of vibro-load appliance on the tool for pipes widening.* Impulse processes in mechanics of continuous medium. Materials of IX international scientific conference. Nikolayev: 329-332.

Dzoz, N.A. & Zhulay, Y.A. 2008. *Initiation of water wells by way of cavitation hydro-dynamic influence*. Mining information-analytical bulletin. Moscow: MGGU, 3: 345-350.

Patent 87038 of Ukraine, МКИ E21F 5/02. *Unit for hydro-impulsive influence on coal seam* / L.M. Vasilyev, Zhulay Y.A., V.V. Zberovskiy, P.Y. Moiseenko, N.Y. Trokhimets; claimer and patent owner is IGTM of NAS of Ukraine. # a 2007 10209/9822; appl.13.09.07; publ. 10.06.09, bul. #11.

Zhulay, Y.A. & Zberovskiy, V.V. 2010. *Solution of conceptual tasks of hydro-impulsive loosening of coal seams prone to outbursts in mode of periodically stalling cavitation.* Collection of scientific works of NMU of Ukraine. Dnipropetrovs'k: NGA, 35. Vol. 2: 246-253.

Geomechanical Processes During Underground Mining – Pivnyak, Bondarenko, Kovalevs'ka & Illiashov (eds)
© 2012 Taylor & Francis Group, London, ISBN 978-0-415-66174-4

Economic aspects of development strategy of mining companies

J. Kudelko
KGHM CUPRUM Ltd – Research and Development Centre, Wroclaw, Poland

ABSTRACT: The demand and supply of raw materials have an influence on mining companies activity. The investment needs a new look on the mining projects market and requires an adequate strategy. The paper is focused on development strategies and their evaluation. A special attention was paid to the economic aspects and its measures. Cost effectiveness and risk of consolidation projects can be described as a comparison of a real net value of integrated and basic companies.

1 INTRODUCTION

During last several years the intensive activity of big mining companies aimed on integration through mergers or acquisitions, as a direction of development strategy, has been noticeable. Trend for companies consolidation and for creating the global giants caused breaking the successive records of the most valuable transactions.

Mergers called the consolidation processes include three types of activity:

- horizontal integration, what means joining two companies operating in the same business, at the same stage of production. Generally the third company is the result of the connection.
- vertical integration, in which participate at least two companies with the similar production profile but dealing with the different stages of the production cycle,
- conglomerate consolidation, which joins the companies having totally different sphere of activity.

Subsidiaries may still have a form of separate corporate entity forming the holding or may join together into one corporation, which will have divisions in different branches.

New companies, including the former competitors or partners (in case of vertical integration) are mostly the result of merger. The major purpose of merger is the so called synergy effect i.e. obtaining the higher value of the company after the connection, than the sum of the parts forming the new entity. The shareholders of companies being jointed, instead of previous interests take the shares of new created company. In case of conglomerate consolidation, the additional result is the effect of branch risk diversification or the fact that company profits are becoming independent from the fluctuation caused by production seasonality.

The company take-over (acquisition) is not very precise definition and consists in the trial of achieving, by the taking-over company, the control over the company being taken-over. Taking-over process may be under the approval of the former company bodies (board and stakeholders) – amicable takeover or under the objection – hostile takeover. The transfer may happen through:

- redemption the shares of the taken-over company on the stock exchange,
- redemption the shares outside the stock exchange or redemption of interests,
- collection of the adequate number of authorities from shareholders or partners,
- taking-over of the company's assets.

Take-over can make for liquidation of the taken-over company, its including into the capital group on which the taking-over company has the control, or incorporation i.e. including the taken-over company into the structures of taking-over entity.

Finishing the stage of negotiations and preparing for take-over means the necessity of providing the adequate funds for the transaction completing. Those resources do not have to come from the company (take-over may be financed by, for example credit or emission of bonds) and do not have to be of monetary nature (payment may be done, for example, by the shares of taking-over company).

Although the presented procedure concern only the taking-over process, the quite similar pattern will be in case of joining the companies. It must be, however, paid attention on the differences between those two kinds of companies integration.

2 MERGERS AND ACQUISITIONS

External method of development strategy implementation, consists, first of all, in connecting the companies, *mergers and acquisitions* and in opposite actions i.e. dividing the companies or/and selling the part of company (*divestements*). Those two opponent action are mostly undertaken at the same time or in specific sequence, for example after dividing the company, the acquisition or merger with another company takes place. Merger, dividing and selling processes, being a method of company development are interrelated and should be analyzed jointly. As far as the large scale mergers have been taking place since a long time, the large scale dividing processes occurred only in eighties.

Company mergers may be divided into several types using the different criteria. Basic classification is obtained using the criterion of type of synergy, on which the merger is based, what in turn is related with the integration degree and legal status of jointed company.

Strategic criteria are best reflected by dividing on mergers and acquisitions. The first create the integrated company, implementing the effects of technological, production and market synergy, while the second are related with finance strategies and are based on finance synergy (Figure 1).

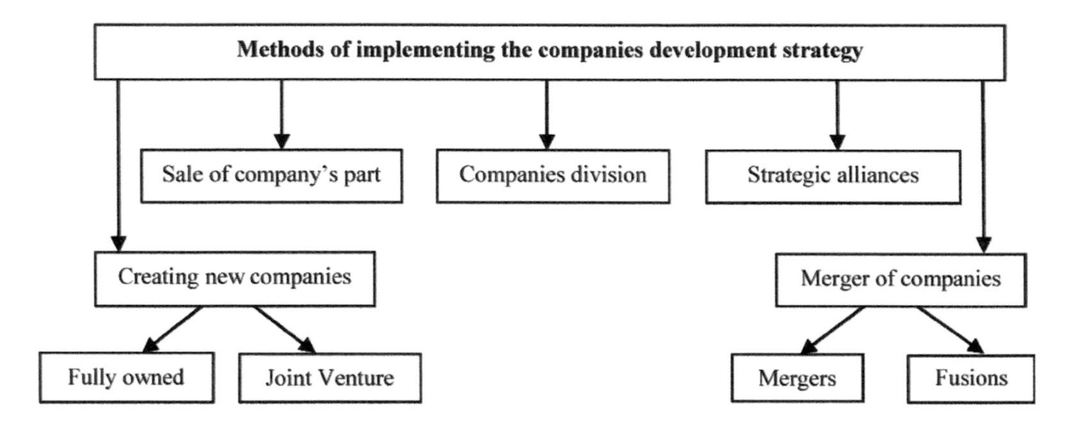

Figure 1. Methods of implementing development strategies and types of connections.

Acquisition (take-over) may concern the company in good or bad financial condition. In the second case the company is taken-over to be restructured and for its possible further sale after rehabilitation.

Classical role of mergers and acquisitions is:

- taking over the competitive assets,
- taking over the customers,
- reduction of competition,
- adapting the production capacity of sectors to the demand.

It may be said that M&A are the tool for quick and aggressive growth of the company (to enter the circle of big players), implementation of business strategy consisting in buying a company to sell it after or without restructuring. Those actions play a critical role in the development of internationalization and globalization of companies and in creating the so called "new economy".

3 EFFECTS OF CONSOLIDATION

Positive and negative effects of consolidation of big companies come out on macro and micro scale (Figure 2). The first, **macroeconomic** ones, concern the consumers, sector and the economy as a whole. Positive effects for clients are price reductions, dynamic in innovations, new products, better service.

Holding the prosperity in the sector entering the saturation may be the result of consolidation. Thus the consolidation is the best method for the sector development correction, especially with regard of the phase where the production capacity surplus is encountered.

Microeconomic effects result from the company size growth and from the increase of market share as well as reduction of competition and improvement of resources competitiveness of jointed company.

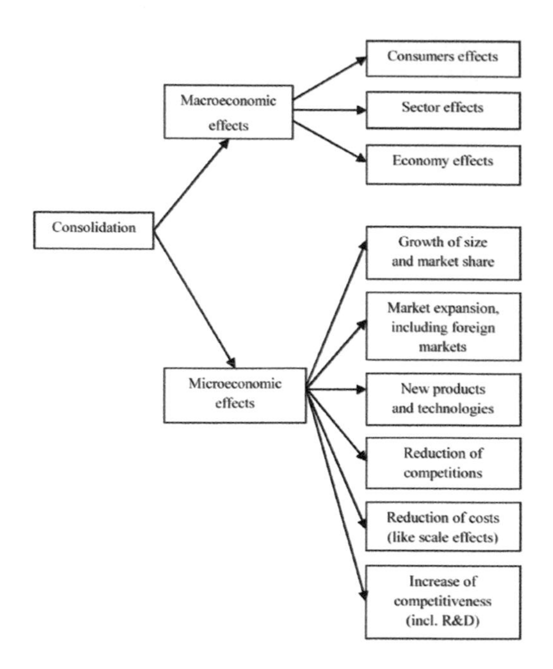

Figure 2. Positive effects of consolidation.

Effect of mergers and acquisitions for the companies being connected are therefore the same as effects of internal growth. Their specificity is the considerable rate of growth and synergy of resources from different entities being connected. It is aimed to improve the market position of the company, and, in special case, reaching the leadership or even the position of dominant.

In 2009 the transactions of mergers and acquisitions were focused on consolidation of smaller companies, selling illiquid assets and foreign expansion of Chinese investments. In 2010 activity of big mining companies and units under the governmental control was concentrated on using the financial reserves for taking-over and completing the projects with perspective resources, being at the final stage of development or the *minesite projects*. Number of declared acquisitions increased significantly in the second half of 2010, mainly with regard of gold and base metals projects. Results of Price Waterhouse Coopers study for 2010, quoted by Wilburn D.R. at al., say about 2 693 mergers and acquisitions having the value of 113.0 billion. Next PWC report concerns 2011, and gives the number of over 2 600 transactions amounting 149.0 billion.

In 2011 (data cover the period from the beginning of January until the end of December) 92 transactions concerning the non-ferrous metals were concluded (Figure 3). Almost half of them referred to the copper.

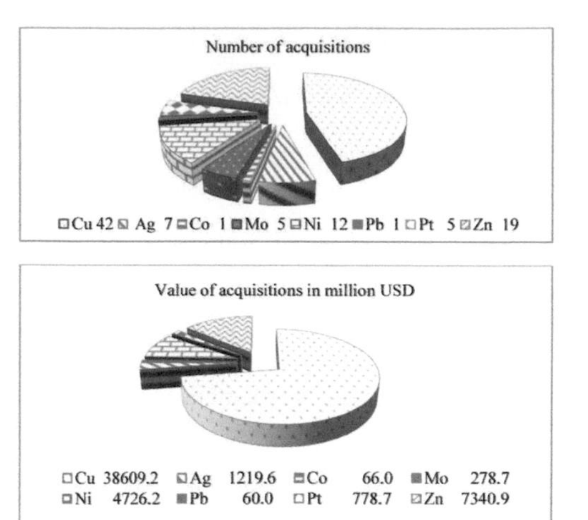

Figure 3. Number and value of acquisitions concerning non-ferrous metals in 2011.

Value of the transaction reached the sum of 53 079.3 billion USD. As it is visible, almost 75% of costs in the analyzed period, concerns the copper projects.

4 VERTICAL INTEGRATION

Cost-effectiveness of mining company integration with supplying or buying firm may be analyzed using the method which combines the strategic and economic evaluation of the venture in the following four steps:

1. Strategic analyze of integration advisability.
2. Comparing the economic effectiveness of base company and integrated company.
3. Comparing the integration risk.
4. Calculating the economic effects of integration.

First step refers to the strategic, long-term plan of company's development within the domain, markets, targets and competitive edge and verifies the integration advisability. Detailed analyze of integration advisability must be adjusted to the company's specificity. The example factors are:

- in backward integration, towards suppliers: tender force of suppliers and distributors, quality, reliability and punctuality of deliveries, competences of integrated company, financial condition of integrated company;
- in forward integration, towards buyers: tender force of buyers, share in market of the integrated company, market of integrated company, competences of integrated company, financial condition of integrated company.

During evaluation of integration strategic analyze methods, such as sector analyze, developed by M.E. Porter and/or SWOT/TOWS analyzes, are employed.

Since the strategic analyze is a multi-criterion evaluation, the aggregate feasibility (advisability) rate is calculated using range-point method or quotient normalization (Stabryla 2000; Penc-Pietrzak 2003 & Butra et al. 2009).

The second stage of this method is economic evaluation of the integration. For the needs of this evaluation the concept of *economic rate of return*, which has not been used until now, both in literature and in practice (Kudelko 2008), was introduced. *Economic rate of return* is defined as relation of economic added value and invested capital. Economic added value is the operation profit after tax, minus total capital cost, and for the first time was formulated by Stern and Steward. Invested capital is the corrected value of company's assets. Both categories concern the basic company before integration and integrated company. The integration is chosen when the integrated company has the highest economic rate of return.

In economic evaluation of investment projects static and dynamic methods are used as a standard (Butra et al. 2004; Rogowski 2004; Sierpinska & Jachna 2005; Wirth et al. 2000). Static methods are based on annual incomes and costs while the dynamic methods base on receipts and expenditures, whereat the time factor in the cash value is taken into consideration through the discount rate (Koźmiński A.K. & Piotrowski 2005; Struzycki 2002; Butra et al. 2004; Kudelko 2007; Kudelko & Nowak 2006; Ehrbar 2000).

About 25% of companies takes the rate of return basing on profit as a single evaluation criterion. Proposed rate of return based on economic added value includes own cost and cost of debt of the project.

During the third stage of evaluation the project risk is estimated under the assumption that different scenarios may happen in future i.e. differential values of variables defining the economic rate of return. As a measure of risk the semi standard deviation from expected rate of return and variation coefficient were taken. Semi standard deviation is a measure of negative deviations from expected value, while the variation coefficient is defined by semi standard deviation vs. the expected value.

In the fourth stage the economic effects of integration, which are the difference between the economic added value before and after integration, were calculated.

The company making the decision about vertical integration changes its chain of the added value. The biggest scope of integration the longest chain of values.

5 ECONOMIC EVALUATION OF VERTICAL INTEGRATION

Integration projects are usually investment ones – material or financial. Standard methods for such project evaluation are known. They are dynamic and static methods describing the transaction cost of vertical integration in mining (Joskow 1985, 2003; Stuckey 1983 & Kudelko 2009). Dynamic methods take into consideration the time factor or cash value in time, while static methods base on average annual economic results. Among popular dynamic methods are: net current value and internal rate of return, and lately also the option value of the project. These methods use category of cash flows i.e. revenues and expenditures, which sometimes are incorrectly identified as incomes and costs.

Amongst widespread simple methods are: operation profit and net profit, periods and rates of return of investment expenditures and breakeven points. Capital rate of return or capital yield is generally described as annual „profits" or net annual cash flows versus invested capital. In practice different methods for rate of return calculation are used, expressing "profit" through net profit, if necessary enlarged by interest or depreciation and interest, while capital is included in total expenditures or financed by own funds.

At present return on investment, return on equity as well as economic rate of return are recognized among standard rates of return.

Return on investment is calculated from the formula:

$$ROI = \frac{EBIT}{IC},$$

where ROI – return on investment; $EBIT$ – profit from operational activity, USD / year; IC – total investment expenditures or company assets, USD.

Return on equity measures the effectiveness of equity capital of the company. This rate is calculated from the formula:

$$ROE = \frac{NI}{E},$$

where ROE – return on equity; NI – net profit, USD / year; E – company equity capital invested, USD.

Economic rate of return (ESZ) is the relation between economic added value and the company assets (Kudelko 2009). For such approach speak all those arguments, which are quoted for using the economic added value as a measure of financial results of a company. The paper presents the method of evaluation the vertical integration of the mining companies and risk of using it. Economic added value is calculated from the formula:

$$EWD_t = EBIT_t(1 - CT_t) - WACC_t \cdot IC_{t-1},$$

where EAV_t – economic added value in t year, USD /year; $EBIT_t$ – profit from operational activity in t year, USD / year; CT_t – tax rate in t year; $WACC_t$ – average weighted capital cost in t year; IC_{t-1} – total investment expenditures at the end of $t-1$ year, USD.

If we divide EAV equation by IC_{t-1} we obtain:

$$\frac{EWD_t}{IC_{t-1}} = \frac{EBIT_t(1 - CT_t)}{IC_{t-1}} - WACC_t.$$

Left side of equation may be interpreted as an economic rate of return:

$$ESZ = \frac{EWD_t}{IC_{t-1}}.$$

After substituting EWD_t into ESZ we obtain:

$$ESZ = \frac{EBIT_t(1 - CT_t)}{IC_{t-1}} - WACC_t.$$

Integration is considered as successful, when the following condition is met:

$$ESZ_{int} \geq ESZ,$$

where ESZ_{int} – economic rate of return of integrated company, USD / year; ESZ – economic rate of return of basic company, USD / year.

Economic effects of integration (EE) are calculated from the formula:

$$EE = ESZ_{int} - ESZ.$$

Supportive criterion, while making a decision with regard of integration, is variation coefficient, being the semi standard deviation versus rate of return ratio. Integration due to the incorporated risk is expedient, when:

$$CV_{int} < CV.$$

where CV_{int} – variation coefficient of integrated company; CV – variation coefficient of basic company.

Semi standard deviation and variation coefficient is calculated for different strategic options:

$$ss = \sqrt{\sum_{i=1}^{n} p_i \cdot d_i^2}.$$

$$d_i = \begin{cases} ESZ_i - E(ESZ) & when\ ESZ_i - E(ESZ) < 0; \\ 0 & when\ ESZ_i - E(ESZ) > 0. \end{cases}$$

$$CV = \frac{ss}{E(ESZ)}.$$

where ss – semi standard deviation; CV – variation coefficient; p_i – probability of i scenario of events occurrence.

Expected value of rates of return is calculated from:

$$E(ESZ) = \sum_{i=1}^{n} ESZ_i \cdot p_i,$$

where $E(ERR)_{int}$ – expected value of economic rate of return for integrated company; ERR_{int} – economic rate of return for integrated company in i scenario of events; p_i – probability of i scenario of events occurrence; $E(ERR)$ – expected value of economic rate of return for basic company; ERR_i – economic rate of return in i scenario of events.

6 CONCLUSIONS

Mergers and acquisitions it is a multistage and multidimensional process, which in many cases may end with defeat expressed for example by the fall in shares value. In order to avoid that problem, the strategic reasonability of consolidation should be considered and then the whole consolidation process must be managed properly, starting from choosing the adequate consolidation candidate, through competent consolidation procedure, ending on effective joint of previously independent organizations.

After choosing the candidate, evaluating its value, estimating the value of synergy effects, establishing the bonus which is possible to accept and the choice of methods and sources of financing, the subject initiating the merger or acquisition transaction face with the process of effective connecting integration. Negligence at any stage of the consolidation process may result in the transaction failure.

Making a decision about the most adequate, for the company, scope of vertical integration is a complex process requiring the right investments, sometimes bonding the substantial financial resources and giving the specific economic effects. Therefore implementation of such strategy must have the strong economic grounds.

Selection of strategic option is made as follows:

– if $ERR_{int} > ERR$ and $CV_{int} > CV$, integration option is chosen;

– if $ERR_{int} > ERR$ and $CV_{int} < CV$, integration option is chosen;

– if $ERR_{int} < ERR$ and $CV_{int} < CV$, option without integration is chosen;

– if $ERR_{int} < ERR$ and $CV_{int} > CV$, option without integration is chosen.

In order to calculate the risk, the describing variables are presented in several scenarios of variables, usually optimistic, pessimistic and basic. Scenarios of variables are developed after deep analyze, consulting, for example, with the experts.

The presents method of evaluating the vertical integration, based on economic rate of return, is a very good strategic tool, which enables making the decision about implementing or dropping the venture. It complies with all relevant standards and requirements of the due diligence. Used together with other evaluation methods, enables completing the safe integration and substantial reduction of the related risk.

REFERENCES

Butra, J., Kicki, J., Kudelko, J., Wanielista, W. & Wirth, H. 2009. *Basics of economic calculation in mining companies.* Monography. IGSMiE PAN Publishing, Krakow.

Butra, J., Kicki, J., Kudelko, J., Wanielista, K. & Wirth, H. 2004. *Economy of geological and mining projects.* Monography. CBPM CUPRUM Publishing, Wroclaw.

Ehrbar, A. 2000. *EVA – strategy of creating the company's value.* Company's Finance Publishing, Warszawa.

Kozminski, A. K. & Piotrowski, W. 2005. *Management. Theory and practice.* PWN, Warsaw.

Kudelko, J. 2008. *Economic rate of return as a measure of capital yield.* KGHM CUPRUM Publishing, 3.

Kudelko, J. 2007. *Strategies of mining companies development.* Ores and metals, 12.

Kudelko, J. & Nowak, J. 2006. *Characteristic of selected elements of geological and mining projects evaluation.* Minerals, 4.

Penc-Pietrzak, I. 2003. *Strategic analyze in company's management. Concept and employment.* CH BECK, Warsaw.

Rogowski, W. 2004. *Efficiency calculation of investment projects.* Economic Printing House Publishing, Krakow.

Sierpinska, M. & Jachna, T. 2005. *Evaluation of company under world standards.* PWN, Warszawa.

Stabryla, A. 2000. *Strategic management in company's theory and practice.* PWN, Warsaw-Krakow.

Struzycki, M. 2002. *Business management.* Difin, Warsaw.

Wilburn, D.R., Vasil, R.L. & Nolting, A. – Exploration Reviev, U.S. Geological Survey, 2010.

Wirth, H., Wanielista, K., Butra, J. & Kicki, J. 2000. *Strategic and economic evaluation of industrial investment projects.* IGSMiE PAN Publishing, Krakow.

Geomechanical Processes During Underground Mining – Pivnyak, Bondarenko, Kovalevs'ka & Illiashov (eds)
© 2012 Taylor & Francis Group, London, ISBN 978-0-415-66174-4

About management of processes of deformation and destruction of rocks around of working of deep mines

O. Novikov & Y. Petrenko
State Higher Educational Institution "Donetsk National Technical University", Ukraine

ABSTRACT: On the basis of the analysis of results of analytical researches the basic rock-geological parameters which are influencing for the size of a zone of destruction of rocks (ZDR) and giving in management are allocated. Are carried out laboratory researches of influence of scheme of spatial location of bolting on size of these parameters. The new construction procedure of parameters roof bolting on the basis of the developed concept of management by process of formation around of working of deep mines ZDR is offered.

One of the major problems of rock geomechanics is providing of stability of working. In states of deep mines dynamics and character of the geomechanical processes occurring in a rock massif, containing working changes.

The analysis of existing concepts about the mechanism of display of rock stress in an area of supported working's shows, that formation and progress ZDR in many respects defines their state. During too time of law of the processes occurring inside ZDR while are studied insufficiently.

Results of known analytical calculations have significant divergences with data of mine supervision. Sizes of ZDR turn out significantly the smaller size and displacement on perimeter of working have uniform character that mismatches the validity. The reason for divergences calculation and the fact data describing size of ZDR, in our opinion, is discrepancy of dynamic character of a solved problem to its static statement. In opinion of G. Litvinskiy, this discrepancy is reflected that the problem by definition of size of ZDR, the majority of researchers is solved at disposable "salvo" acceptance of boundary terms (Litvinsky 1974). Such statement of a problem does not consider the redistribution of stress connected with moving of front of destructions in depth of a massif. The account of this circumstance is possible at the step-by-step solution to the problem when boundary terms for each subsequent step strike root in view of the variations connected with realization of the previous step.

Displacement of a contour of working results from the complex geomechanical processes occurring in the surrounding rock massif. Therefore attempts of researchers unequivocally (functionally) to establish dependence between the parameters describing terms of formation ZDR in time, behavior of rocks inside of destroyed area and displacement of a contour of working lead to significant idealization of the mechanism of display of rock stress in which the physical essence of geomechanical processes is not reflected.

The developed analytical method of the prognosis of size of ZDR, based on use of positions of mechanics of fragile destruction and the step-by-step solution to the problem in view of variation strength and reological parameters of a rock massif, concerning their characteristics received by results of rocks in the sample, has allowed to receive dependence of relative size of ZDR on lifetime of working (Maintenance and carrying ... 2005).

For a background of ways of efficient control a state of a rock formation around of mines it is necessary to set the degree of influence of major coefficients on size of ZDR. To these coefficients concern: depth location up workings (H), strength of containing rocks (R), residual strength of rocks within the limits of ZDR (R_0), rebuff support in working (P), angle of internal friction of rocks (φ), coefficient of cohesion of rocks (C).

With these objective analytical calculations of size of ZDR in which value of one of the above-stated parameters changed at the fixed average values of the others have been executed. The quantitative assessment of a degree of influence of each parameter was made through parameter K_r, which was defined:

$$K_r = r_p^m / r_p,$$

where r_p^m, r_p – accordingly, ZDR, appropriating current (changeable) and average values of parameter interesting us, m.

On Figure 1 charts of dependence of parameter K_r from values of influencing coefficients are shown. The analysis of charts shows, that on a degree of influence on size ZDR coefficients settle down in following sequence: C, φ, H, R, R_0, P.

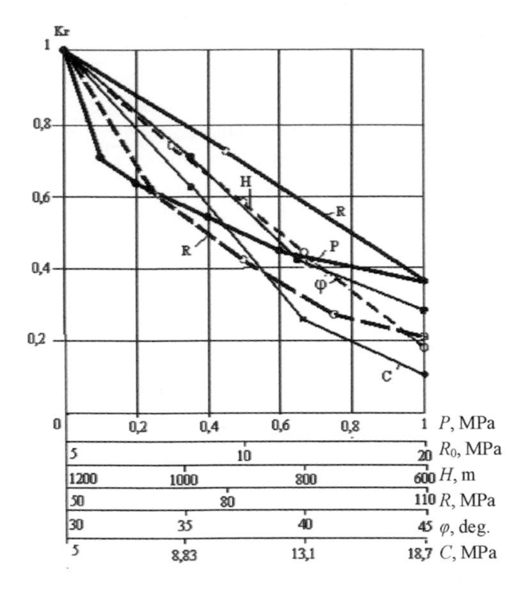

Figure 1. Variation of coefficient of a degree of influence on size of ZDR K_r depending on major coefficients: P – rebuff support in working, MPa; R_0 – residual strength of rocks within the limits of ZDR, MPa; H – depth location up working, m; R – strength of containing rocks, MPa; φ – angle of internal friction of rocks; C – coefficient of cohesion of rocks, MPa.

The greatest influence on size of forming ZDR renders angle of internal friction of the destroyed rocks (φ) and coefficient of cohesion of rocks (C). So, the increase with 30 up to 45 degree leads to reduction ZDR by 72%, and an increase of coefficient of cohesion of rocks with 9 up to 13 MPa – on 39%.

Residual strength of the destroyed rocks characterizes their possibility to save bearing ability in some volume as constructions. By researches IPRF NAS of Ukraine it is proved, that variation of a type of the intense state of the destroyed rocks (from monoaxial to plan and three-dimensional) it is possible to raise essentially their residual strength (Vynogradov 1989).

The analysis of existing means and methods of supporting of mines has shown that effectively and technologically to operate these coefficients it is possible by means of roof bolting.

In our opinion, the mechanism of work of roof bolting consists not only in concept about anchors, as about a bearing construction of type of a frame, and as about the elements, changing structure of a massif and preventing its destruction, i.e. to formation around of working of a zone of not elastic deformations.

Proceeding from the given concept, the radial scheme of location of anchors applied now is not the most rational, since area of influence of anchors on a massif in this case minimal. In this context, developed and tested in DonNTU spatial location of bolting of a massif allow to make the most at a minimum quantity of anchors of bearing ability of a rock massif (Kasyan, Petrenko & Novikovn 2010.).

Such spatial location of roof bolting allows the use of it not only as the strength element, preventing stratification of rocks and their displacement in a cavity of working, but also the element providing communication between separate fragments of destroyed rocks in all directions (radial, tangential and along axis workings). It provides substantial increase bearing ability anchored shell of the destroyed rocks due to increase of their residual strength.

For an assessment of influence of roof bolting on processes of deformation and destruction of rocks around of working have been carried out laboratory researches the limit and out of limit intense-deformed state of the rock massif reinforced spatial anchor by systems (Novikov 2011).

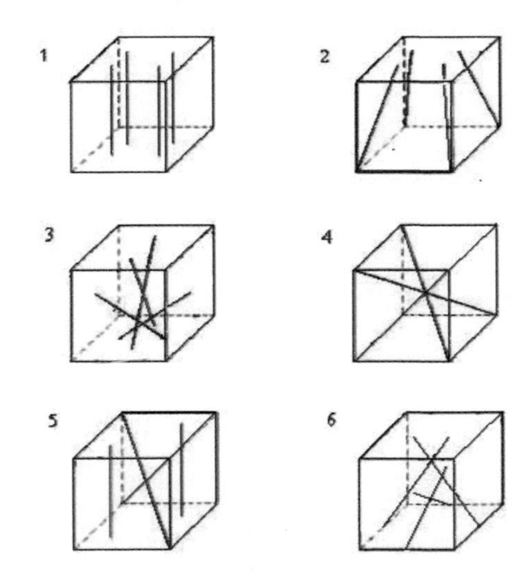

Figure 2. Schemes of spatial location of anchors.

The investigations were carried out at the setting for triaxial compression tests of the rock in IPRF NAS of Ukraine. Cubic samples were made of cement-sandy mixtures with the size of an edge 55 mm which modeled a area of rock massif in volume in 1 m^3 with strength 30-50 MPa. Samples were reinforced under various spatial locations (Figure 2). The strengthening effect from anchors was provided with forces of cohesion of a cement-sandy mixture with a bar of an anchor. Tests on plucking of anchors from the hardened mixture have shown, that process of extraction of a bar (extension from the sample) begins at effort from 93 up to 121 H , that provides compliance with a state of power similarity.

Originally in samples were created the hydrostatic field of stress simulating depth of location up 800, 1200 and 1600 m. Then was created, in a direction of action of stress σ_3 (in a direction of bolting) full unloading with maintenance of size of stress σ_1 and σ_2 at an initial level was made. Further, under specially developed programme the average size of operating stress, average relative deformations, size of residual strength, limiting and residual relative deformations in a direction of unloading, relative variation of volume, energy of variation of the form and volume, modules of deformation, elasticity, shift and recession paid off. For an assessment of character of destruction of the rocks reinforced by various spatial locations of bolting, criterion Nodai-Lode was used.

The limiting state of samples was defined on one of four below listed criteria:

– achievement of the maximal value by tangents by stress;

– excess of limiting relative deformations at unloading the sample;

– sharp decrease in size of the module of shift;

– excess of the chart describing variation of relative volume of the sample ($\Delta V / V$).

Results of researches are presented on Figure 3-12.

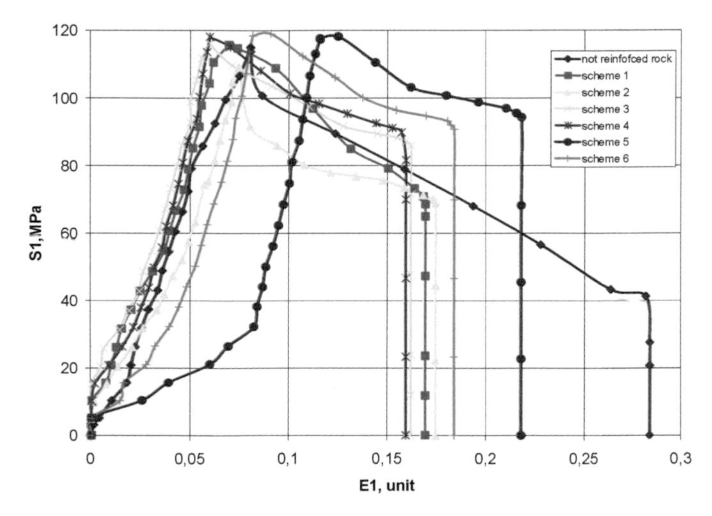

Figure 3. Diagrams "stress-deformation" for rock-anchor constructions reinforced under various schemes, at modeling strength of containing rocks 30 MPa and depth location up working 1200 m.

As you can see from (Figure 3), all charts $\sigma_1 = f(\varepsilon_1)$, received from tests of rock-anchor constructions for the generalized tension, have three characteristic plots: a plot of uniform comprehensive compression up to the stress simulating set depth, a plot of unloading in a direction of bolting up to some value at which there is a destruction and a plot before full unloading from stress σ_3 (out of limit deformation). If the parameters describing intense-deformed state it is rock-anchor constructions and rocks without insert at loading on the first plot essentially do not differ, at unloading, values of the parameters describing limiting and especially out of limit state (residual strength) are broken a set up to 4.5 times. It is necessary to note, that the limiting state is rock-anchor constructions (a point of an excess of charts $(\Delta V/V) = f(\sigma_3)$ on Figure 4) comes at unloading on axis 3 (in a direction of bolting) on size of from 59 up to 89% from initial, that is essential more, than to not reinforced rock (37%). Thus, for rock-anchor constructions, value of normal stress in 3.5 times, and the maximal tangents of stress – in 3.6 times above, than for not reinforced rock. Limiting relative deformations for the constructions reinforced under locations 3-6, up to 2.1

times above, than for not reinforced rock (for the constructions reinforced under locations 1 and 2, they on 26% below). Values of modules of volumetric deformation, shift and elasticity for rock-anchor constructions in a limiting state decrease up to 1.7 times. In a limiting state dilatancy it is rock-anchor constructions on 40% less, than for not reinforced rocks. The attitude of size of energy deformation to energy of variation of volume for rock-anchor constructions in a limiting state increases in 18.7 times, in comparison with not reinforced rock that speaks about greater power capacity of their destruction and explain effective composite action of rock and anchors on reaction to destruction.

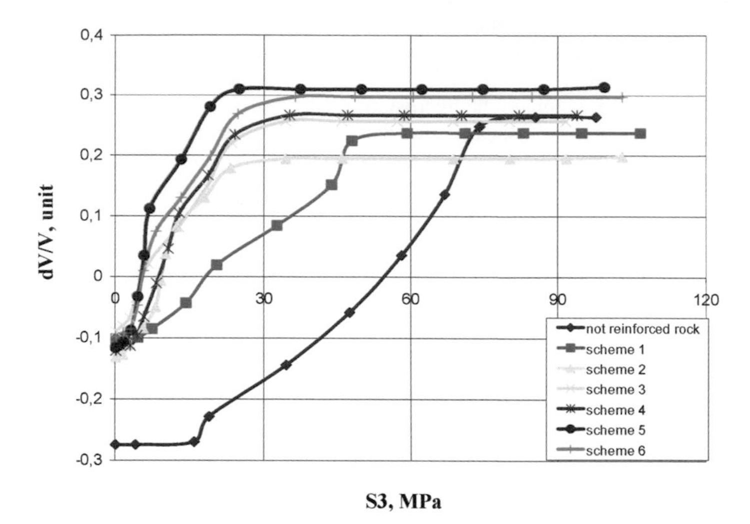

Figure 4. Charts of variation of relative volumetric deformations (ΔV/V) from stress σ_3 at unloading is rock-anchor constructions reinforced under various schemes, in the field of out of limit deformation at modeling strength of containing rocks 30 MPa and depth location up working 1200 m.

About character of destruction and the subsequent out of limit deformation it is rock-anchor constructions and not reinforced rock speak values of coefficient Poisson (coefficient of cross deformations). You can see from the charts presented on Figure 5, values of coefficients of cross deformation for the constructions reinforced under various locations, in a limiting state change from 0.45 up to 0.48 (0.46 for not reinforced rock), that testifies about fragile character of destruction. Unlike not reinforced rock for which the coefficient of cross deformation remains practically to constants up to an instant of full unloading from stress σ_3, rock-anchor constructions after achievement of a limiting state quickly pass in area of plastic deformation.

So, at unloading from initial value of a stress on 62-84%, all rock-anchor constructions, irrespective of the location of reinforcing, are plastically deformed down to full unloading from stress σ_3 that proves to be true in size of coefficient of cross deformation from 0.99 up to 1.61 (less than 0.5 is characteristic for fragile the destruction not reinforced rock). About plastic character of deformation at rock-anchor constructions in out of limit area speak also charts of dependence of the attitude of energy of forming to full energy of deformation (A_f/A_F) from stress σ_3 at the unloading, presented on Figure 6.

For not reinforced rock the domination of the fragile deformation connected with gradual accumulation crack of tearing away. As far as approaching to the area of the complete unloading from stress σ_3, there is an speed-up replacement of fragile deformation on plastic. Thus, in not reinforced rock cracks of complex shift are formed, that significantly increases its degree of dilatancy. For rock-anchor constructions reinforced under various locations, already at an initial stage of unloading from stress σ_3 a speed-up growth parts of plastic deformations in cumulative dynamics of their progress is characteristic. This process proceeds with practically constant intensity down to full unloading constructions from stress σ_3. Thus, relative residual deformations in a direction of unloading up to 1.7 times is more, than for not reinforced rock. Besides reinforcing has allowed to decrease on 70% relative variation of volume on an out of limit plot of deformation, an event from a proceeding destruction (scarification) in comparison with not reinforced rock. It speaks about efficiency of reaction of the rocks connected by anchors in an out of limit state to their further destruction.

It is established, that out of limit deformation rock-anchor constructions reinforced under schemes 1, 2, 5 and 6 and not reinforced rock, occurs in the field of extension with a shear, and the constructions reinforced under scheme 3 and 4-in area of a stretching. In process of unloading from stress σ_3 in out of limit area of deformation there is a stress reduction of rocks and rock-anchor constructions. Despite of about identical gradient of decrease in stress σ_1 at reduction of stress σ_3 for not reinforced rock and rock-anchor constructions reinforced under various schemes, due to fundamental importance of limiting stress residual strengths of constructions from 3.1 up to 4.5 times more, than for rock. So, residual strength for the best construction reinforced under the scheme 5 makes 0.81 from initial. Residual unit values normal and tangents of stress for rock-anchor constructions accordingly in 4.5 and 4.6 times above, than for rock.

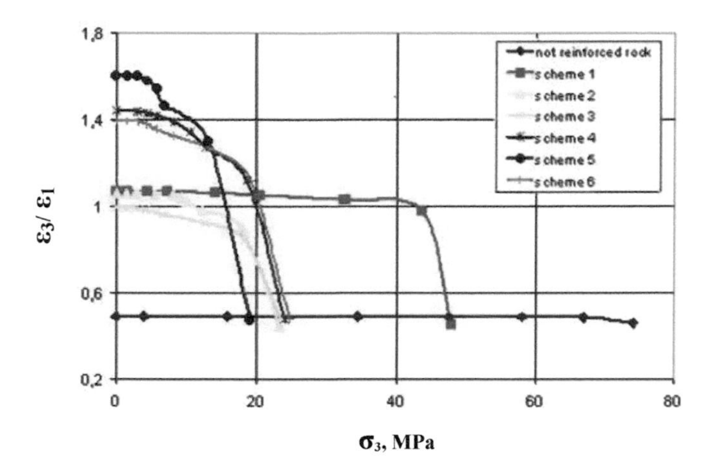

Figure 5. Charts of dependence of coefficient of cross deformations ($\varepsilon_3/\varepsilon_1$) from stress σ_3 at unloading rock-anchor constructions reinforced under various schemes, in the field of out of limit deformation at modeling strength of containing rocks 30 MPa and depth location up working 1200 m.

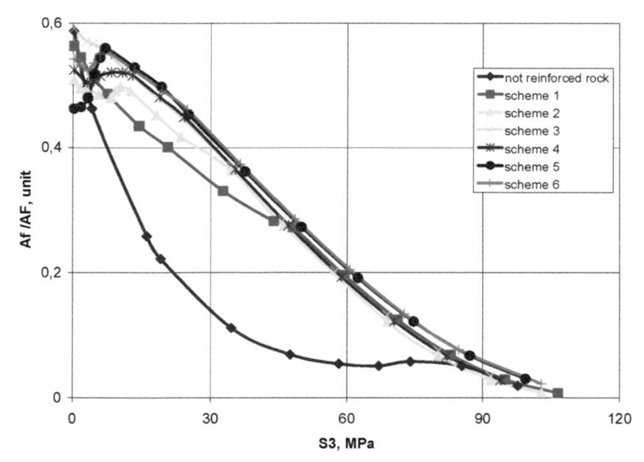

Figure 6. Charts of dependence of the attitude of energy deformation to full energy of deformation (A_f/A_F) from stress σ_3 at unloading rock-anchor constructions reinforced under various schemes, at modeling strength of containing rocks 30 MPa and depth location up working 1200 m.

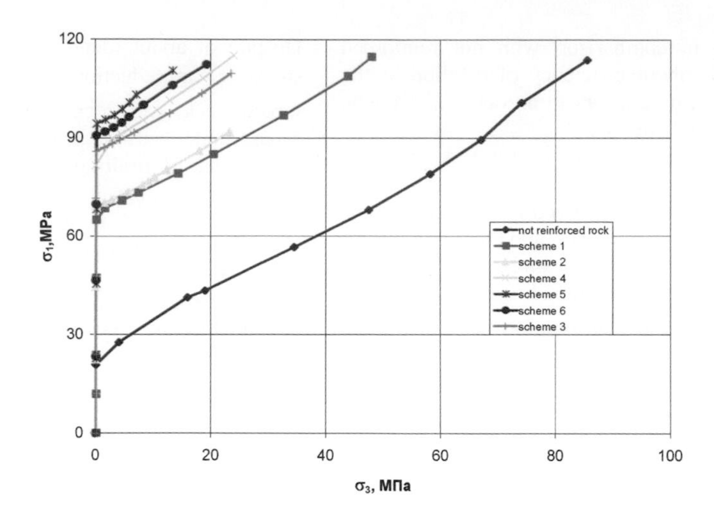

Figure 7. Charts of dependence of the main stress σ_1 from stress σ_3 at unloading rock-anchor the constructions reinforced under various schemes, in the field of outrageous deformation at modeling strength of containing rocks 30 MPa and depth location up working 1200 m.

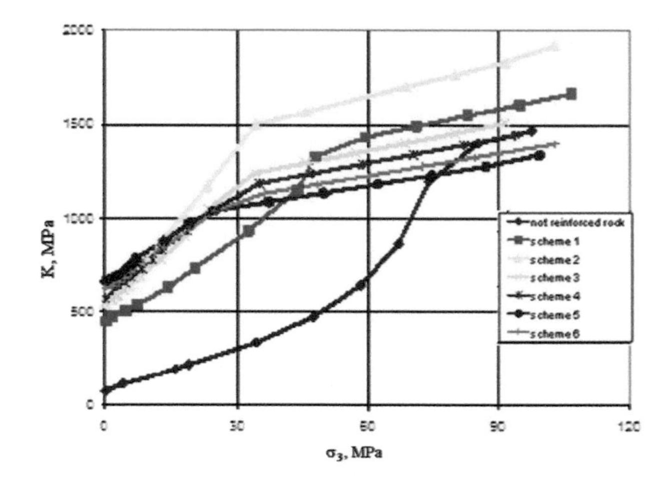

Figure 8. Charts of variation of the module of volumetric deformation (K) from stress σ_3 at unloading rock-anchor constructions reinforced under various schemes, in the field of out of limit deformation at modeling strength of containing rocks 30 MPa and depth location up working 1200 m.

Apparently from the charts presented on Figure 8, at unloading not reinforced rock and rock-anchor constructions in the field of out of limit deformation (from a point of an excess of charts and down to full unloading from stress tension σ_3) there is an accelerated decrease in the module of volumetric deformation that characterizes an increase of their rigidity and decrease in residual strength. More intensively this process goes for the constructions reinforced under schemes 1-3, more slowly – for reinforced under schemes 4-6. Residual values of the module of volumetric deformation for rock-anchor from 5.7 up to 8.4 times it is more than constructions, than for not reinforced rock. Similarly there is a decrease in values of the module of shift for not reinforced rock and are rock-anchor constructions in the field of out of limit deformation – Figure 9. It characterizes process of slackening and rolling of internal structure of a material of the construction, accompanied by decrease in its strength to action of tangents of stress and increase of a tribute of plastic deformations.

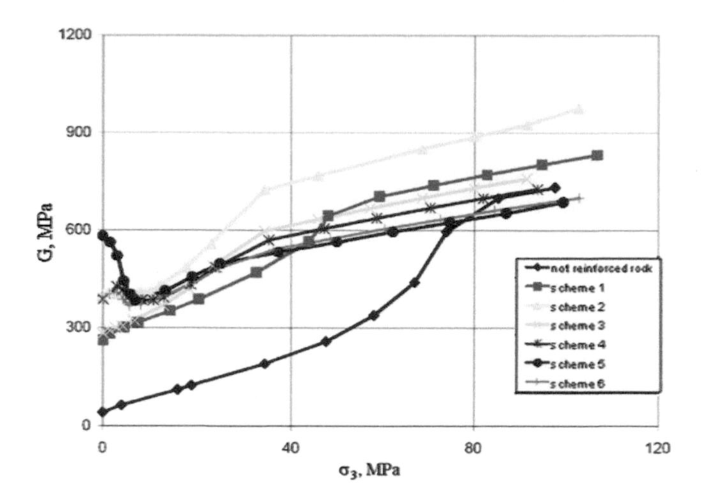

Figure 9. Charts of variation of the module of shift (G) from stress σ_3 at unloading rock-anchor constructions reinforced under various schemes, in the field of out of limit deformation at modeling strength of containing rocks 30 MPa and depth location up working 1200 m.

For rock-anchor the constructions, reinforced under the schemes 5, on a plot of out of limit deformation, there is an increase of the module of shift in 1.4 times in comparison to limiting value. For all it is rock-anchor constructions residual values of the module of shift from 6.2 up to 13.9 times more, than for not reinforced rock, that speaks about high efficiency of strength of constructions to action of shift loads due to inclined to a plane of unloading of anchors. Thus, in a material of constructions cracks separation extend, that is accompanied minimal it scarification, thus residual strength decreases slowly, and the module of recession accepts the minimal values. These data well will be coordinated with results of the researches (Alekseev, Revva & Rjazantsev 1989) executed in IPRF NAS of Ukraine.

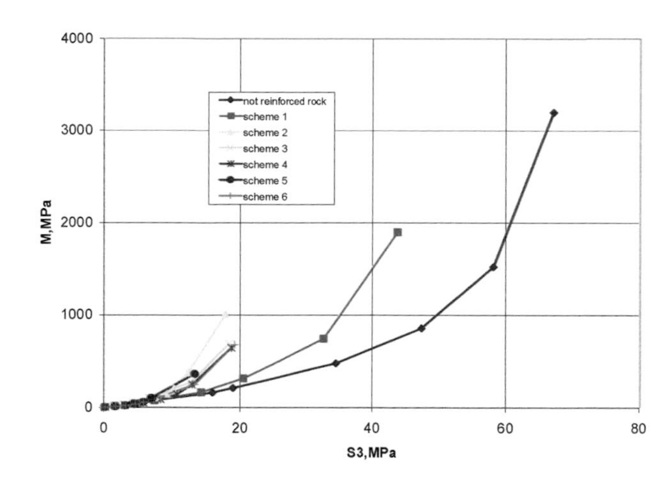

Figure 10. Charts of variation of the module of abatement (M) from stress σ_3 at unloading rock-anchor constructions reinforced under various schemes, at modeling strength of containing rocks 30 MPa and depth location up working 1200 m.

It is established, that at unloading it is rock-anchor constructions from stress σ_3, in out of limit area, at a constant lateral load $\sigma_1 = \sigma_2$, be observable transition from fragile to more viscous (quasi-plastic) to character of the deformation, characterized by accelerated decrease of the module of recession in comparison with not reinforced rock (see Figure 9), by increasing out of limit deformations (Figure 11) and conservation of high residual strength (see Figure 3) is observed, that is characteristic for plastic rocks. Apparently from the charts presented on im.10, value of the module of recession for rock-anchor constructions from 1.9 up to 4.0 times is less, and relative deformations in a direction of unloading ε_3 (see Figure 11) from 11 up to 65% is more, than for not reinforced rock, that determine on quasi- plastic character of their deformation in out of limit area.

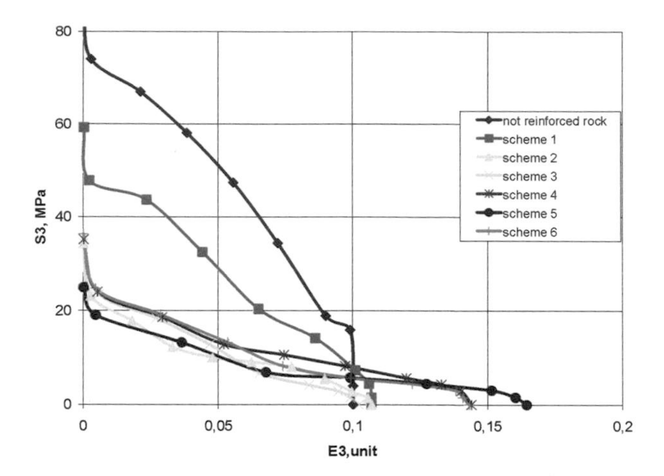

Figure 11. Charts of variation of stress σ_3 at unloading, on an outrageous site of deformation, from relative deformations ε_3 for rock-anchor constructions reinforced under various schemes, at modeling strength of containing rocks 30 MPa and depth location up working 1200 m.

At the unloading not reinforced rock from stress σ_3 on an out of limit plot of deformation, there is a formation of cracks of complex shift, that significantly increases a degree dilatancy a material at practically constant value of coefficient of cross deformations. Planes of destruction, thus, as a rule, are parallel to a plane of unloading. For rock-anchor the constructions reinforced under various schemes, at unloading in out of limit area, propagation of cracks tensile is typical, thus occurs minimal disintegration (dilatancy) a material, value of coefficient of cross deformations and relative deformations in a direction of unloading increase, thus residual strength decreases slowly, and the module of recession accepts the minimal values.

Variation of a type of a deformation state is rock-anchor constructions at unloading in the field of out of limit deformation and value of the module of recession is shown on Figure 12.

The deformation state of not reinforced rocks and rock-anchor constructions, reinforced under the scheme 1, at unloading from stress σ_3 in out of limit area, changes in not significant limits (a stretching with shift). It is connected with progress in them of cracks of complex shift, and reinforcing of 1 anchor radially scheme, minimally resist to this process. For others rock-anchor constructions reinforced under schemes 2-6, wider range of variation of a deformation state (from compression with shift up to a stretching with shift) is characteristic, that is caused by effective involving in composite action on reaction of deformation angled the located anchors and rock.

Besides reinforcing under schemes 4 and 5 allows to use to the greatest degree supporting ability of rock-anchor construction due to more recent variation of a favorable mode of its work (from a deformation state "compression with shift" up to "stretching with shift") at unloading.

It is established, that the effect quasi-plastic (fast decrease in size of the module of recession at unloading construction in comparison with not reinforced rock) in a greater degree is shown in less stable rocks (30 MPa) and decreases in process of an increase of strength.

At an increase of depth location the level of the initial volumetric intense state of rock-anchor constructions that affects character of their out of limit deformation at unloading raises.

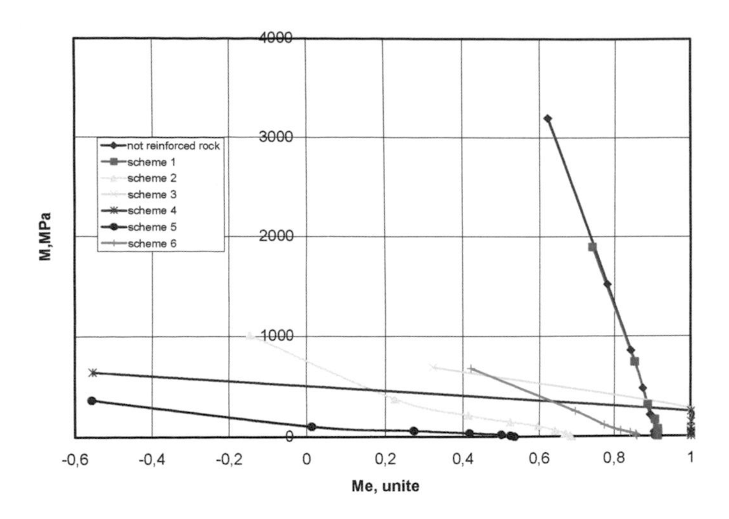

Figure 12. Charts of dependence of the module of recession (M) from a type of a deformation state (μ_ε) rock-anchor constructions reinforced under various schemes, in the field of out of limit deformation at modeling strength of containing rocks 30 MPa and depth location up working 1200 m.

The out of limit plot of deformation begins with growth of depth at fundamental importance of a stress σ_3 (so at smaller removal of a face of working from an installation plot anchor systems), that in turn causes slower decrease in the module of recession at unloading from stress σ_3 and more recent transition of a construction in area quasi-plastic deformations (intensive decrease in the module of recession and increase of out of limit deformations are observed at small values σ_3, characteristic for practically full unloading). In turn, fundamental importance of the module in the beginning of a plot of out of limit deformation, right after achievements by a construction of a limiting state, cause a greater degree of infringement of its initial internal structure that increases a degree dilatancy constructions in the field of out of limit deformation and reduces residual strength. So, with an increase of strength of rocks with 30 MPa up to 50 MPa, a residual strength rock-anchor construction decreases on 8%. It is connected with an increase of a degree dilatancy in constructions from more durable rocks on a plot of out of limit deformation. Thus, relative residual deformations of a construction in a direction of unloading have increased for 17% and value of coefficient of cross deformations – in 1.5 times.

The increase of modeled depth location up rock-anchor constructions leads to reduction of their residual strength. So, at an increase of depth location up rock-anchor construction reinforced under the location 5, about 800 m up to 1600 m, its residual strength has with other things being equal decreased with 0.85 up to 0.77 from initial. It is connected by that the anchors having on two orders higher elastic properties, than rock, at a stage before limiting deformation of a construction represent concentrators of stress which influences within the limits of the areas on a massif, strengthen destruction (a degree scarification) rocks. At an increase of modeled depth location up constructions, influence of this coefficient amplifies that in turn and leads to decrease in residual strength of a construction.

Similar results have been received at tests of rock-anchor construction reinforced under other schemes, at modeling strengths of rock 40 both 50 MPa and depths of disposition arrangement up 800, 1200 and 1600 m. They allow to draw following conclusions.

It is established, that constructions reach a limiting state at unloading from radial stress on size of from 59 up to 91% from initial (for not reinforced rock at unloading on 25-47%). Thus, values of modules of volumetric deformation, shift and elasticity decreased in 2.8 times in comparison with values to unloading (for not reinforced rocks decrease made in 1.8 times), remaining on size of up to 2.3 times it is less, than for not reinforced rock. Limiting values of normal and maximal tangents of stress were in 4.8 times above, and limiting relative deformations of constructions in 1,3 times it is less, than for not reinforced rock. Relative variation of volume for constructions from rocks with strength up to 30 MPa in a limiting state up to 1.7 times is less, and from rocks strength 40 MPa and more – up to 2.4 times above, than for not reinforced rock. The attitude of size of energy deformation and energy of variation of volume for constructions in a limiting state in 30 times more, than for not reinforced rock that testifies to greater power strength of their destruction and speaks effective composite action of rock and anchors on reaction to it.

It is established, that bolting changes the mechanism of destruction of rocks. At unloading, under angles up to 30 degree (theoretical value of angle $45 \pm \varphi / 2$) to a plane of unloading, planes of destruction on which it results from tensile with shift or tensile are formed.

At the further unloading constructions (an out of limit plot), there is their fast transition in area of quasi-plastic deformation to coefficient of cross deformation from 0.7 up to 2.2, decrease in values of the module of deformation and the module of shift to 2.2 times in comparison with limiting values (except for the constructions reinforced under schemes 5, 6 and 7), and their residual (at full unloading) values up to 22.2 times above, than for not reinforced rock. For rock-anchor constructions reinforced under schemes 5, 6, 7 and 8, on a plot of out of limit deformation, there is an increase of the module of shift up to 6.8 times in comparison to limiting value that speaks about increase of strength of a construction to action of shift loads due to inclined to a plane of unloading of anchors. Residual values normal and tangents of stress for constructions up to 4.7 times above, and residual strength up to 5 times above, than at not reinforced rock also makes from 0.31 up to 0.85 from initial. Relative residual deformations it is rock-anchor constructions in a direction of unloading reach 0.23 (up to 2.5 times more, than for not reinforced rock). Out of limit deformation of all is rock-anchor constructions and not reinforced rocks occurs in the field of a stretching to shift, thus, relative variations of volume from a proceeding destruction (scarification) rocks decrease up to 97% in comparison with not reinforced rock. It speaks about efficiency of reaction of the rocks connected by anchors in an out of limit state to their further destruction.

Similar results have been received at the tests of real samples of clay slate selected from a adjacent roof of a seam m_5^{1h} of mine "Dobropolskaya", which have shown their good convergence with results of the researches lead on models (a divergence does not exceed 20%).

Thus, as a result of tests it has been established, that accommodation in rocks of spatial set of reinforcing elements allows changing size of the parameters, describing structurally-mechanical and deformation properties of a rock massif, creating hold up to destruction, afford an opportunity to operate processes of deformation and destruction of rocks.

REFERENCES

Litvinsky, G.G. 1974. *Kinetics of a fragile destruction of a rock massif in a mine working area*. FTPRPI, 5: 15-22.

Maintenance and carrying out of working of deep mines of Donbass: monograph. 2005. Under the general ed. S.S. Grebenkin. Donetsk: Kashtan: 256.

Vynogradov, V.V. 1989. *Geomechanic of management of state of rock massif in the area of rock working*. Kyiv: Scientific thought: 192.

Kasyan, M.M., Petrenko, Y.A. & Novikov, O.O. 2010. *Technique of definition of parametres of anchor rock-reinforcing systems for maintenance of stability of rock working: STP (02070826) (26319481)*. Donetsk-Dobropillja, 27.

Novikov, O.O. 2011. *Deformation and destruction of rocks, reinforced by spatial schemes*. News of Donetsk Mine Institute. – Donetsk: DonNTU, 1: 45-51.

Alekseev, A.D., Revva, V.N. & Rjazantsev, N.A. 1989. *Destruction of rocks in a volumetric floor compressing voltage*". Kyiv: Scientific thought, 168.

Geomechanical Processes During Underground Mining – Pivnyak, Bondarenko, Kovalevs'ka & Illiashov (eds)
© 2012 Taylor & Francis Group, London, ISBN 978-0-415-66174-4

The formation of the finite-element model of the system "undermined massif-support of stope"

I. Kovalevs'ka, M. Illiashov & V. Fomychov
National Mining University, Dnipropetrovs'k, Ukraine

V. Chervatuk
"DTEK Pavlogradugol" Pavlograd, Ukraine

ABSTRACT: The features of the numerical modeling was research by finite element method taking into account the impact on the development of stress-strain state of the "undermined massif-support of stope", of the time and the destruction of its individual elements.

1 INTRODUCTION

Calculation of the finite element method for modeling of nonlinear media, under the influence of complex stress, is a nontrivial task. Most of tasks of geomechanics refer precisely to this type. In the course of their solution to perform each time to the correcting-calculation model on several parameters, roughly speaking runs multivariate optimization in previously unknown space of available solutions. The simplest way to correcting the calculation model is its maximum simplification. This approach, in some cases, ensures quick results of computational experiment. However, one can't assert, that the results are fully adequate for the qualitative and quantitative indicators of the real object.

First of all, the simplification of the computational model in tasks of geomechanics associated with the exception of factors and nonlinearity of the medium with a change in SSS (stress-straight state) system in time. Such calculations using FEM (finite element method) requires a deep understanding of the laws of behavior of materials under external load and significant computational resources. As a result, the behavior of the "undermined massif-support of stope" is regarded as a one-stage model of a physical object. However, under real operating conditions of extraction workings stresses and strains in the system are not immediately fixed. On the contrary, during the whole process of exploitation of the mining work geometric and mechanical properties are changing, so the distribution of stresses and strains should be considered as a process. Correcting of such calculating models requires considerable resources of time and computing power. Become necessary to use different methods of describing the complicated physical phenomena in the description of materials and the interaction of individual objects within a single computational model.

At this stage of development of the finite element method there are a whole number of software products, which include computational systems that implement by variety of mathematical methods the wide range of physical problems. These complexes have different degrees of flexibility and focus on different groups of applied tasks. For solve the tasks of geomechanics, in various productions, usually used SolidWorks Simulation (COSMOS/M), FLAC 2D/3D, ABAQUS and ANSYS.

Currently, the most ample opportunities for the modeling of materials, conditions of interaction of objects, the solution of nonlinear problems and tasks of mechanics destruction has ANSYS (Chigarev 2004). Also this software product has a high level implementation of numerical algorithms for different conditions of media behavior and other characteristics of the modeling. Therefore, implementation features described below correspond to the tasks of geomechanics approaches implemented within a computer system ANSYS.

2 SHOTCRETE AND BLASTING

The basis of adequate and mathematically correct solution of any problem with the help of the numerical grid method is to construct a high-quality computational grid. In FEM – it is creation of the finite element model. During this stage it is necessary to provide two main criteria: the geometric accuracy of the description of objects and choice of physically reasonable parameters of finite elements.

Geometric accuracy is achieved by selecting the

right combination of geometric shapes and locations of points of finite elements. In today's software products, the process of building a finite element grid to varying degrees is automated. However, almost all of them used a minimal set of initial conditions, consisting of: linear dimension of a finite element and accuracy description of the geometry. In fact, these parameters represent the average linear dimension of the finite element in the chosen coordinate system and the maximum permissible deviation calculated from the surface of the point or edge of the model. For complex geometries encountered in problems of geomechanics, the manipulation of these parameters may not always lead to the construction of the grid (Basov 2002). In the construction of the grid in the area of contact frame lining and contour of mining work with a large amount of linear elements and a large tolerance can't get an accurate description of the geometry of the frame, but with a small amount of linear elements and the small dimension of the admission of the whole task may exceed the available computing resources.

Therefore, the number of computing systems that implement the finite element method, using an approach that is associated with a change in the size of finite elements in various zones of modeling object. Thus, objects that have small linear dimensions (Figure 1) (anchors, elements of frame lining, crossbar, pit props, powered roof supports, etc.) are described by finite elements of small size and peripheral elements of the computational area is described by the large size. This allows us to obtain a satisfactory description of the geometry and reduce the overall dimension of the problem.

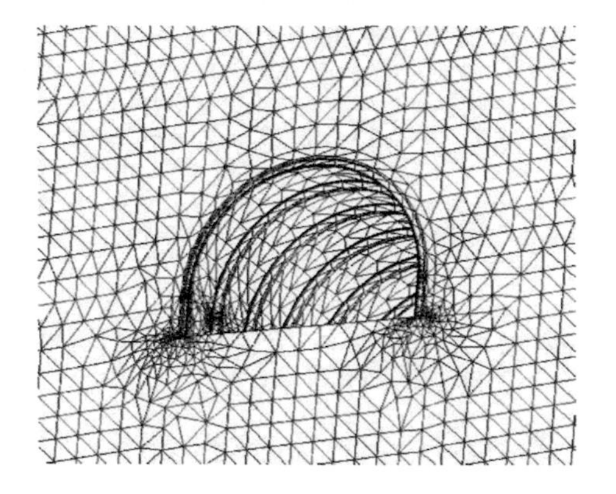

Figure 1. Modeling of the grid for mining working, including a frame support.

Use of tetrahedra and parallelepipeds as the finite element within a finite element model can most adequately realize the stress field under complicated loading. The elements in the form of parallelepipeds are highly resistant to the computation process and, due to the coverage of relatively large size of the computational area, are ensured savings of computational resources. At the same time, the tetrahedra are good as final elements in areas with possible high stress gradient zones of fracturing and the solution of problems of mechanics of destruction.

Another feature of a finite element model – is a task of singular points in the computational area (Basov 2002). These points indicate where you want to set the point of finite element. This approach is used to a high degree of accuracy to determine the SSS in particular, critically important point of the modeling object or to change the terms for grid construction of defined element of computational area. In both cases, well-chosen singular point can substantially improve the adequacy of the results obtained at different steps of the solution.

The physical parameters of a finite element should be assigned by the order and type of material, described in it (Chigarev 2004). The choice of the order of the finite element allows not only with a high degree of accuracy to describe the geometric surface of the higher orders, but also to make appropriate calculations of zones of high gradients in the stresses and strains arising in the lining cell expression and processing in the rocks adjacent to its contour.

As you know, all descriptions of physical medium in finite element method are concentrated in the finite element. In the course of computational experiments to select the most appropriate for the specific calculation of the physical medium. On the basis of made choice to describe the desired properties of the material or materials which, in consequence, to bind to specific finite element generated for the computational model. The greatest breadth of capability in this direction is the editor of materials of ANSYS (Chigarev 2004).

Versatile software system of finite-element analysis ANSYS, being one of the world leaders in the field of computer engineering (CAE, Computer-Aided Engineering), ensure the possibility of solving linear and nonlinear, stationary and non-stationary spatial tasks of mechanics of solid deformation and mechanics of construction. At the same time ANSYS has broad capabilities in dealing with non-stationary geometrically and physically nonlinear problems of contact interaction of the elements of computing area.

The implementation of ANSYS to describe the response or reaction to the impact of a complex system of different physical nature allows you to use the same model for solving such problems of bound, like strength of thermal stress or the influence of magnetic fields on the construction strength.

The rapid growth of computer technology in the early 70s of the twentieth century has allowed a considerable degree of empower ANSYS. The systems made a large number of changes were added to the nonlinearity of different nature, have introduced the possibility of using the method of subconstruction, greatly expanded the library of finite elements. In solving problems of geomechanics, the program offers a wide range of computational tools that allow you to:

– to consider a variety of structural non-linearity;

– to solve the most general case of contact interaction for three-dimensional bodies of complex configuration;

– allow the presence of large (finite) deformations, displacements and rotation angles;

– to perform multi-parameter optimization in interactive mode;

– and much more, along with parametric modeling, adaptive grid re-formation, using the elements and extensive opportunities to create macros using parametric design language of ANSYS – APDL.

All functions performed by the program ANSYS, combined into groups called processors. The software package has one preprocessor, one processor of solutions, two post-processors and several supports' processors, with the optimizer. ANSYS preprocessor is used for creating finite element models and options for implementation of the solutions process. The processor executes the application solution loads and boundary conditions, and then determines the response of the computational model. Using a postprocessor have access to the results of the solution and evaluate the behavior of computational model, as well as carry out additional calculations.

In ANSYS for the entire set of information relating to the model and the results of the solution, used one, central database (Figure 2). Model information is stored in the database at the stage of preparation preprocessor. Load and results of solutions are recording by solutions processor. Data obtained on the basis of the decision at their Postprocessing, written in the form describes the level of post-processor. The information entered by one of the processors are available, if necessary, for other processors.

When constructing a grid of high-quality CAD-models ANSYS uses many means of quality control of the grid. In the software package provides four ways to generate the grid: to use the method of extrusion, creation of an orderly grid, creation of arbitrary grid and adaptive construction.

The generators of the arbitrary network have a wide range of internal and external operating options of the grid quality. For example, the algorithm is implemented reasonable choice of the finite element size, allowing building the grid of elements, with the curvature of a surface of the model and the best approximation – displaying its real geometry. For a simple model commonly used areas of hexahedral elements, and for the rest – tetrahedron.

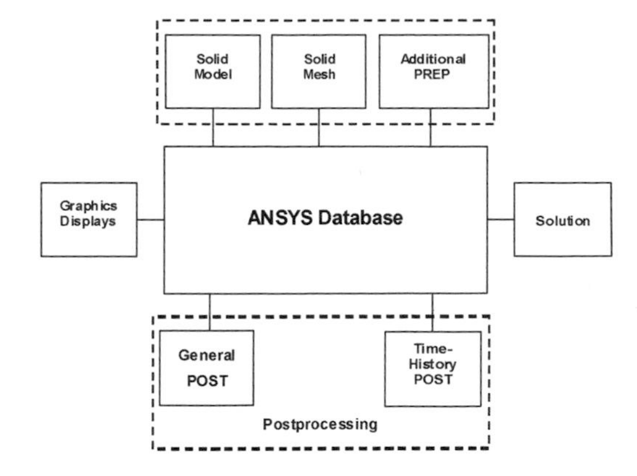

Figure 2. Connection diagram of a central database ANSYS.

The transition from the hexagonal grid to a tetrahedral with pyramidal elements is a valuable tool for modeling the geometry at the junction of areas with different grid. The researcher has the ability to automatically stitch the areas without need to enter the conditions-limitations or skip the median node points of elements and avoid the mathematical gaps in sought-for functions.

Adaptive grid construction is that after creating the solid model and the boundary conditions, ANSYS generates a finite element grid and performs calculations based on it. Then, estimates the error introduced into the finite element model and change the grid size from one solution to the other solution as long as the estimated error is less than a predetermined value or will reach the limited of iterations.

With the help of parametric projecting language (APDL) in the solid state module simulation of ANSYS runs parameterization of the model. Parameters are used as objects from which the performance of certain procedures is depend on, they can control the geometrical dimensions of the model, providing an opportunity, in the course of computing experiment, change the size in the following analysis. Attributes of parameterization are entered into the log-file in ASCII format, which contains all information entered during the work. The parameters in this file can be changed, then enter new data into the ANSYS program to rebuild the model with the correct size. Parametric log-file that uses solid modeling tools of ANSYS software for calculating models creation is especially useful during optimizing. By setting the size of the computational area with parameters in the optimization process can change the solid-state and finite-element model. In this case, the boundary conditions, are applying for a new computational model automatically.

Any material creates by selecting from a structured list of the required total characteristics and assigning them specific values. All possible models of physical medium behavior or characteristics of materials are divided into logical groups, which are selected by, or on the basis of the exclusion, or on the basis of superposition, it all depends on the compatibility of selected physical parameters. In the result is a series of spreadsheets to the principle of an inverted tree. The main table is a summary list of selected characteristic of a specific material.

The combination of specific relations for the fluidity condition, flaw of flow and law of hardening defines a particular model of the plastic material behavior. In ANSYS modeled the following types of plastic behavior: a classical linear kinematic hardening, polygonal kinematic hardening, linear isotropic hardening, polygonal isotropic hardening, the anisotropic behavior, materials models of Drucker-Prager and Ananda. Besides that, the user can define own version of the plastic model.

The model of classical linear kinematic hardening describes the behavior of conventional metallic materials, schematic diagram of an elastic deformation which has a plot and a plot of linear hardening. This model is applicable to most common, initially isotropic structural metals at the area of low strains. Used by the induced-modified by Mises' fluidity condition and associated law of flow. Manifestation of kinematic hardening is the Bauschinger effect.

Polygonal model of kinematic hardening also applies to metals, but are more applicable to those of them, that has figure more than one linear section of hardening. This model employs overlay, or Besselinga scheme to describe the complex behavior of polygonal material by combining the individual responses received on the basis of a simple dependency "stress-strain". Used a modified by Mises yield condition and associated flow law. Manifestation of kinematic hardening is the Bauschinger effect.

The model of a linear isotropic hardening refers to the common, widely used metallic materials with linear hardening. It is applicable to isotropic materials and in-state deformation significantly preferable to a model with kinematic hardening. Mises fluidity criterion is used in conjunction with the equations of the theory of Prandtl-Reuss. Bauschinger effect is not taking into account.

Polygonal model of isotropic hardening describes the behavior of conventional materials, hardening with increasing strain, and describes more accurately for large deformations. Fluidity condition of Mises is used; the Bauschinger effect is not modeled.

The model of anisotropic behavior describes of materials that behave differently in the stretch and compression or differently deform in different directions. The use of isotropic hardening allows using this model to determine the work of hardening in anisotropic material. Use a modified fluidity condition of Mises associated law of flow.

Drucker-Prager model is applicable to such grainy, granular materials such as rocks, concrete or soil. Use Mises fluidity condition, depending on the medium pressure to modeling the increase in the limit of fluidity of stress of the material with the full pressure. Law of flow can be associated or no associated. Hardening is missing.

Anand model describes the behavior of metals at elevated temperatures, but can be used at lower too. It is a model of an isotropic material hardening with increasing loading rate; a model that is usually sets by parameters of state, and not by using the curve "stress-strain". Anand model uses Mises fluidity condition with associated law of flow.

User's model can be used to specify any actual non-linearity behavior of the material. Subprogram written by user in the language FORTRAN, is introduced to the program ANSYS, and user's model can be used with the other.

Select the type of calculation determines which of the characteristics of the material to be used in construction of particular finite element matrix.

3 NUMERICAL MODELS

In some cases, the solution of nonlinear task of mechanics of solid media cannot be performed using one type of computational model that is we are talking about performing the task in several interrelated stages (Morozov 2010). There are two stages in task of geomechanics, if not taken into account the factors of the original tectonic stress, water production, etc.

The first step is the determination of stresses and strains of the static load, showing its condition at the time of ideal contact of set support and contour of mining opening, formed by rocks of the massif. The second stage is to analyze the processes occurring in the rock massif during the time. These processes include the rheology of rocks.

The behavior of the elements of the frame support (Figure 3) depends on the conditions of the problem. When modeling has elastic formulation, response of the lock setting is only possible in recording to dynamic components, which lead immediately to a significant complication of the task, because it requires changes in accounting of contact area, and, consequently, frictional force acting in the lock. In the elastic-plastic formulation of the task is possible to model the behavior of the lock, even without large displacement, to model the behavior of the lock by using the so-called "synthetic" approach. The essence of such a solution is to replace the roof support element, in this case -the lock, with geometric objects with physical parameters that allow the most complete display required for the modeling of the real characteristics of the object.

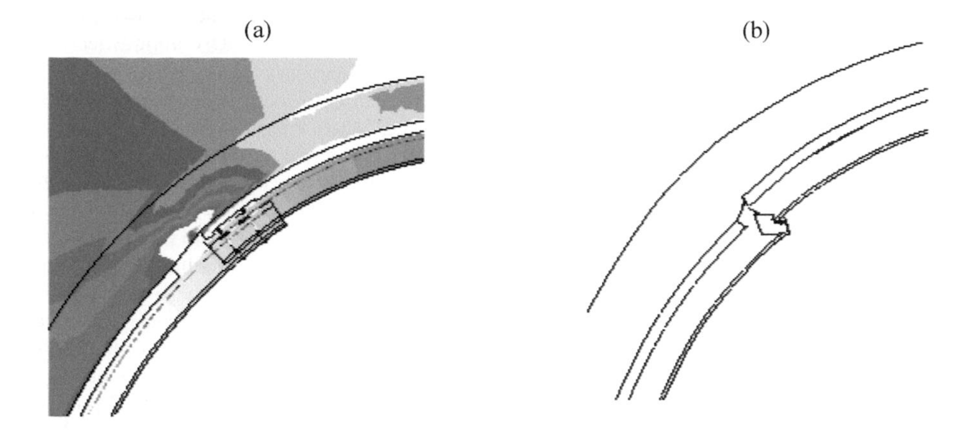

Figure 3. Modeling of the node unit of suppleness of the frame support: (a) geometrically accurate model, and (b) model of the node unit, providing its mobility without taking an account the dynamics.

In the calculations in the elastic formulation of the distribution of stresses and deformations in a layered rock massif is usually significantly different from the elastic-plastic recording to rheology. As a rule, it is connected with the effect of slip at the boundaries of lithological differences (Figure 4). As on the contacting surfaces of the rock layers are modeled two symmetrical non-connected node units, then they move relative to each other a lot more in the modeling of the nonlinear environment than for the deformable environment with small movements. Thus, it becomes possible mediately modeling the local dumped rock into the vesicle of mining opening, without any "synthetic" approaches.

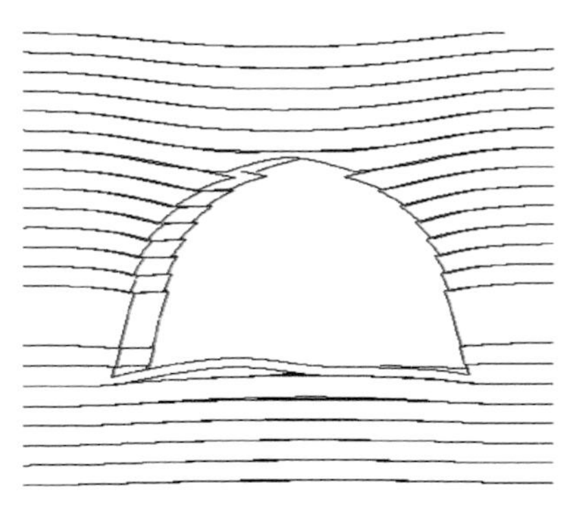

Figure 4. Relative slippage of rock layers in the task solution of nonlinear definition.

Changing the calculation model, conclude in choosing other rheological characteristics for materials used in the computational experiment, can greatly affect the development process of changing SSS system in time (Figure 5). As shown in the figure presented by the final movement of distribution diagram of replacement of the circular section contour of the mining opening, passed on the coal seam, and surrounded by sandstone, has a different geometry depending on the application of the same rheological characteristics, only applied to different objects of the same computational areas. With an increasing number of rock layers and complication of the geometry of the calculating area, influence of the rheological properties of materials is only increase, which together with time factor brings to significant changes in the distribution pattern of stresses and strains. So, the localization of zones of possible occurrence of cracks can be changed.

The actual physical processes occurring in the immediately adjacent to the mining opening contour processing rock massif, include softening and following complete destruction of the rocks. As noted above, these processes occur over time and require consideration of nonlinearity of the medium used in the modeling. That is, the analysis of main cracks becomes the third final step of solutions to describe the behavior of the rock massif in the vicinity of mining opening (Morozov 2002).

The implementation of this step is fundamentally

different for the development workings and working faces. The basis for these differences is the following factors: time of operation, methods of maintaining the original contour of mining opening, the degree of mutual influence and the geometry of the section.

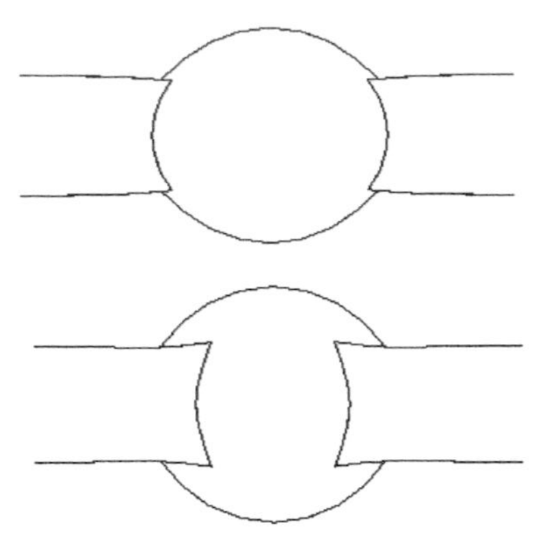

Figure 5. Displacement of mining opening contour with different ratios of rheological factors of coal and rocks.

The formation of cracks in the roof, for example, the drift with circular cross-section is usually a relatively long physical processes taking place in several stages, from which you want to distinguish two main ones: the development of cracks to the zone of influence of sewage treatment works and the development of cracks in this zone. In the result of research by SSS of rock massif around the mining opening should develop in stages to consider the development of cracks in its roof. In the first stage main cracks formed along the axis of the mining opening of the first and second type. Some of these cracks cease to grow. And some that are in the areas of maximum gradient of the strain continued to grow until the meeting in the rock massif, above the roof of mining opening. In such circumstances, the load to frame support of mining opening begins to increase rapidly, which leads to trigger off support's locks.

After entering these cracks in the zone of influence of sewage treatment works in the crack begins to dominate the third type. Moreover, cracks are formed along the cross section of the mining opening, or at small angles to this section. The result is cracking system, it is extremely difficult give in to detailed modeling. Modeling such a system of cracks requires a certain ranging with increasing sampling within a single computational area. This leads to the use in the calculation of "synthetic" approach to the description of geometric cracking, investigated in the course of solving the problem. That in turn means an additional phase of the study with a modified geometry of the computational area.

Cracks, prior to the collapse of the roof in the excavation face, initially developed by the second type, starting to grow on the surface of the roof working face, but with the growth of deformations on a vertically oriented cleaved rocks formed in the top of the development of the crack, confirm to the first type. Joint development of these cracks leads to the formation of packs of loose rocks, within which, in consequence, when crossing the front of the crack unit formed rocks, collapsing in the goaf. Thus, in calculating the behavior of roof of excavation face involving mechanics of fracture, task is solved by using two types of analysis of cracks in the rock massif.

As a result of the calculations performed can be considered as a separate phase of the quasi-dynamic modeling with a parametric variation of the lengths of face drift and the magnitude of bottom-hole drifts out space in the plane of the working face.

4 CONCLUSIONS

The main feature of numerical modeling application by method of finite element of geomechanics tasks includes in ability to display with high degree of confidence, during the experiment a large number of features of the behavior of real objects of study. Described above approaches in dealing with individual applied task of mechanics of solid environment and fracture mechanics have a wide range of applied possibilities and allow to modeling the development of SSS system "massif-support" according to the influence to the time and the destruction of its individual elements.

At the same geometrical features and terms of physical processes can be described within a multistep solution to the task. Each of the steps carried out related to the previous and following by instantaneous state of stress and strain of fields, which allows you to perform an analysis of the results obtained with the smallest possible set of assumptions and constraints that invariably arise in a phased solution of boundary value problems in mechanics.

The use of different finite elements allows optimizing the dimension of solving tasks and describing various local features of the objects. The solution of spatial task of geomechanics is the surest approach to obtain adequate results when analyzing the state of mining openings and their mutual influence in the area of sewage treatment works. In this case, the use of second-order tetrahedra as final elements in the areas of greatest stress gradients in combination with parallelepipeds on the periphery of the computational area and macro elements in the areas of destruction of the material makes it possible to obtain results with high initial accuracy of computation.

Taking into account of fracturing, and development of cracks during the growth of external loads in the solution of applied tasks, associated with finding the best approaches to solving each particular case. Themselves within a single crack in the computational area can be formed and grow in different patterns. The most common of these are implemented in a numerical form for the finite element method. Thus, when analyzing the stability of the connection of mining opening and working face by the method of finite element can be obtained mathematically accurate and physically adequate stress-strain state, taking into account the time factor and the possible processes of destruction of rock massif.

REFERENCES

Chigarev, A.V., Kravchuk, A.S. & Smalyuk, A.F. 2004. *NSYS for engineers: inquiry book*. Moscow: Mashinostroenie-1: 512.

Basov, K.A. 2002. *ANSYS examples and tasks*. Moscow: ComputerPress: 224.

Morozov, E.M., Muzeymnek, A.J. & Shadsky, A.S. 2010. *NSYS in the hands of the engineer: fracture mechanics*. Moscow: LENAND: 456.

Geomechanical Processes During Underground Mining – Pivnyak, Bondarenko, Kovalevs'ka & Illiashov (eds)
© 2012 Taylor & Francis Group, London, ISBN 978-0-415-66174-4

Software development for the automatic control system of deep-water hydrohoist

Y. Kyrychenko, V. Samusia & V. Kyrychenko
National Mining University, Dnipropetrovs'k, Ukraine

ABSTRACT: The article is devoted to software creation for the transients modes automatic control systems of the deep-water hydrohoists. The purpose is reached by means of mathematical simulation on meta-, macro- and micro- levels of the most typical transient processes in the basic technological units of underwater equipment. The primary result of the research is the mathematical formulation of hydraulic hoisting basic elements: metering device, supply and elevating pipes, mixer, pneumatic pipeline, compressor and receiver. The developed mathematical support describes distribution of shock and kinematic waves in hydraulic hoist's pneumatic-hydraulic paths as well as transient processes in the power equipment. It formed the basis for software development allowing the basic transient processes' parameters defining for installations with different productivity and mining depths.

1 INTRODUCTION

Over the past few decades, the international community is showing more interest in the development of mineral deposits of the oceans, where there are rich deposits of polymetallic ores, the number of which greatly exceed reserves on land. To date, the most promising in terms of industrial development, are deposits of polymetallic nodules, polymetallic sulphide ores, cobalt-rich ferromanganese crusts, hydrates and phosphorites. The largest companies in the U.S., UK, Canada, Germany are actively working to develop technical ways and means of extraction of polymetallic ores from the sea bottom.

2 FORMULATING THE PROBLEM

This article focuses on the actual problem – the development of efficient ways and means of mining ore deposits of the oceans.

According to the experts today the most promising way of transportation of mined minerals on the basic floating craft is a deep-water airlift hydrolift due to its high rates of reliability in difficult conditions at great depths.

Below is a fragmental mathematical description of the most typical process in basic technological units of underwater equipment.

3 MATHEMATICAL DESCRIPTIONS OF PROCESSES IN THE BASIC UNITS OF AIRLIFT HYDROHOIST

1. "Metering device's drive" unit is described as follows. The dependence between the output parameter – the frequency of shaft rotation $n_m(t)$ and the input setpoint $u_{is}(t)$ is determined by the following differential equation solvation (Kyrychenko, Shvorak, Kyrychenko & Yevteev 2010):

$$T_{tm}\frac{dn_m(t)}{dt}+n_m(t)=K_{tm}u_{is}(t),$$

as

$$n_m(t)=K_{tm}(1-e^{-\frac{t}{T_{tm}}})u_{is}(t),$$

where T_{tm} – determined by the curves of transient process of the metering device's drive by the changing $u_{is}(t)$ from the rated value u_{is}^{nom} to $1.1u_{is}^{nom}$;

$$K_{tm}=\frac{n_m^{nom}}{u_{is}^{nom}},$$

where n_m^{nom} – nominal metering device's shaft rotation frequency.

2. "Metering device" unit has the following description. Output consumption $Q_{md}(t)$ is determined by the formula (Kyrychenko, Shvorak, Kyrychenko & Yevteev 2010):

$$Q_{md}(t) = K_{d1} \cdot n_d(t - \tau_{is}),$$

where K_{d1} – coefficient depending on characteristics of solid minerals; τ_{is} – delay time when moving the solids from the bunker into the transport pipeline, value of which is determined by the design parameters of the metering unit and its shaft rotation frequency. The input size of the solids d_e^d is determined by the formula

$$d_e^d = K_{red}^d \cdot d_e,$$

where d_e – average dimension of the solids in the bunker; K_{red}^d – coefficient of the particle size reduction by the metering unit.

3. "Supplying pipe" unit. The input parameters for this unit are the output parameters of the "metering device" unit. Taking into account the technological scheme of the hydrohoist, supply pipe should be divided into two sections. The first section of pipe transports pulp from the level of the metering device to mixer. Fast wave processes associated with the possible water hammer on this site, are described with the transient model (Goman, Kyrychenko & Kyrychenko 2008):

$$(1-C_1)\left(\frac{\partial p}{\partial t} + V_0\frac{\partial p}{\partial x}\right) - \rho_0 a_0^2\left(\frac{\partial C_1}{\partial t} + V_0\frac{\partial C_1}{\partial x}\right) + \\ + \rho_0 a_0^2 (1-C_1)\frac{\partial V_0}{\partial x} = 0. \quad (1)$$

$$C_1\left(\frac{\partial p}{\partial t} + V_1\frac{\partial p}{\partial x}\right) - \rho_1 a_1^2\left(\frac{\partial C_1}{\partial t} + V_1\frac{\partial C_1}{\partial x}\right) + \\ + \rho_1 a_1^2 C_1\frac{\partial V_1}{\partial x} = 0. \quad (2)$$

$$\left(1+\frac{C_{1k1}}{2}\right)\left(\frac{\partial V_0}{\partial t} + V_0\frac{\partial V_0}{\partial x}\right) - \frac{C_{1k1}}{2}\times \\ \times\left(\frac{\partial V_1}{\partial t} + V_1\frac{\partial V_1}{\partial x}\right) + \frac{(1-C_1)}{\rho_0}\frac{\partial p}{\partial x} = -(1-C_1)g\, sin\alpha - \\ - \frac{\lambda_n}{2D_t}\frac{\rho_n}{\rho_0}|V_n|V_n - -\frac{3}{8}\left[\frac{C_1 C_{x1}}{R_1}|V_0 - V_1|(V_0 - V_1)\right]. \quad (3)$$

$$\left(\frac{\rho_1}{\rho_0} + \frac{k_1}{2}\right)\left(\frac{\partial V_1}{\partial t} + V_1\frac{\partial V_1}{\partial x}\right) - \left(1+\frac{k_1}{2}\right)\times \\ \times\left(\frac{\partial V_0}{\partial t} + V_0\frac{\partial V_0}{\partial x}\right) + \frac{1}{\rho_0}\frac{\partial p}{\partial x} = -\frac{\rho_1}{\rho_0}g\, sin\alpha + \\ + \frac{3}{8}\frac{C_{x1}}{R_1}|V_0 - V_1|(V_0 - V_1) \quad (4)$$

Indications of the flow parameters, used in equations (1)-(4) are deciphered below in order to avoid repetitions.

The system of equations (1)-(4) is closed with respect to unknown p, V_0, V_1, C_1. The coefficient of hydraulic resistance during the motion of the pulp λ_n is determined from empirical formulas (Kyrychenko, Shvorak, Kyrychenko & Yevteev 2010). Distribution of solid phase concentration kinematic waves in the first part of the supply pipe is described by a quasi-stationary model (Kyrychenko 2009):

$$(\rho_1 - \rho_0)\frac{\partial C_1}{\partial t} - \frac{\partial}{\partial x}\left[C_1 V_1 \rho_1 + (1-C_1)V_0\rho_0\right] = 0,$$

$$\frac{\partial p}{\partial x} = g\left[C_1\rho_1 + (1-C_1)\rho_0\right]sin\alpha + \\ + \frac{\lambda_n}{2D_t}\left[C_1\rho_1 V_1^2 + (1-C_1)\rho_0 V_0^2\right] - C_1\rho_1 V_1\frac{\partial V_1}{\partial x} - \\ -(1-C_1)\rho_0 V_0\frac{\partial V_0}{\partial x} = 0,$$

$$V_0 - V_1 = V_\infty(1-C_1)^{n-1}, \quad \frac{\partial C_1}{\partial t} + V_W\frac{\partial C_1}{\partial x} = f,$$

$$V_W = C_1 V_1 + (1-C_1)V_0 + \frac{\partial}{\partial C_1}\left[V_0(1-C_1)C_1\right],$$

$$\lambda_n = \frac{1}{(1.8 lg\, Re_n - \delta)^2},$$

$$Re_n = \frac{4(Q_1 + Q_0)}{\pi D_t}\left(\frac{C_1}{v_1} + \frac{1-C_1}{v_0}\right),$$

δ – absolute roughness of pipe; Q_i – volumetric flow rate of components; v_i – kinematic coefficient of viscosity of the components.

The movement of sea water in the second section of the inlet section of the supply pipe to the level of the metering auger is described by the following equations:

$$\left(\frac{\partial p}{\partial t} + V_0\frac{\partial p}{\partial x}\right) + \rho_0 a_0^2\frac{\partial V_0}{\partial x} = 0,$$

$$\left(\frac{\partial V_0}{\partial t} + V_0\frac{\partial V_0}{\partial x}\right) + \frac{1}{\rho_0}\frac{\partial p}{\partial x} = -g\, sin\alpha - \frac{\lambda_0}{2D_t}|V_0|V_0.$$

4. "Mixer" unit. The input parameters are the output parameters of the unit "supply pipe", namely, the flow rate and pulp density as well as air flow at the output of air line. At the mixer's output pressure $p(t)$, flow rate

of the pulp $Q_p(t)$ and density of the pulp $\rho_p(t)$ are formed. In the general case (Kyrychenko, Shvorak, Kyrychenko & Yevteev 2010)

$$p(t) = p^{st}(t) + \Delta p(t),$$

where $p^{st}(t)$ – static component, calculated by the method, described in (Kyrychenko, Shvorak, Kyrychenko & Yevteev 2010). $\Delta p(t)$ – oscillatory component. By $q(t) > 1.1 q_{opt}$ and $\Delta p(t) = \Delta p^{max}(t) sin(kt)$, where q_{opt} – calculated optimal specific air consumption; k – oscillation frequency corresponding to the shell flow structure (Kyrychenko, Yevteev & Romanyukov 2007; Yevteev & Kyrychenko 2009). By $q(t) > 1.1 q_{opt}$ oscillatory component in the mixer has a decaying nature:

$$\Delta p(t) = \Delta p^{max}(t) \cdot e^{-nt} \cdot sin\left(\sqrt{k^2 - n^2} \cdot t\right),$$

where n – decrement. It should be noted that the definition of the oscillatory component of pressure in the mixer, for the case when different flow modes are consecutively changed in the elevating pipe, is an independent research (Kyrychenko 2009).

5. "Elevating pipe" unit. The input parameters are the same as the output ones of the mixer. Fast wave processes during the movement of the pulp in the elevating pipe are described by the non-stationary model (Goman, Kyrychenko & Kyrychenko 2008):

$$(1 - C_1 - C_2)\left(\frac{\partial p}{\partial t} + V_0 \frac{\partial p}{\partial x}\right) - \rho_0 a_0^2 \left(\frac{\partial C_1}{\partial t} + V_0 \frac{\partial C_1}{\partial x}\right) - \rho_0 a_0^2 \left(\frac{\partial C_2}{\partial t} + V_0 \frac{\partial C_2}{\partial x}\right) + \rho_0 a_0^2 (1 - C_1 - C_2) \frac{\partial V_0}{\partial x} = 0. \quad (5)$$

$$C_1\left(\frac{\partial p}{\partial t} + V_1 \frac{\partial p}{\partial x}\right) + \rho_1 a_1^2 \left(\frac{\partial C_1}{\partial t} + V_1 \frac{\partial C_1}{\partial x}\right) + \rho_1 a_1^2 C_1 \frac{\partial V_1}{\partial x} = 0. \quad (6)$$

$$C_2\left(\frac{\partial p}{\partial t} + V_2 \frac{\partial p}{\partial x}\right) + \rho_2 a_2^2 \left(\frac{\partial C_2}{\partial t} + V_2 \frac{\partial C_2}{\partial x}\right) + \rho_2 a_2^2 C_2 \frac{\partial V_2}{\partial x} = 0. \quad (7)$$

$$\left(1 + \frac{C_1 k_1}{2} + \frac{C_2 k_2}{2}\right)\left(\frac{\partial V_0}{\partial t} + V_0 \frac{\partial V_0}{\partial x}\right) - \frac{C_1 k_1}{2}\left(\frac{\partial V_1}{\partial t} + V_1 \frac{\partial V_1}{\partial x}\right) - \frac{C_2 k_2}{2}\left(\frac{\partial V_2}{\partial t} + V_2 \frac{\partial V_2}{\partial x}\right) + \frac{(1 - C_1 - C_2)}{\rho_0}\frac{\partial p}{\partial x} = \varphi_0, \quad (8)$$

$$\left(\frac{\rho_1}{\rho_0} + \frac{k_1}{2}\right)\left(\frac{\partial V_1}{\partial t} + V_1 \frac{\partial V_1}{\partial x}\right) - \left(1 + \frac{k_1}{2}\right)\left(\frac{\partial V_0}{\partial t} + V_0 \frac{\partial V_0}{\partial x}\right) + \frac{1}{\rho_0}\frac{\partial p}{\partial x} = \varphi_1, \quad (9)$$

$$\left(\frac{\rho_2}{\rho_0} + \frac{k_2}{2}\right)\left(\frac{\partial V_2}{\partial t} + V_2 \frac{\partial V_2}{\partial x}\right) - \left(1 + \frac{k_2}{2}\right)\left(\frac{\partial V_0}{\partial t} + V_0 \frac{\partial V_0}{\partial x}\right) + \frac{1}{\rho_0}\frac{\partial p}{\partial x} = \varphi_2, \quad (10)$$

where

$$\varphi_0 = -(1 - C_1 - C_2) g\, sin\alpha - \frac{\lambda}{2D_t}\frac{\rho_p}{\rho_0}|V_p|V_p - \frac{3}{8}\left[\frac{C_1 C_{x1}}{R_1}|V_0 - V_1|(V_0 - V_1) + \frac{C_2 C_{x2}}{R_2}|V_0 - V_2|(V_0 - V_2)\right],$$

$$\varphi_1 = -\frac{\rho_1}{\rho_0} g\, sin\alpha + \frac{3}{8}\frac{C_{x1}}{R_1}|V_0 - V_1|(V_0 - V_1),$$

$$\varphi_2 = -\frac{\rho_2}{\rho_0} g\, sin\alpha + \frac{3}{8}\frac{C_{x2}}{R_2}|V_0 - V_2|(V_0 - V_2),$$

$$\frac{1}{a_1^2} = \frac{\rho_1}{K_1} + \frac{\rho_1}{F}\left(\frac{\partial F}{\partial p}\right), \quad \frac{1}{a_2^2} = \frac{\rho_2}{K_2} + \frac{\rho_2}{F}\left(\frac{\partial F}{\partial p}\right),$$

$$\frac{1}{a_0^2} = \frac{1}{a_l^2} + \frac{\rho_0}{F}\left(\frac{\partial F}{\partial p}\right), \quad a_l^2 = \frac{K_l}{\rho_0},$$

$$K_1 = \frac{E_1}{3(1 - 2\theta_1)}, \quad \frac{1}{K_2} = \frac{1}{\rho_2}\left(\frac{\partial \rho_2}{\partial p}\right),$$

$$\rho_p = \rho_0^* + \rho_1^* + \rho_2^* = (1 - C_1 - C_2)\rho_0 + C_1\rho_1 + C_2\rho_2,$$

$$V_p = \frac{1}{\rho_{см}}\left(\rho_0^* V_0 + \rho_1^* V_1 + \rho_2^* V_2\right).$$

K_1, E_1, θ_1 – the bulk modulus, Young's modulus and Poisson's ratio of solids; K_2 – the bulk modulus of gas bubbles; K_l – the bulk modulus of elasticity of the fluid; a_l – the speed of sound in the pure unbounded liquid; R_1, R_2 – the equivalent radius of solids and gas bubbles; k_1, k_2 – coefficients taking into account the effect of non-sphericity and the concentration of solids and air bubbles at the connected mass; g – acceleration of gravity; a – the pipe angle; D_t – the diameter of the pipeline; λ – the coefficient of Darcy; t – time; C_{xs}, C_{xa} – the resistance coefficients of solids and air bubbles; C_i – phase volume concentration; p – the pressure; ρ_i – the true density of the phase; ρ_i^* – the reduced density of phase; V_i – the phase velocity; x – longitudinal coordinate. The subscripts denote: "0" – water, "1" – solids; "2" – gas bubbles, "p" – three-phase water-air mixture of groundwater. F – area of open pipeline section.

The six equations system (5)-(10) is closed related to the unknown functions p, V_0, V_1, V_2, C_1 and C_2 and allows to study the non-stationary processes while motion of a three-phase mixture in the lift pipe.

Distribution of the kinematic wave of air phase concentration (see the second stage of the launch) is described by the quasi-steady model (Kyrychenko 2009):

$$(\rho_0-\rho_2)\frac{\partial C_2}{\partial t}-\frac{\partial}{\partial x}\left[C_2V_2\rho_2+(1-C_2)V_0\rho_0\right]=0,$$

$$\frac{\partial p}{\partial x}=g\left[C_2\rho_2+(1-C_2)\rho_0\right]sin\alpha+$$

$$+\frac{\lambda_p}{2D_t}\left[C_2\rho_2V_2^2+(1-C_2)\rho_0V_0^2\right]-$$

$$-C_2\rho_2V_2\frac{\partial V_2}{\partial x}-(1-C_2)\rho_0V_0\frac{\partial V_0}{\partial x}=0\,,$$

$$V_2-V_0=V_\infty(1-C_2)^{n-1},\ \frac{\partial C_2}{\partial t}+V_W\frac{\partial C_2}{\partial x}=f\,,$$

$$V_W=C_2V_2+(1-C_2)V_0+\frac{\partial}{\partial C_2}\left[V_2(1-C_2)C_2\right],$$

$$f=-\frac{C_2}{(pV_2)}\cdot\frac{d(pV_2)}{dt},$$

$$\lambda_c=\left\{2\,lg\left[\left(\frac{6.81}{Re_c}-\frac{\delta}{3.7D_t}\right)\right]\right\}^{-2},$$

$$Re_c=\frac{4(Q_2+Q_0)}{\pi D_t}\left(\frac{C_2}{\nu_2}+\frac{1-C_2}{\nu_0}\right),$$

where the m subscript denotes the parameters of water-air mixture.

6. "Air line" unit. The input parameters are the pressure and air flow at the outlet of the receiver compressor station, and output – the pressure and the air flow at the mixer inlet.

The water displacement process from the vertical air-line by compressed air (first stage of the launch) is described by the following equation (Kyrychenko 2009):

$$\frac{1}{g}\left\{\left[1+\left(\frac{d}{D_t}\right)^2\right]\cdot h_i-x_1\right\}\cdot\frac{d^2x_1}{dt^2}+\frac{1}{2g}(K_1-K_2)\left(\frac{dx_1}{dt}\right)^2+$$

$$+x_1+\frac{p_1-p_a}{\rho g}=0,$$

where

$$K_1=\varsigma_1-1+\lambda_{pn}\frac{h_i}{d}+\left(1+\frac{h_i}{d}\lambda_0\right)\left(\frac{d}{D_t}\right)^4,\ K_2=\frac{\lambda_{pn}}{D_t}x_1,$$

where d – diameter of air line; h_i – mixer immersion; x_1 – the vertical coordinate whose origin coincides with the input section of air line, and positive direction coincides with the direction of air movement in the air line; P_a – atmospheric pressure; ς_1 – the coefficient of local mixer resistance; λ_{pn}, λ_0 – Darcy's coefficients for the air line and lift pipe filled with sea water.

The entire air line will be filled with compressed air at the end of the first stage of launching. Non-stationary isothermal movement can be described by the following system of equations (Kyrychenko, Shvorak, Kyrychenko & Yevteev 2010):

$$\left(\frac{\partial p}{\partial t}+V_2\frac{\partial p}{\partial x}\right)+\rho_2a_2^2\frac{\partial V_2}{\partial x}=0\,,$$

$$\left(\frac{\partial V_2}{\partial t}+V_2\frac{\partial V_2}{\partial x}\right)+\frac{1}{\rho_2}\frac{\partial p}{\partial x}=-g\,sin\alpha-\frac{\lambda_2}{2D_t}|V_2|V_2\,.$$

The conventional assumption of process isothermality is hardly acceptable due to the specifics of supplying air system. Therefore, the steady flow in the air line can be described by the following equations considering the temperature variation of compressed air (Kyrychenko 1988):

$$\frac{dp}{dx_1}\left(1-\frac{16M_2k_hRT}{\pi^2d^4p^2}\right)=\lambda_{nu}\frac{8M_2^2k_hRT}{\pi^2d^5p}-\frac{gp}{k_hRT}-\frac{16M_2^2R}{\pi^2d^4p}\frac{dT}{dx_1},$$

$$k_m(T-T_0)dx_1=C_pM_2dT+gM_2dx_1-\frac{16M_2^3(k_hRT)^2}{\pi^2d^4}\frac{dp}{p^3},$$

where k_m – the linear (per unit length of the tube) coefficient of air to water heat transfer; M_2 – mass flow of gas; T_0 – absolute temperature of the washing line water; k_h – coefficient that takes into account the properties differences between the high pressure compressed air and the properties of an ideal gas; R – gas constant; C_p – mass isobar heat capacity of air. The mechanism which considers the change of compressed air temperature which influences the dynamics of the air line is described in (Kyrychenko 2009).

Metamathematical description of transient processes in all levels of underwater equipment does not fit into the limited scope of this article. Therefore, the authors have to only provide the references to their works, which are describing the following transitional regimes in detail (for example, for launching stage):

– acceleration of water in the inlet pipe (Kyrychenko, Shvorak, Kyrychenko & Yevteev 2010);

– the lifting of the first portion of the solid material to the output section of the lift tube (Kyrychenko, 1998);

– the "pump – flexible pipes" unit (Kyrychenko, Shvorak, Kyrychenko & Yevteev 2010).

7. A standard accumulation unit as the modulation for a compressible gas form is used for the receiver. The receiver is described by the first order linearized equation (Kyrychenko, Shvorak, Kyrychenko & Yevteev 2010):

$$T_r \frac{d\Delta Q_r}{dt} + \Delta Q_r = \Delta Q_r ,$$

where T_r – time constant, which depends on the initial regime. The transfer function of the receiver is:

$$W(p) = \frac{1}{T_r p + 1} = \frac{Q_r(p)}{Q_c(p)},$$

where $Q_r(p)$ – output flow of the receiver; $Q_c(p)$ – productivity of compressor.

8. "Station of Compressor" unit description. The input parameter – the controlling action $u_{cs}(t)$. Output parameter – the air flow at the inlet of air-conduction $Q_{vk}(t)$. Communication is described by the differential equation (Kyrychenko, Shvorak, Kyrychenko, Yevteev 2010):

$$T_c \frac{dQ_{vk}(t)}{dt} + Q_{vk}(t) = K_{tc} \cdot u_{cs}(t),$$

where T_c – time constant of the compressor (as determined by the transient characteristics); K_{tc} – transfer coefficient, calculated by the nominal data of compressor station:

$$K_{kc} = \frac{Q_{vk}^{nom}}{u_{cs}^{nom}},$$

where Q_{vk}^{nom} – air flow at the output of the compressor station at the nominal discharge pressure and the nominal temperature; u_{cs}^{nom} – nominal setpoint-term of compressor station.

4 CONCLUSIONS

The software for the proposed method of automatic control of deep hydraulic hoisting is developed. It was used for calculation of basic transient processes duration in the core elements of hydraulics system. The calculations were performed in a wide range of plant productivity and the depths of field development areas.

The developed software "HydroWorks 3p" is designed to calculate the dynamics of three-phase flows. The software is compatible with SolidWorks 2010/2011 and supports MS Windows Vista (x32, x64) and Windows 7 (x32, x64) operating systems. There are two installation modes: add-in and stand-alone. The software includes the following units:

– calculation unit – dynamic link library (dll), conducting the calculation method;

– add-in unit – Solid Works integrated dll;

– stand-alone executable unit;

– isualization unit.

The next researching step should be the development of software for calculation of the transition processes in the elements of underwater mining equipment (collection module with energy unit, a pump unit with a flexible pipe, tank spout with a screw feeder). The goal of the product is the harmonization and synchronization of parameters of transient modes with time-varying characteristics of the bottom block machines.

REFERENCES

Kyrychenko, Y., Shvorak, V., Kyrychenko, V. & Yevteev, V. 2010. *Dynamics of the deep-water hydrohoists in the ocean mining*. Dnipropetrovs'k: State Higher Educational Institution National Mining University.

Goman, O., Kyrychenko, Y. & Kyrychenko, V. 2008. *Determining the propagation velocity of pressure waves in the elements of deep-water airlift hydrohoist*. State

Higher Educational Institution National Mining University. Dnipropetrovs'k: Research bulletin of SHEI NMU, 9.

Kyrychenko, Y. 2009. *Mechanics of the deep-water hydrotransport systems in the ocean mining*. Dnipropetrovs'k: State Higher Educational Institution National Mining University.

Kyrychenko, Y., Yevteev, V. & Romanyukov, A. 2007. *The research of shell flow mode parameters in the elevating pipe of the deep-water airlift*. State Higher Educational Institution National Mining University. Dnipropetrovs'k: Research bulletin of SHEI NMU, 9.

Yevteev, V. & Kyrychenko, Y. 2009. *Experimental researches of parameters of vertical flow of three-component mixture in airlift hydrohoist*. State Higher Educational Institution National Mining University. Dnipropetrovs'k: Research bulletin of SHEI NMU, 1.

Kyrychenko, Y. 1988. *Analysis of compressed air flow in the pneumatic system of the airlift*. Dnipropetrovs'k: State University. Mathematical modeling of heat and mass transfer.

Kyrychenko, Y. 1998. *Numerical simulation of transient processes in deep-water airlift*. Dnipropetrovs'k: Mining mechanics and automatics,1.

 ISBN 978-0-415-66174-4

Degassing systems rational parameters selection at coal mines

N. Kremenchutskiy, O. Muha & I. Pugach
National Mining University, Dnipropetrovs'k, Ukraine

ABSTRACT: the substantiation is given to coal deposits degassing efficiency increase by means of calculation method development and optimization of degassing systems basic parameters at methane-abundant coal mines. Selection of boreholes rational number is carried out based on maximum methane debit provision, observing limit-permissible concentration of natural gas in degassing system. Total cost of degassing pipeline and consumed electro energy by vacuum-pump with specified debit of methane-air mix (MAM) is accepted as optimization criterion of degassing system parameters.

1 INTRODUCTION

In coal industry there is an acute problem of occupational safety provision. Special attention is paid to mines with high gas inflow where high methane concentration repeatedly leads to blow-ups that become the reason of large-scale accidents. The basic requirement for their prevention is safety rules observation during mining operations. But with increase of working depth and methane-inflow of coal seams it is impossible to ensure permissible gas concentration in the air. Under such conditions degassing of gas-bearing coal seams, bearing rocks and worked-out areas is an important technological process contributing to mining operations safety increase and face output increase.

The aim of the work is to consider and substantiate questions connected with coal deposits degassing efficiency increase, development of new and improving of already existing calculation methods and optimization of basic parameters of degassing systems:

– to perform natural researches of degassing systems defining dynamics of methane capture value and methane-air mix consumption that come into degassing network;

– to establish correlation dependences of degassing parameters and equation of regression based on natural observations and statistic materials depending on mining-technical conditions;

– to improve calculation methodology and boreholes number optimization, and distance between them providing maximal total methane debit captured at production, unit taking into account permissible methane concentration in gas pipeline;

– to develop methodology focused on pipeline diameter optimization based on economic indices for all parts of the pipeline that are included in the degassing network.

2 RESEARCH OBJECT

The "Almaznaya" mine of a production complex "Dopropol'eugol'" has become the object of experimental researches of coal seams degassing parameters.

The field of mentioned mine is located in north-western part of Krasnoarmeysk coal region of Donbass at the territory of Donetsk oblast in Ukraine. Based on gas-inflow volume, the mine is considered to be extra-dangerous having relative methane-bearing capacity equal to 32 m^3 / t of a daily output, by coal dust explosiveness - dangerous. At present, the mine has been developing l_3 and m_5 seams with thickness 1.45-2.5 and 1.0-1.1 respectively. The seams are not inclined to self-ignition, and they are not dangerous by bursts of coal and gas. Bedding of the seams is flat having dip angle of 8-12°. Biggest part of the host rocks is clay and sand shales with middle strength and stability values. Continuous prediction of outburst hazard is carried out during development of l_3 seam at depth of 600 m.

Development method of the mine field is panel-wise, mining method – by long pillars along the strike. Ventilation scheme of the mine is central-flank, ventilation method – exhaustive. Degassing of underworked coal seams and worked-out area is implemented during development of l_3 seam at "Almaznaya" mine.

Natural observations are carried out at the extraction area of the 5th north longwall face of southern stone incline (SSI) of l_3 seam at horizon of 550 m. In order to decrease methane release at this area the degassing of underworked coal seams is provided by means of boreholes drilling into the l_4^H seam. The boreholes are drilled from the filled ventilation drift after the longwall advance and with turning toward longwall face. Length of the degassing boreholes makes up 34 m. Turning angle between perpendicular that is drawn to the drift axis and borehole projection on the horizontal plane is $\varphi = 65°$; borehole inclination angle to the horizon is $\beta = 15°$. Distance between the boreholes is 8 m.

Methane-air mixture coming into the degassing boreholes is pumped out with help of surface vacuum-pump station equipped with four pumps of VVN-150/2 type. As the boreholes collars are underworked by the longwall face they get disconnected from gas pipeline and shut by chump plugs.

Area gas pipeline consists of pipes with diameter of 325 mm except for the adjoining to the longwall area having length of 100 m, where gas pipeline diameter is equal to 100 mm. Parallel to this area a pipeline with diameter of 100 mm is laid and it is designed for gas suction from the worked out area thanks to degassing column left in the caved ventilation working. There is a perforated vertical pipe outgrowth ("candle") with diameter of 250 mm and height – 2 m installed at the end of the pipeline for gas suction from the worked-out area. "Candles" are installed every 16-24 m.

Together with underworked coal seam degassing, isolated capture of methane from the ventilation drift dead-end is implemented. Capture of methane-air mixture is carried out along metallic pipeline with 800 mm diameter. From the worked-out area isolated from the working by chump plug, methane is sucked out by gas-suction unit equipped with VMCG-7 fan. The fan is located in special chamber located at the 4th northern conveyor drift SSI (southern rock incline) of 550 m horizon of l_3 seam between an incline and northern foot-way.

Methane-air mixture is supplied from the fan into the mixing chamber located at northern foot-way of southern rock incline above the 4th conveyor drift.

Degassing scheme of an extraction area is shown on Figure 1.

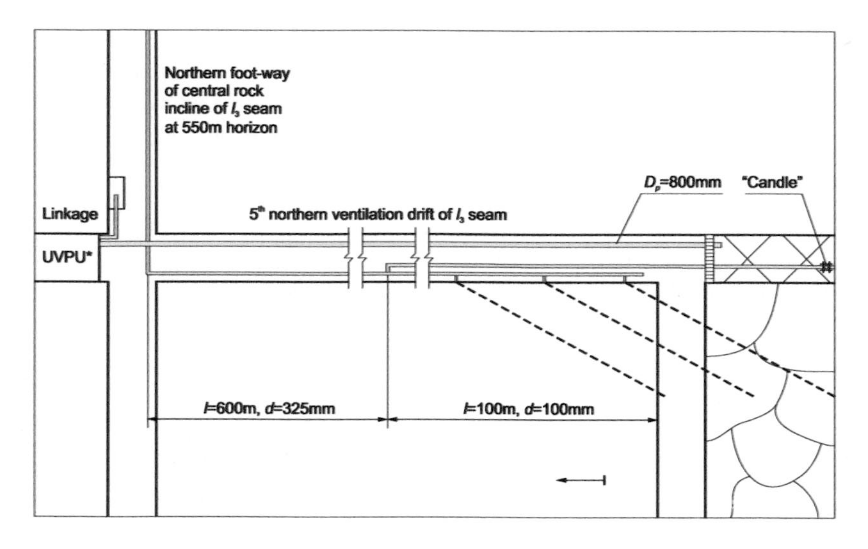

Figure 1. Degassing scheme of the 5th northern longwall of SSI of l_3 seam at "Almaznaya" mine. *UVPU – underground vacuum-pump unit.

3 EXPERIMENTAL RESEARCHES

In given mining-geological and mining-technical conditions the faces of degassing boreholes do not reach full unloading zone of an adjoining seam from rock pressure, where the spot with maximum methane release is located. In this case, methane debit captured by i-th degassing borehole is most reliably described by the following exponential dependence

$$I_i = a_1 e^{a_2 L_i} + a_3 e^{a_4 P_i},$$

where a_1, a_2, a_3, a_4 – regression coefficients.

After experimental observations processing, the following values of correlation coefficient and value of reliability are gained for the following dependences:

– $P = f(L) - r_{PL} = 0.85$ at $\mu = 20.8$;

– $c = f(L) - r_{cL} = 0.97$ at $\mu = 92.8$;

- $Q = f(L) - r_{QL} = 0.73$ at $\mu = 11.1$;
- $I = f(L) - r_{IL} = 0.95$ at $\mu = 52.7$;
- $Q = f(I) - r_{QI} = 0.23$ at $\mu = 1.8$;
- $I = f(P) - r_{IP} = 0.71$ at $\mu = 9.4$;
- $Q = f(P) - r_{QP} = 0.98$ at $\mu = 191.7$;
- $I = f(L,P) - R = 0.968$;
- $Q = f(L,P) - R = 0.999$.

Thus, the above-mentioned dependences have close correlation connection. The exception is $Q = f(I)$ dependence at which value of μ is less than the required one ($\mu \geq 2.6$). This is substantiated by a wide scatter of points toward different directions along axis of ordinates at small values of abscissa.

Under conditions of an extraction area of the 5th northern longwall of SSI of l_3 seam the regression equation looks like

$$I_i = 0.4665e^{-0.0956P_i}.$$

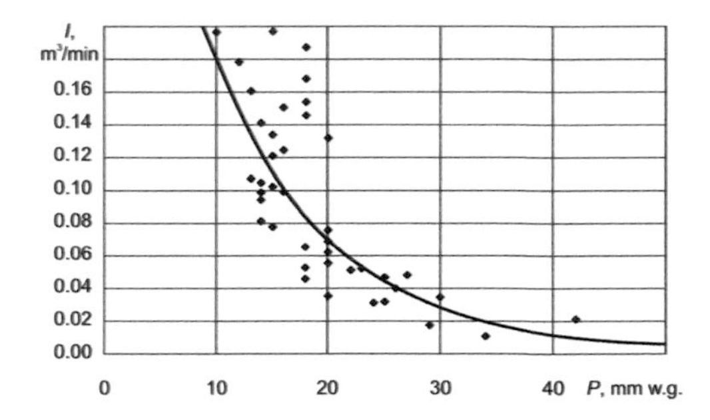

Figure 2. Dependence diagram of methane consumption on depressurizing in the collar of degassing borehole.

In order to solve the set tasks, the evaluation criterion of an approximation permissible error is accepted as ε = 15-20%. Average approximation error for this equation makes up 32.3%. Relatively high calculations error is substantiated by a considerable influence of the factors not taken into consideration by the given dependence. One of these factors is the factor of an inter-location of the borehole collar and longwall face line.

Dependence of methane debit captured by i-th borehole, from the borehole collar to the longwall is presented by the following exponential equation

$$I_i = b_0 e^{b_1 L_i}.$$

For the conditions of an extraction area of the 5th northern longwall of SSI of the l_3 seam the last equation will have a look

$$I_i = 0.1947e^{-0.0743L_i}.$$

Graphical illustration of the given dependence is shown on Figure 3. An average approximation error is 14.9%.

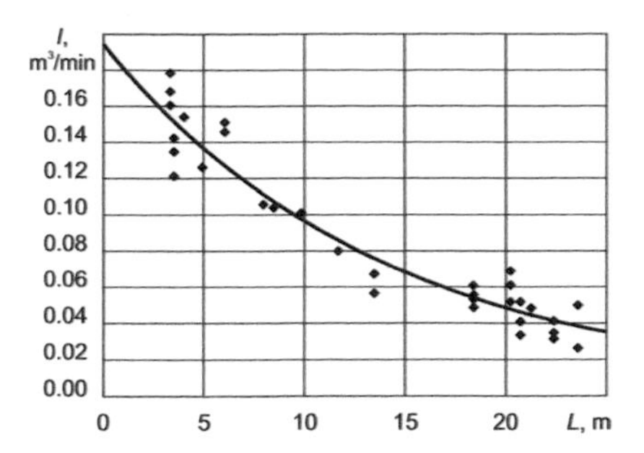

Figure 3. Dependence diagram of methane consumption on the distance between the borehole collar and longwall.

When defining dependence of the debit of methane captured by the borehole on the distance from borehole collar to longwall in the shape of polynomials of the 2nd and 3rd degrees for the conditions of the 5th northern longwall, the following regression equations are received:

$$I_i = 0.0002L_i^2 - 0.0121L_i + 0.1911;$$

$$I_i = -7 \cdot 10^{-6} L_i^3 + 0.0005L_i^2 - 0.0154L_i + 0.2001.$$

The average error of approximation for these equations is 23.7 and 16.4% correspondingly. Received results testify about the possibility to use required type of description of an equation depending on its practical usage.

After initial data processing, the following equations are established:

1. Depression in the borehole collar, methane concentration and methane-air mixture consumption dependence on the distance between degassing borehole collar and longwall (Figure 4).

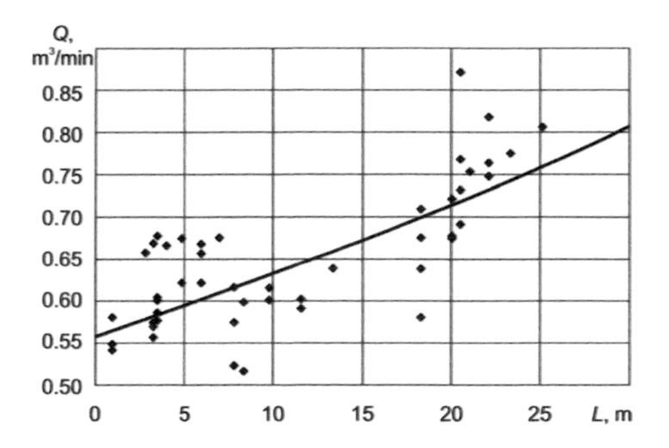

Figure 4. Dependence diagram of methane-air mixture consumption on distance between borehole collar and the longwall.

2. Methane-air mixture dependence on methane debit in a borehole collar.

3. Methane-air mixture consumption dependence on depression value in the borehole collar (Figure 5).

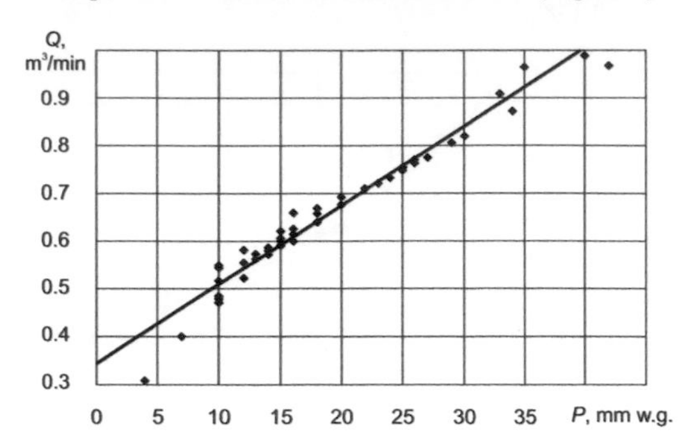

Figure 5. Dependence diagram of methane-air mixture consumption on depression in degassing borehole collar.

Regression equation of the above-mentioned dependences are presented below:

– $P = 12.258e^{0.0307L}$;

– $c = 33.827e^{-0.087L}$;

– $Q = 0.5583e^{0.0123L}$;

– $Q = -0.3534I + 0.6633$;

– $Q = 0.0165P + 0.3433$.

Average error of approximation for given equations makes up 12.9; 14.9; 7.2; 14.0 and 3.2% correspondingly.

Debit of methane captured by degassing borehole considering distance change from the borehole collar to longwall and depression in it under conditions of the 5th northern longwall of SSI of l_3 seam is described by the following regression equation

$$I = 0.1799e^{-0.07102L - 0.00178P}.$$

An average error of approximation for this equation is equal to 21%.

Dependence of methane-air mixture consumption on borehole collar distance to longwall and depression in degassing borehole:

$$Q = 0.226834I + 0.016657P + 0.316527.$$

Average error of approximation – 2.2%.

4 SELECTION OF RATIONAL BOREHOLES NUMBER AND DISTANCE BETWEEN THEM

Solution of the task focused on rational boreholes number selection that work simultaneously at an extraction area and distance between them foresees gaining of maximal total debit of the methane captured at the area. There are limiting conditions that affect this task:

a) methane concentration in degassing pipeline is below the permissible one;

б) methane concentration in the borehole collar is lower than the set value.

Scanning method is used during this task solving. The result of calculations – values of l_{opt} and n_{opt} at which the condition $\sum_{j=1}^{S} I_j = max$ is provided.

Based on the above-mentioned methodology, calculations have been made for conditions of the 5th northern longwall of southern rock incline of l_3 seam at "Almaznaya" mine.

Value of degassing efficiency coefficient of adjacent seams during borehole drilling from the mine working that is backfilled after the stope advance, can reach 0.4 (Degassing... 2004).

Based on the mining-geological and mining-technical conditions at an extraction area, area length L is accepted as such that characterizes maximal distance of collar from the area along movement of borehole methane-air mixture from the stope line (for conditions of the 5th northern longwall with $L = 25m$). After that, possible number of simultaneously working boreholes n_b is accepted and distance between the boreholes is calculated based upon the following formula

$$l_d = \frac{L}{n_b}, \text{ m}.$$

For above-mentioned area the boreholes number changes from 2 to 5 at distance between them equal to 12.5; 8.3; 6.3 and 5.0 m correspondingly.

After individual methane debit determination for each borehole, total methane consumption captured by simultaneously working boreholes is calculated. In connection with the fact that when the boreholes collars approach stope line, the value of $\sum_{i=1}^{n} I_i$ increases and during calculation total methane debit for two positions is defined:

a) collar of the first borehole based on the methane-air mixture movement direction is located at a distance l_d from longwall $\sum_{i=1}^{n} I_i$;

б) collar of the first borehole is located at the beginning of coordinates, i.e. at the level of stope line – $\sum_{i=1}^{n} I_i^*$.

Final result of total methane debit for the given borehole number is an average value of total methane debit calculated according to the next formula

$$\sum_{i=1}^{n} I_i \,_{av} = \frac{\sum_{i=1}^{n} I_i + \sum_{i=1}^{n} I_i^*}{2}, \text{ m}^3/\text{min}.$$

The calculations results for conditions of the 5th northern longwall of SSI of l_3 seam at "Almaznaya" mine are shown in Table 1.

Table 1. Results of degassing parameters calculations at an extraction area.

Parameters of degassing network	Number of boreholes			
	2	3	4	5
	Distance between the boreholes, m			
	12.5	8.3	6.3	5.0
I_0, m³/min	2.0	2.0	2.0	2.0
I_1, m³/min	0.8	1.1	1.3	1.4
I_2, m³/min	0.3	0.6	0.8	1.0
I_3, m³/min	–	0.3	0.5	0.7
I_4, m³/min	–	–	0.3	0.5
I_5, m³/min	–	–	–	0.3
Q_0, m³/min	3.5	3.5	3.5	3.5
Q_1, m³/min	4.0	3.8	3.8	3.7
Q_2, m³/min	4.7	4.2	4.0	3.9
Q_3, m³/min	–	4.7	4.3	4.2
Q_4, m³/min	–	–	4.7	4.4
Q_5, m³/min	–	–	–	4.7
c_0, %	58.2	58.2	58.2	58.2
c_1, %	20.4	28.9	34.5	38.3
c_2, %	7.2	14.4	20.4	25.2
c_3, %	–	7.2	12.1	16.6
c_4, %	–	–	7.2	10.9
c_5, %	–	–	–	7.2
$c_{pl.1}$, %	20.4	28.9	34.5	38.3
$c_{pl.2}$, %	13.3	21.3	27.2	31.5
$c_{pl.3}$, %	–	16.1	21.8	26.2
$c_{pl.4}$, %	–	–	17.7	22.1
$c_{pl.5}$, %	–	–	–	18.7
$\sum_{i=1}^{n} I_i$, m³/min	1.2	2.1	3.0	3.9
$\sum_{i=1}^{n} I_i^*$, m³/min	2.9	3.8	4.7	5.6
Average total debit	2.0	2.9	3.8	4.8

Following the gained results: as the boreholes number increases, total methane debit grows by linear dependence and methane concentration in degassing pipeline – by logarithmic dependence.

5 DEGASSING BOREHOLE DIAMETER OPTIMIZATION

In order to increase income of an enterprise at production prime cost reduction, expenses for purchase and maintenance of exploited equipment must be minimized. During coal seam degassing the selection of optimal degassing pipeline diameter is the necessary condition for creation of efficient degassing process and provision of minimal expenditures for purchase of pipeline and electrical energy needed for vacuum-pump in order to move methane-air mixture along the pipeline.

As an optimization criterion, total cost of degassing pipeline and consumed by the vacuum-pump electro-energy is accepted with specified debit of methane-air mixture.

Electro-energy consumption needed to move gas-air mixture along degassing pipeline is defined by formula (1) and its cost by formula (2).

$$E_i = \frac{Q_{b.i}(\Delta P_i)^2}{10\eta} \cdot 24 \cdot 365 \cdot t_i \text{, Kw·h,} \quad (1)$$

$$K_{e.i} = 454 \cdot \frac{c_i^{-0.17} d_i^{-5.35} L_i t_i Q_{b.i}^3}{\eta} K_k \text{, UAH,} \quad (2)$$

where E_i – electro-energy consumption required to move methane-air mixture along an i-th area for all service life of the pipeline, Kw·h; $Q_{b.i}$ – average-geometrical air consumption at an i-th area of the pipeline taking into account air inflows, m^3/s; $(\Delta P_i)^2$ – square-law losses of pressure at an i-th area of the pipeline, GPa; t_i – pipes service life at an i-th area of the pipeline, years; 24 – number of continuous working hours of vacuum-pumps a day; 365 – number of days in a year; 10 – correction coefficient considering measuring units of used values; η – efficiency of vacuum-pump; c_i – methane content in a drawn off mixture at an i-th area of the pipeline, %; d_i – diameter of an i-th area of the pipeline, m; L_i – length of an i-th area of the pipeline, m; K_k – cost of 1 Kw·h of electro-energy, UAH.

Determination of square losses of pressure in pipeline and average-geometrical air consumption at an i-th area of degassing network is calculated by the following equations:

$$(\Delta P_i)^2 = \alpha_i L_i Q_{b.i}^2 \text{, GPa;}$$

$$\alpha_i = 0.52 c_i^{-0.17} d_i^{-5.35} \text{, GPa·s}^2/\text{m}^7;$$

$$Q_{b.i} = \sqrt{Q_i Q_{i+1}} \text{, m}^3/\text{s,}$$

where α_i – specific aerodynamic resistance of an i-th area of pipeline, GPa·s^2/m^7; Q_i, Q_{i+1} – methane-air mixture consumption at initial and end points of an i-th area correspondingly, m^3/s.

Methane-air mixture consumption at the end point of an i-th area of pipeline is defined by the following formula:

$$Q_{i+1} = Q_i + aL_i \text{, m}^3/\text{s,}$$

where a – specific value of air inflow into degassing pipeline, m^3/s, that, according to the manual (Degassing... 2004) is accepted to be equal to $1.67 \cdot 10^{-5}$ m^3/s. With known factual consumptions of methane-air mixture, specific value of inflow at an i-th area is defined by the following formula

$$a_i = \frac{Q_{i+1} - Q_i}{L_i} \text{, m}^3/\text{s.}$$

The authors of this paper have gained linear dependence of gas pipes cost on their diameter in a shape of the following equation: $k_m = f + bd_i$. This equation is fair at diameter d_i change in range of [0; 0.5]. At big diameter values, the total pipes cost of an i-th area of degassing pipeline should be calculated based on the following equation

$$K_{pl.i} = L_i\left(b' d_i^2 + f' d_i + k'\right) \text{, UAH,}$$

where b', f', k' – empirical coefficients.

When defining optimal pipe diameter of gas pipeline, taking into account cost of the pipeline and electro-energy consumed by vacuum-pump, the aim function will have the following look:

$$\sum_{i=1}^{n}\left(K_{e.i} + K_{pl.i}\right) \rightarrow min,$$

where n – number of pipeline areas in degassing network of a mine.

In order to introduce some restrictions the task is considered with respect to specific degassing system (Figure 6).

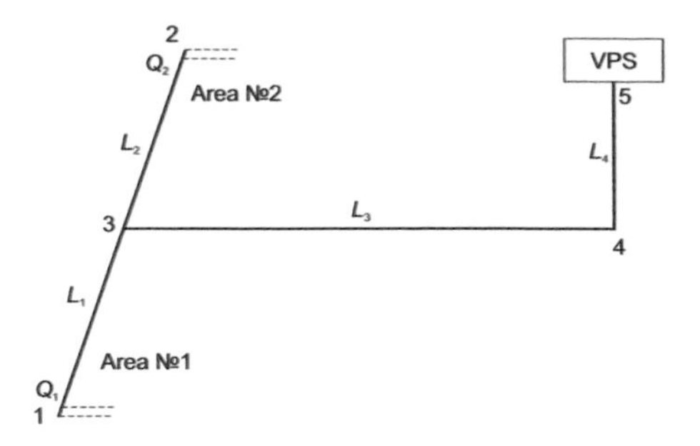

Figure 6. Scheme of the mine degassing network.

As the restrictions during solving of an optimization task, the following system of equations are accepted

$$\begin{cases} \left(P_{w_1} - P_{c_1}\right)^2 - \alpha_1 L_1 Q_{b_1}^2 - \alpha_3 L_3 Q_{b3}^2 - \alpha_4 L_4 Q_{b4}^2 \le P_{per}; \\ \left(P_{w_1} - P_{c_1}\right)^2 - \alpha_1 L_1 Q_{b1}^2 = \left(P_{w_2} - P_{c_2}\right)^2 - \alpha_2 L_2 Q_{b2}^2. \end{cases}$$

The set task of the gas pipes diameter optimization is solved by method of Lagrange. Lagrangian function in expanded form:

$$L(d_1, d_2, d_3, d_4, \lambda_1, \lambda_2) = \sum_{i=1}^{4}\left[L_i\left(bd_i^2 + fd_i + k\right) + 454\frac{c_i^{-0.17} d_i^{-5.35} L_i t_i Q_{b.i}^3}{\eta} K_k \right] + \lambda_1\left[\left(P_{w_1} - P_{c_1}\right)^2 - 0.52 c_1^{-0.17} d_1^{-5.35} L_1 Q_{b_1}^2 - \right.$$

$$\left. - 0.52 c_3^{-0.17} d_3^{-5.35} L_3 Q_{b_3}^2 - 0.52 c_4^{-0.17} d_4^{-5.35} L_4 Q_{b_4}^2 - P_{per}\right] + \lambda_2\left[\left(P_{w_1} - P_{c_1}\right)^2 - 0.52 c_1^{-0.17} d_1^{-5.35} \times \right.$$

$$\left. \times L_1 Q_{b_1}^2 - \left(P_{w_2} - P_{c_2}\right)^2 + 0.52 c_2^{-0.17} d_2^{-5.35} L_2 Q_{b_2}^2 \right]. \qquad (3)$$

During solving the task focused on pipeline diameter optimization, the equation (3) will be presented by the following system of equations:

$$\begin{cases} \dfrac{\partial L}{\partial d_i} = 0; \quad i = 1 - 4; \\ \dfrac{\partial L}{\partial \lambda_j} = 0; \quad j = 1; 2. \end{cases} \qquad (4)$$

The roots of received system of equations define stationary point (d_{1opt}, d_{2opt}, d_{3opt}, d_{4opt}).

After differentiation and conversions, the system of equations (4) is solved by one of the calculation methods, as a result of which it is possible to determine optimal pipes diameters at the gas pipeline areas (d_{1opt}, d_{2opt}, d_{3opt}, d_{4opt}).

Versatility of the developed methodology lies in the following: as the number of extraction areas, at which degassing is carried out, grows, the algorithm of the task solving remains the same. In this case, only number of equations that describe degassing system changes and are subject to joint solution during optimal diameters determination at degassing pipeline areas.

6 CONCLUSIONS

Natural researches of degassing systems are performed with determination of methane and methane-air capturing value. Based on natural observations in conditions of coal mines extraction areas, dependences of degassing parameters on mining-technical conditions are defined, their reliability and adequacy evaluation is done, and also close correlation connection is established.

The authors have developed original methodology of rational boreholes number determination that provides maximal total methane debit when disconnected from boreholes degassing network in which (boreholes) methane concentration in the mixture is less than the permissible value.

Also the result of researches is the developed methodology of degassing system parameters optimization taking into account costs of the pipes and electro-energy consumed by vacuum-pumps in order to move methane-air mixture along the pipeline.

REFERENCES

Degassing of coal mines. Requirements for methods and schemes of degassing. SOU 10.1.00174088.001-2004. 2004. Kyiv: Mintopenergo of Ukraine: 162.

Geomechanical Processes During Underground Mining – Pivnyak, Bondarenko, Kovalevs'ka & Illiashov (eds)
© 2012 Taylor & Francis Group, London, ISBN 978-0-415-66174-4

Analysis of stress-strain state of rock mass while mining chain pillars by chambers

V. Buzylo & A. Yavorsk'yy
National Mining University, Dnipropetrovs'k, Ukraine

V. Yavorsk'yy
Union of coal industry Dnipropetrovs'k territorial organization, Pavlograd, Ukraine

ABSTRACT: In order to analyze stress – strain state of the "roof-pillars-ground" system calculation algorithm was developed, which is based on boundary element method and analytical solution of Kelvin task. This geomechanical problem is solved in additional stresses. Non-linear multiple correlation analysis of results allowed to find out the regression equation for the maximum values of roof and ground convergence in percentage of seam thickness.

1 TOPICALITY OF THE WORK

Ukraine's power strategy up to 2030 foresees considerable increase of coal percentage in Ukraine's fuel and energy complex. In this connection extraction of thin seams and chain pillars with the thickness of 0.6…1.2 m where the exploited coalfields contain up to 70% of conditioned coal deposits has become one of the most urgent problems. The complexity of solving it at the mines of West Donbass Region is explained by specific mining and geological conditions. The coal in this region is solid enough (strength coefficient according to M.M. Protodyakonov is $f = 2.0...3.5$) and bearing strata – argillites, siltstones and sandstones are weakly stable ($f = 0.8...2.5$). Therefore, previously used augers in these conditions turned out to be of little effect: cutting units often did not cope with solid coals and due to weak stability of rocks it was necessary to insert wide pillars between the stalls. The situation somewhat improved after transferring to setting BZM-1M use with the improved cutting units but coal loss remained to be considerable.

Coal extraction of row or block of chambers with retaining relatively narrow inter- chambers pillars and retaining quite wide support pillars between the blocks could be more expedient under the present conditions. With the purpose of reducing losses it is expedient to minimise the size of the inter-chamber pillars allowing their crushing after the block removal.

In these conditions it is possible to use more efficient equipment, for instance, "Dnepr" coal-cutting combiner, in case of supporting stall roofing with individual bonding.

In all the cases of stipulating parameters of the accepted development systems thorough research of geological and mechanical processes, occurring in the rocks massif around the block of chambers and coal pillars, are necessary.

2 LEVEL OF THE ISSUE DEVELOPMENT

Various aspects of coal seam destruction and weak enclosing rocks on the periphery of breakage heading are considered in the works of O.V. Kolokolov, A.F. Kurnosov, N.A. Lubenets, A.N. Zorin, V.G. Kolesnikov and others.

And with it in most of the modern methods the pillars are calculated without taking into account the time factor and separately from the roof and ground. It is considered that they are loaded evenly with the weight of strata up to the surface, which is incorrect in case there are pillars of various width. The hypothesis of bridging of M.M. Protodyakonov (A.A. Borysov, G.Ye. Gulevych, S.V. Vetrov, P.P. Korzh and others) is also applied but for determining the size of roof failure the data obtained by the means of natural observations are required.

Roofing is mainly calculated according to the scheme of the beam the size of the span of which depends on the width of the chambers block and the angle of failure of the roofing rocks (Yu. A. Modestov, S.G. Borysenko, Ye.I. Kamskiy, Yu.B. Gubenin and others). It also requires conducting mining or laboratory research.

For the considered mining-geological conditions it is impossible to use analytical ways of solving flat problems of geomechanics on massifs, weakened by finite number or infinite number of planes of various delineations, as all of them are received for homogenous environments not taking into account rheological properties of rocks (G.N. Savyn, L.D. Shevyakov, D.I. Sherman, A.S. Kosmodamyanskiy, V.V. Rahymov, L.I. Ilshtein and others).

Although numerous ways of solving problems carried out by well-known numerical methods of finite and boundary elements relate to nonuniform environments, they take into account common roof, pillars and ground deformation, including time parameters, relate to concrete technological schemes, concrete mining and geological conditions (Zh.S. Yerzhanov, T.D. Karymbayiv, Yu.M. Liberman, A.V. Usatenko, L.V. Novikova, Ye.A. Sdvyzhkova and others) and also cannot be applied for stipulating parameters of heading-and-chamber method in case of availability of solid coals and soft rocks.

Another very important aspect the researchers have not paid much attention to is probabilistic nature of changing physical and mechanical qualities of heterogeneous massif of soft rocks. The present factor is determining not only in the process of evaluating stress-strain state of rocks massif itself but also while solving problems, connected with reliability evaluation of equipment and technology of carrying out second working (K.F. Sapytskiy, S.M. Lypkovych, N.N. Lebedev and others).

3 THE MAIN PART

In the present work for increasing reliability of the projected chamber system of developing probabilistic nature of physical and mathematical characteristics of rocks was taken into account, which was stipulated by heterogeneity of the massif and their changing in course of time.

Increasing the sizes of inter-chambers pillars rapidly enhances exploitation losses of the natural resource. It stipulates the necessity of either introducing the systems of development with laying of the worked-out area or advancing constructive schemes and parameters of chamber system.

In both cases it is necessary to conduct thorough analysis of stress-strain state of "roof-pillars-ground" system.

For this purpose the authors developed calculating algorithm, grounding on method of boundary elements and analytical solving of Kelvin's problems (Krauch & Starfield 1987; Buzylo & Yavorskiy 2011). The initial equations systems for determining stresses and displacements in the researched area of the massif on the periphery of stopes and inter- chambers pillars row is formed according to the prescribed conditions on the contours of the stopes. The calculation is carried out on the action of forces of rocks weight. The model of linear creeping is used with Abel's kernel. The set task of geomechanics is solved with additional stresses. Complete stresses are placed as equal to the sum of the initials contained in the massif till the beginning of working out (at any point and at infinity in vertical direction it is $(\sigma_{yy})_0^\infty = \gamma H$ and in horizontal position it is $(\sigma_{xx})_0^\infty = \lambda\gamma H$; bulk density of rock is t/m^3; $\lambda = \nu/(1-\nu)$ is coefficient of lateral bearing reaction; ν is Poison's rock coefficient) and additional ones stipulated by formation of cells and inter-chamber pillars. As full stresses constitute zero on the chambers edges, additional stresses are known on these chambers edges, which is taken into account while creating initial system of equations.

The main stages of calculation are:

– approximation of the limits of the researched area with finite elements and setting boundary conditions in each of them;

– introduction in each finite element unknown fictitious normal P_y and tangential P_x uniformly distributed loads;

– conveying stresses and displacements in each finite element through unknown "fictitious" loads P_x and P_y with the help of analytical solving of the problem of Kelvin, calculation of influence coefficients, formation of the system of equations for determining P_x and P_y ;

– solving the obtained system of algebraic (in the developed algorithm the method of Gauss is used), calculation of loads P_x and P_y , providing for performing of the set boundary conditions along with the real load from the weight of the rocks;

– calculation of stresses and displacement on the edges and at any inner point of the researched area with the help of the determined fictitious loads according to the method of superposition.

Thus, the initial system of equations is formed with the help of the following correlations:

– for navigation in vertical and horizontal directions:

$$u_y^i = \sum_{j=1}^{N} B_{yx}^{ij} P_x^j + \sum_{j=1}^{N} B_{yy}^{ij} P_y^j ,$$

$$u_x^i = \sum_{j=1}^{N} B_{xx}^{ij} P_x^j + \sum_{j=1}^{N} B_{xy}^{ij} P_y^j ; \qquad (1)$$

– for the corresponding stresses:

$$\sigma_y^i = \sum_{j=1}^{N} A_{yx}^{ij} P_x^j + \sum_{j=1}^{N} A_{yy}^{ij} P_y^j ,$$

$$\sigma_x^i = \sum_{j=1}^{N} A_{xx}^{ij} P_x^j + \sum_{j=1}^{N} A_{xy}^{ij} P_y^j . \quad (2)$$

In correlations (1) and (2) N – is the number of finite elements; B_{yx}^{ij}, B_{yy}^{ij}, B_{xx}^{ij}, B_{xy}^{ij} – navigation influence coefficients; A_{yx}^{ij}, A_{yy}^{ij}, A_{xx}^{ij}, A_{xy}^{ij} – stress influence coefficients.

The formula used for determining influence coefficients obtained with the use of the above mentioned analytical method of solving, are included in monograph (Novikova & Ponomarenko 1997).

The researched area of the rock massif includes three stopes, two support and two inter-chamber pillars. Mining and geological conditions of PLC "DTEK Pavlogradugol" mines are considered. Deposition of coal strata are presented in this minefield with argillites and siltstones, locally with sandstone and numerous coal seams. The position of rocks and coal seam is flat. Deposition of covering stratas us represented by sands, clays and loams with the total thickness of 120 м. Coal seams with different thickness ranging from 0.55 m to 1.28 m lie at different depths – from 90 m to 480 m. At various mines roof and ground of the seams are constructed with above listed rocks, breaking point of which for compression $(\sigma_c)_r$ equals the value of the interval of 8...40 MPa, and moduluses of elasticity E_r – to the interval of $0.1 \cdot 10^4 ... 2.26 \cdot 10^4$ MPa. Mean value of volume density and the coefficient of Poisson of country rocks constitute correspondingly $\gamma_r = 2.65 \, t/m^3$, $\nu_r = 0.3$.

Coal characteristics with variation coefficient of $V = 0.25$ have mean values of: $\gamma_y = 1.47 \, t/m^3$, $E_y = 3.5 \cdot 10^3$ MPa, $(\sigma_c)_r = 35$ MPa, $\nu_y = 0.4$.

The width of the processed pole is determined according to the formula: $A = na + (n-1)b$, where b – width of inter-chamber pillar, m; a – the width of the chamber, m; n – the number of chamber in the allocated block.

Parameter A had the following value in calculations: 4.8...8.7 m, and the width of the supporting pillar b_1 was assumed as equal to $2b$.

Thus, initial data to calculation are physical-mechanical E_y / E_r , $\gamma_r H / (\sigma_c)_y$ and geometrical a/b and b/m parameters are of probabilistic nature. Therefore calculation according to the developed algorithm was carried out with different meaning of the indicated values:

$E_y / E_r \in [0.15; \ 0.39; \ 1.25]$, $\gamma_r H / (\sigma_c)_y \in [0.05; \ 0.11; \ 0.23; \ 0.36]$, $a/b \in [1.55; \ 1.75; \ 2.33; \ 3.50; \ 4.66]$, $b/m \in [0.4; \ 0.5; \ 0.6; \ 0.8; \ 0.9; \ 1.0; \ 1.2; \ 1.5; \ 1.6]$.

The height of chambers and pillars was equal to the thickness m of the extracted seam.

Moduluses of elasticity of rocks according to the accepted deformation model were determined through the known creep function, taking into account the fact that full technological cycle of processing one chamber with the width of 2.1 m and length of 30 m with an auger constitutes 6 hours. According to the data of VNIMI the parameter of coal and rocks creeping α is assumed as equal to 0.7, and the parameters δ of argillite, siltstone, sandstone and coal correspondingly had the following values:

$1.17 \cdot 10^{-2} \cdot c^{-0.3}$, $5.54 \cdot 10^{-3} \cdot c^{-0.3}$, $3.28 \cdot 10^{-3} \cdot c^{-0.3}$, $2.32 \cdot 10^{-3} \cdot c^{-0.3}$.

The considered combinations of values of the above listed parameters from possible intervals of their varieties constituted 180 calculative variants.

Nonlinear multiple correlated analysis of the obtained data performed according to the known method (Lvovskiy 1982) allowed to determine equations of regression foe maximal values of approach of roof and ground in percentage of the seam thickness:

$$\frac{\Delta u_y}{m} = -0.965 + 6.739 \frac{\gamma_r H}{(\sigma_c)_y} - \frac{E_y}{E_r} \left(1.044 - 0.640 \frac{a}{b} - 1.195 \frac{b}{m} \right), \quad (3)$$

as well as for correlation of maximal stretching stresses σ_{xx} in the roof to $\gamma_r H$:

$$\frac{(\sigma_{xx})_{roof}}{\gamma_r H} = 0.284 + 0.116 \frac{a}{b} + \frac{b}{m} \left(0.00874 \frac{E_y}{E_r} - 0.0934 \right). \quad (4)$$

For inter-chamber pillars the following correlations have been obtained:

$$\frac{(\sigma_{yy})_{max}}{\gamma_r H} = -3.555 + \frac{E_y}{E_r} \left(1.736 - 1.130 \frac{a}{b} \right) + 0.523 \frac{b}{m} ;$$

$$\frac{(\sigma_{eq})_{max}}{\gamma_r H} = 1.344 +$$

$$+\frac{a}{b}\left(0.296\frac{E_y}{E_r} + 0.591\right) - 0.665\frac{b}{m}, \qquad (5)$$

where $(\sigma_{yy})_{max}$ and $(\sigma_{eq})_{max}$ – normal σ_{yy} and equivalent σ_{eq} stress in the dangerous section of the pillar. Equivalent stress was determined according to the criterion of P.P. Balandin:

$$\sigma_{eq} = \frac{(1-\psi)(\sigma_1 + \sigma_3)}{2\psi} +,$$

$$+\frac{\sqrt{(1-\psi)^2(\sigma_1+\sigma_3)^2 + 4\psi(\sigma_1^2 + \sigma_3^2 - \sigma_1\sigma_3)}}{2\psi},$$

where σ_1 and σ_3 – maximal and minimal main stresses; $\psi = \sigma_p / \sigma_c$; σ_p and σ_c – limits of rock strength for extension and compression.

For the considered rocks $\psi = 0.1$.

For the support pillars there is the following correlation:

$$\frac{(\sigma_{yy})_{max}}{\gamma_r H} = -1.757 -$$

$$-\frac{a}{b}\left(0.436\frac{E_y}{E_r} + 0.854\right) + 0.509\frac{b}{m}. \qquad (6)$$

4 CONCLUSIONS

The obtained correlations (3)-(6) are aimed at identifying parameters of the projected chamber systems of sloping coal seams development in soft rocks. In particular, according to them rational parameters of coal augering in the conditions of West Donbass Region mines ($m = 0.8$ m; $H = 160$ m; $a = 2.1$ m; $b_1 = 1.2$ m; $b = 0.6$ m).

Industrial testing at the PLC "DTEK Pavlogradugol" has proved their efficiency.

REFERENCES

Krauch, S. & Starfield, A. 1987. *Methods of finite elements in mechanics of solids*. Moscow: Mir: 328.

Novikova, L., Ponomarenko, P., Prykhodko, V. & Morozov, I. 1997. *Methods of finite elements in problems of mining geomechanics*. Dnipropetrovs'k: Science and education: 178.

Recommendations on calculation of edge displacement and loads on the mine working support according to experimental indicators of rock deformation beyond the breaking point. 1982. Leningrad: VNIMI: 36.

Lvovskiy, Ye. 1982. *Statistical method of creating empirical formula*. Moscow: Vyschaya Shkola: 223.

Buzylo, V., Yavorskiy, V., Koshka, O. & other authors. 2011. *Technology of processing coal pillars with chambers in the conditions of West Donbass Region mines*: monograph. Dnipropetrovs'k: National Mining University: 95.

Geomechanical Processes During Underground Mining – Pivnyak, Bondarenko, Kovalevs'ka & Illiashov (eds)
© 2012 Taylor & Francis Group, London, ISBN 978-0-415-66174-4

Identification the cutting machine rational feed rate according to the working area stability factor

S. Vlasov & O. Sidelnikov
National Mining University, Dnipropetrovs'k, Ukraine

ABSTRACT: Methodology of identification the cutting machine rational feed rate according to the working area stability factor is offered in the article. Its implementation allows to exclude incidents engendered by fall of roof rocks exposures into longwall working area and, in certain cases, gives the chance to choose the rational cutting machine operating mode that, by-turn, allows to prove-out process of coal seam sloughing. Generally, prognostication stability of the mining seam and roof rocks in the working face plane depending on cutting machine feed rate is an important component of technological designing the production unit operation, and feed rate – important technological parameter of the coal mining technology.

1 INTRODUCTION

The result of mining workings driving is rock mass discontinuity that implies the redistribution rocks stress and strain state and change the initial stress field in the mass. Process of continual stress redistribution in the mass is characteristic for mining workings which change the position in time and space, i.e. for operating longwalls and development faces. The artificial cavities are driven in the rock mass, cause change of stresses in any limited area of mass that leads to formation so-called zone of mining working influence on this mass that is characterized anomalous stress field in comparison with the initial one. The specified zone moves synchronous with working face advancing, i.e. in process of face advance all new, earlier virgin parts of the rock mass get to zone of mining working influence.

2 FORMULATING THE PROBLEM

In the research works (Skipochka 2006; Porcevsky 2004 & Kuklin 2003) was repeatedly noticed that there is a feedback between duration of operating the increased stress in the mass and rocks strength properties. Take into consideration this phenomenon, it is possible to compute a rational face advance rate at which strength of the rocks surrounding directly a plane of exposure, taking into account time factor, will be more or equal to operating stresses value in the working face plane, taking into account the relaxation factor.

3 PURPOSE OF THE ARTICLE

Researching the longwall face space stability depending on the cutting machine feed rate.

4 RESULTS

The problem above specified was solved taking into account following positions and assumptions:

– the stress in any point of the mass forms of an initial stress (underground pressure in a virgin mass) and additional which results from mining operations (longwall advance);

– elastic deformations extend in a mass instantly;

– at once after elastic deformations redistribution in the mass processes of the stress relaxation begin;

– plastic rock deformations for a considered time interval is accepted greater infinitesimal order in comparison with elastic deformations, therefore it is considered that, relaxation process occurs at constant relative deformation;

– eventually elastic deformations turn (transform) into the plastic;

– plastic deformations – irreversible deformations which result from displacement of rock grains from each other, cracking on micro- and macrolevel and other irreversible processes which lead to the rock strength decrease;

– the elasticity module of rock is considered constant and weakly dependent on time and stresses operating in the mass;

– the form of relaxation stresses curves weakly depends on value of the operating stress, i.e. analytical function which describes process of the stress relaxation for concrete rock is fair for all range of initial stresses variation.

For stope accident-free operation it is necessary, that value operating in the coal seam and in the immediate roof stresses along the longwall, within a web width of a shearer cutting head, was less or equaled considered rocks strength, taking into account the time factor. This condition fulfillment allows to avoid uncontrolled coal seam sloughing and rock fall of the immediate roof at once after shearer cutting head pass. On the assumption of it, as control points for which computation will be made, necessary take the most remote points, located on boundary of the stope influence zone ahead the longwall, accordingly in the coal seam plane and in the plane of immediate roof rocks.

Stress distribution dependence ahead the longwall in a considered rock seam in section, perpendicular stope line and driving through the chosen control point, is identified by means of three-dimensional computer simulation (Vlasov 2010, 2011 & Sidelnikov 2009). In co-ordinate system in which coordinate origin is superposed with stope position, this dependence will have the following form

$$\sigma^{eq} = \left(F(z;L)^{eq} - P_0^{eq}\right)e^{-kX} + P_0^{eq},$$

$$X \in [0; X_0], \tag{1}$$

where σ^{eq} – equivalent stress, MPa; $F(z;L)^{eq}$ – function of equivalent stress distribution in the stope plane along the longwall length z depend on longwall position along the extracting pillar L, MPa; P_0^{eq} – equivalent stress operating in the virgin rock mass at the given depth of mining, MPa; k – coefficient of the stress attenuation along the extraction pillar, m^{-1}; X – value in meters on the co-ordinate axis directed in a direction of longwall advancing $X \in [0; X_0]$, value $X = 0$ – corresponded to stope plane, $X = X_0$ – control point position; X_0 – coordinate the most remote point, located on boundary of the stope influence zone ahead the longwall, m.

Take into account that in the problem the rock mass three-dimensional stress state is considered, it is expedient to express the equivalent stresses by equivalence formula according to P.P. Balandin criterion of strength. Consequently a complex stress state is reduced to simple one. For cause, when $\sigma_1 = \sigma_2 > \sigma_3$

$$\sigma^{eq} = \frac{(1-\psi)(2\sigma_1+\sigma_3)}{2\psi} + \frac{\sqrt{(1-\psi)^2(2\sigma_1+\sigma_3)^2 + 4\psi(\sigma_3-\sigma_1)^2}}{2\psi}, \tag{2}$$

where ψ – brittleness coefficient, $\psi = \dfrac{\sigma_{ten}}{\sigma_{com}}$; σ_{ten} – ultimate tensile strength; σ_{com} – ultimate compression strength.

The account of stress relaxation can be made on dependence of Boltsman-Volterr-Erzhanov.

$$\sigma(t) = \frac{E\varepsilon(t)}{1 + \dfrac{\delta}{1-\alpha} t^{(1-\alpha)}}, \tag{3}$$

where $\varepsilon(t)$, $\sigma(t)$ – deformation and the stress corresponding to the considered time point t, counted from the point when loading of rock sample was started; E – modulus of elasticity; α, δ – creep parameters, are defined on the basis the long-run tests of rocks on creep.

On the other hand, dependence of rock strength on a loading time can be presented the following formula:

$$\left[\sigma^{eq}(t)\right] = \left[\sigma_{lt}^{eq}\right] + e^{-nt}\left(\left[\sigma_{in}^{eq}\right] - \left[\sigma_{lt}^{eq}\right]\right), \tag{4}$$

where $\left[\sigma^{eq}(t)\right]$ – rock equivalent stress at time point t, counted from the point when loading of rock sample was started; $\left[\sigma_{lt}^{eq}\right]$ – rock long-term equivalent stress at time point $t \to \infty$, counted from the point when loading of rock sample was started; $\left[\sigma_{in}^{eq}\right]$ – rock instantaneous equivalent stress at time point $t = 0$, counted from the point when loading of rock sample was started; n – coefficient of stress attenuation in time, s^{-1}.

According to (Skipochka 2006) the variation range of rocks instantaneous strength is in limits 1.2-1.9 from value of their long-term strength. By results of researches (Skipochka 2006; Kuklin 2003) for rocks in the Western Donbas $\sigma_{in} = (1.3-1.7)\sigma_{lt}$, and relaxation time is about 3-5 days.

One more important factor, influencing on rocks strength characteristic, is rate of loading. For Western Donbas conditions the great experimental research work by identification rock strength, deformation and rheological properties was carried out by Institute of Geotechnical Mechanics. For conditions of Novomoskovsk, Pavlograd and Peteropavlovka geological and industrial region of Donbas dependence of rock strength on loading rate vary under the logarithmic law. At increase of loading rate from 0.01 to 10 MPa/s, the ultimate uniaxial compression strength increases on 40-50%. It means that for face space stability estimation it is necessary to consider not only rheological processes operating in the rock mass, but also rate of stress change in rocks which directly surround the stope. Time of stress operating in the mass and rocks loading rate at the coal extraction are in dependence on cutting machine feed rate.

Generally, the face space stability condition can be presented the following

$$\left[\sigma_{us}^{eq}(v)\right] \geq \sigma^{eq}(v), \tag{5}$$

where $[\sigma_{us}^{eq}(v)]$ – rock equivalent ultimate strength as function of cutting machine feed rate, MPa; $\sigma^{eq}(v)$ – equivalent stress operating at face space as function of cutting machine feed rate, MPa; v – cutting machine feed rate, m / s.

To solve the inequality (5) rather v it is necessary to work out correlations and corresponding equations that connecting cutting machine feed rate with strength and rheological rock properties.

4.1 *Dependence of ultimate rock strength on loading rate*

According to (Kuklin 2003), dependence of ultimate rock strength on loading rate varies under the logarithmic law and can be presented in following form

$$\left[\sigma^{eq}\right] = q\,ln\left(\frac{d\sigma}{dt}\right) + u, \tag{6}$$

where $\frac{d\sigma}{dt}$ – loading rate, MPa / s; q, u – the factors depending on type of rock.

4.2 *Identification the face space loading rate*

Function of elastic stresses distribution in rocks ahead of stope obey (1). Then change of stresses value on a final interval $\Delta r = (r_1 - r_2)$, which represents the shearer cutting head web width is possible to find from following expression:

$$d\sigma = \frac{\partial \sigma^{eq}}{\partial x}\Delta r, \tag{7}$$

Take into account that $\Delta r = -vdt$, the (7) will have following expression:

$$\frac{d\sigma}{dt} = -v\frac{\partial \sigma^{eq}}{\partial x} = kv\left(F(z,L)^{eq} - P_0^{eq}\right)e^{-kX}, \tag{8}$$

where X – point co-ordinate in which the slope ratio to an abscissa axis of tangent line is defined,

$$X = \frac{(r_1 + r_2)}{2}.$$

4.3 *Identification dependence of rock strength on stress operating time and on loading rate*

Accept in the equation (3) $\left[\sigma_{in}^{eq}\right] = w\left[\sigma_{lt}^{eq}\right]$, where w – the relation of instantaneous equivalent stress to long-term one (as already it was marked above $w = 1.2\text{-}1.9$. In this cause (3) will be

$$\left[\sigma^{eq}(t)\right] = \frac{\left[\sigma_{in}^{eq}\right]}{w} + e^{-nt}\left(\left[\left[\sigma_{in}^{eq}\right] - \frac{\left[\sigma_{in}^{eq}\right]}{w}\right]\right) =$$

$$= \left[\sigma_{in}^{eq}\right]\left(\frac{1 + e^{-nt}w - 1}{w}\right). \tag{9}$$

It is obvious that for each particularly taken loading rate, strength $\left[\sigma^{eq}\right]$ will equal $\left[\sigma_{in}^{eq}\right]$, then substitute in (5) $\left[\sigma^{eq}\right] = \left[\sigma_{in}^{eq}\right]$, for (9) it is possible to write

$$\left[\sigma^{eq}(t)\right] = \left(q\,ln\left(\frac{d\sigma}{dt}\right) + u\right)\left(\frac{1 + e^{-nt}(w-1)}{w}\right). \tag{10}$$

Substitute in (10) $\frac{d\sigma}{dt}$ for expression (8), get

$$\left[\sigma^{eq}(t)\right] = \left(q\,ln\left(kv\left(F(z;L)^{eq} - P_0^{eq}\right)e^{\frac{-k(r_1+r_2)}{2}}\right) + u\right) \times$$

$$\times\left(\frac{1 + e^{-nt}(w-1)}{w}\right). \tag{11}$$

4.4 *Identification time the stress relaxation in the mass as function of the cutting machine feed rate*

Time, for which the control point co-ordinate moves from X_0 to 0 (from mass depth onto working face plane), will correspond to time of the stresses relaxation and can be found from expression:

$$t = NT, \tag{12}$$

where N – quantity of the extraction cycles; T –

one cycle time, s.

Time of the stress relaxation in the mass as function of the cutting machine feed rate

$$t=\left(\frac{(L_l-\Sigma L_n)+2L_{sh}}{v}+\frac{(L_l-\Sigma L_n)}{v_{cl}}\right)\frac{X_0}{\Delta r}\varphi, \quad (13)$$

where L_l – longwall length, m; ΣL_n – niche total length; L_{sh} – length of end face area where shearer self-cutting in the coal seam, m; v_{cl} – cutting machine feed rate during cleanup (only for the unilateral scheme of shearer operation, for the shuttle scheme $v_{cl}\to\infty$), m / s; φ – factor of increase in time the longwall advance at the expense of presence repair shift and other planned idles, $\varphi=\frac{t_{day}}{t_{prod}}$, where t_{day} – quantity of hours in a day, t_{prod} – quantity of hours in a day on extraction; Δr – effective shearer cutting head web width, m.

Taking into account (12) and (13), the expression (11) will have following form

$$\left[\sigma_{us}^{eq}(v)\right]=\left(q\ln\left(kv\left(F(z;L)^{eq}-P_0^{eq}\right)e^{\frac{-k(r_1+r_2)}{2}}\right)+u\right)\times$$
$$\times\left(\frac{1+e^{-n\left(\frac{(L_l-\Sigma L_n)+2L_{sh}}{v}+\frac{(L_l-\Sigma L_n)}{v_{cl}}\right)\frac{X_0}{\Delta r}\varphi}(w-1)}{w}\right)k_{att}. \quad (14)$$

where k_{att} – coefficient of the rock mass structural attenuation, $k_{att}=0.5\text{-}0.8$.

In consideration of expressions for the stress relaxation (2) and relaxation time (13) for equivalent stresses actually operating in the stope plane it can be written following formula

$$\sigma^{eq}(v)=\frac{\left(F(z;L)^{eq}-P_0^{eq}\right)e^{-kX}+P_0^{eq}}{1+\frac{\delta}{1-\alpha}\left(\left(\frac{(L_l-\Sigma L_n)+2L_{sh}}{v}+\frac{(L_l-\Sigma L_n)}{v_{cl}}\right)\frac{X_0}{\Delta r}\varphi\right)^{(1-\alpha)}}. \quad (15)$$

By substitution in (4) expressions (14) and (15) it is got inequality with one unknown parameter v. At inequality realization at the moment of the stope approach to the control point there will be no loss of exposure rocks stability.

$$\frac{\left(F(z;L)^{eq}-P_0^{eq}\right)e^{-kX}+P_0^{eq}}{1+\frac{\delta}{1-\alpha}\left(\left(\frac{(L_l-\Sigma L_n)+2L_{sh}}{v}+\frac{(L_l-\Sigma L_n)}{v_{cl}}\right)\frac{X_0}{\Delta r}\varphi\right)^{(1-\alpha)}}\le$$
$$\le\left(q\ln\left(kv\left(F(z;L)^{eq}-P_0^{eq}\right)e^{\frac{-k(r_1+r_2)}{2}}\right)+u\right)\left(\frac{1+e^{-n\left(\frac{(L_l-\Sigma L_n)+2L_{sh}}{v}+\frac{(L_l-\Sigma L_n)}{v_{cl}}\right)\frac{X_0}{\Delta r}\varphi}(w-1)}{w}\right)k_{att}. \quad (16)$$

At solving inequality (16) rather v it is identified cutting machine rational feed rate according to the working area stability factor. It is necessary to notice that the solution (16) cannot be obtained in explicitly expressed symbolical form, but only with application of a graphic method or by means of mathematical software packages MathCad, MathLab and other.

For conditions of mining coal seam C_8^H at mine "Zapadnodonbaskay", numerical values of parameters entering in (16) for the coal seam and the immediate roof, are presented in table 1.

Table 1. Initial data to computation the cutting machine rational feed rate for conditions of mining coal seam C_8^H at mine "Zapadnodonbaskay".

Parameter	Unit of measure	Coal	Siltstone (immediate roof)
$F(z;L)^{eq}$	MPa	17.60	15.54
P_0^{eq}	MPa	5.00	3.56
k	m^{-1}	0.15	0.15
X_0	m	30	30
δ	$s^{-0.3}$	0.005	0.0094
α	$s^{-0.3}$	0.690	0.726
n	s^{-1}	86400	86400
w	–	1.64	1.59
ψ	–	0.13	0.11
q	s	2.14	2.18
u	MPa	31.38	24.25
Δr	m	0.8	0.8
φ	–	1.33	1.33
r_1	m	0	0
r_2	m	0.8	0.8
X	m	0	0
L_l	m	190	190
$\sum L_n$	m	0	0
L_{sh}	m	20	20
v_{cl}	m / s	∞	∞
k_{att}	–	0.7	0.5

Dependence of long-term equivalent strength on loading rate for coal is resulted at Figure 1. Initial experimental data for graphing were are taken from research work (Skipochka 2006). According to (Skipochka 2006), tests for coal were carried out at the side pressure $\sigma_1 = \sigma_2 = -3.0$ MPa; for siltstone – at the side pressure $\sigma_1 = \sigma_2 = -1.5$ MPa.

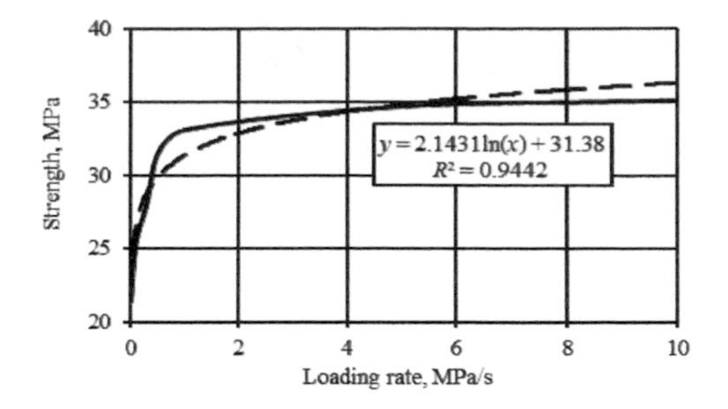

Figure 1. Dependence of coal equivalent strength on loading rate: ——— – experimental curve; – – – – approximating curve.

Graphic solution of the inequality (16) for the coal seam is resulted at Figure 2.

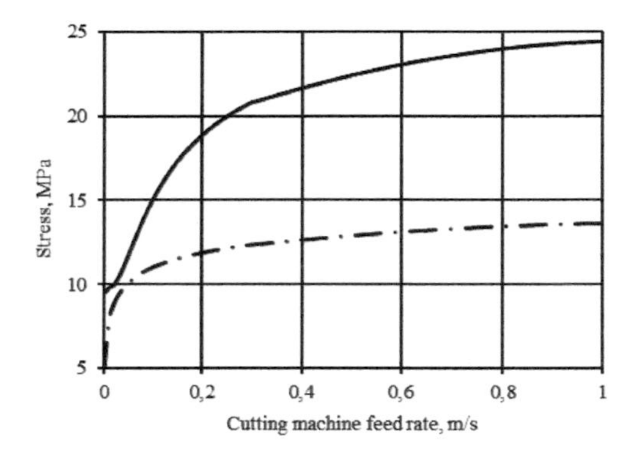

Figure 2. Graphic solution of the inequality (16) for the coal seam: ——— – strength curve; – · – – operating stress curve.

Dependence of long-term equivalent strength on loading rate for siltstone (immediate roof) is resulted at Figure 3.

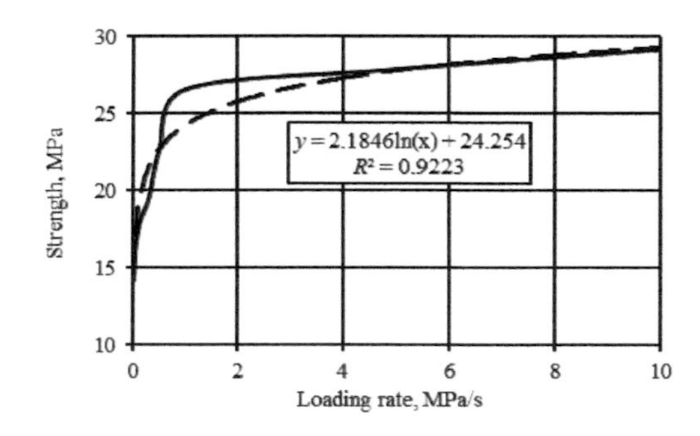

Figure 3. Dependence of siltstone equivalent strength on loading rate: ——— – experimental curve; – – – – approximating curve.

Graphic solution of the inequality (16) for the siltstone (immediate roof) is resulted at Figure 4.

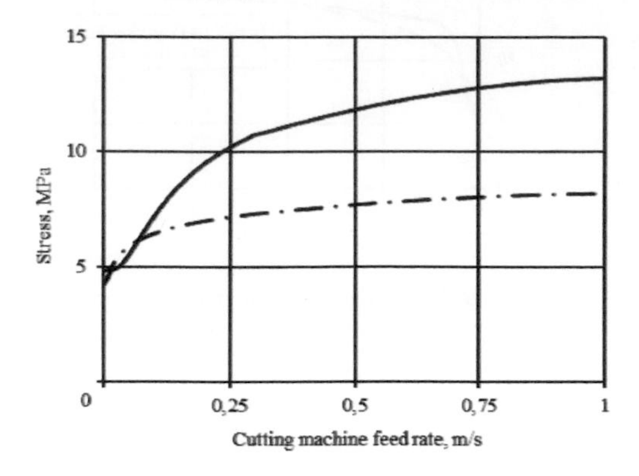

Figure 4. Graphic solution of the inequality (16) for the immediate roof (siltstone): —— – strength curve; – · – – operating stress curve.

5 CONCLUSIONS

The analysis of the obtained solutions for the mining seam (Figure 2) shows that operating stress curve in the working face plane and strength curve are not crossed, and on all number scale the strength curve is above the operating stress curve. It means that at any cutting machine feed rate at the coal seam face space will be no stability loss. The obtained results are well co-ordinated and confirmed by experience of mining longwall #874 at the mine "Zapadnodonbaskay". There during all operation term the coal sloughing and stability loss at the face space was not observed.

For the roof rocks (Figure 4), which is presented by siltstone, in the range of change cutting machine feed rate from 0.015 to 0.067 m / s the stresses operating in working face plane on 7-10% exceed maximum permissible. It makes possible to prognosticate, at the specified cutting machine feed rate, roof rocks stability loss. The experience of mining longwall #874, where cutting machine feed rate was 0.025-0.035 m / s (1.5-2.1 m / min), showed that practically along all extraction pillar at the longwall midsection wedged falls, cavities and cracks in the roof rocks were observed. It confirms operating of the equivalent stresses exceeding equivalent rock strength.

REFERENCES

Kuklin, V.Y. & Bobro, N.T. 2003. *The approached estimation rocks rheological properties*. Dnipropetrovs'k: The geotechnical mechanics, 40: 255-258.

Porcevsky, A.K. & Katkov, G.A. 2004. *Fundamentals of rocks physics, geomechanics and management of a mass state*. Moscow: 160.

Skipochka, S.I., Usachenko, B.M. & Kuklin, V.Y. 2006. *Geomechanics elements of the rock mass at high rates of longwall advance*. Dnipropetrovs'k: 248.

Sidelnikov, A.A. 2009. *Substantiation parameters of the rock mass 3-D simulation around the longwalls and development faces*. Dnipropetrovs'k: The geotechnical mechanics, 82: 77-85.

Vlasov, S.F. & Sidelnikov, O.A. 2010. *Research of the roof fall mechanism applying three-dimensional model of the stratified transversely-isotropic rock mass at a longwall advance*. Dnipropetrovs'k: Naukovy visnyk NMU, 2: 14-17.

Vlasov, S.F. & Sidelnikov, O.A. 2010. *Results of the 3-D simulation of the transversely isotopic rock mass stress state around a longwall*. Dnipropetrovs'k: New techniques and technologies in mining: 145-149.

Vlasov, S.F. & Sidelnikov, O.A. 2011. *The results of convergence researching in the longwall*. Dnipropetrovs'k: Technical and geoinformational systems in mining: 243-246.

Geomechanical Processes During Underground Mining – Pivnyak, Bondarenko, Kovalevs'ka & Illiashov (eds)
© 2012 Taylor & Francis Group, London, ISBN 978-0-415-66174-4

Estimation of reliability and capacity of auxiliary vehicles while preparing coal reserves for stoping

A. Shyrin, V. Rastsvetaev & T. Morozova
National Mining University, Dnipropetrovs'k, Ukraine

ABSTRACT: Factors limiting ground transport capacity are considered. Evaluation of support transport reliability while preparing coal reserves for stoping is given. Technological design of support transport using diesel overhead monorail with high adaptive capacity to develop timely new extraction pillars while mining stimulation was offered.

Timely development of explored reserves is one of the factors determining output increase while mining stimulation. Planned deadline of developing new extraction pillars is determined by the rate of work and capacity of section bedded workings, their maintenance-free service life as well as transport reliability.

Currently, coordination of transport and technological processes at most mines of Western Donbas carrying out coal output in complicated mining and geological conditions is hard to solve. It is due to specific for this region mining and geological, mining and technical and organization factors which retard to some extent mining development and stimulation (Shyrin 2007).

Manifestation of these factors more often takes place spontaneously and it is practically impossible to predict their parameters. So, section transport operation while driving and exploitation extensive watered workings with active soil rock swelling should be considered as system working in condition of uncertainty.

In condition of uncertainty section transport reliability will be characterized by the frequency and degree of failure as well as duration of downtime and adaptive capacity to spontaneous environment change in extreme situations.

To determine the reasons of extreme situations as well as to find out real sources of increasing adaptive capability of section transport while driving section development workings, failure of operative transport and technological design at mines of PJC "DFEC Pavlogradugol" was analyzed. According to the program and methods of complex study all losses of development operations connected with transport failure and necessity to repair and replace damaged units were considered to be accidental ones and were classified as failure within shift – acc/shift.

Interaction of transport and technological processes and operations while driving section development workings was analyzed with the help of 37 estimated figures of nontypical, critical and emergency situations. For initial information seven types of failure within shifts – cases of face outage due to transport facilities which are the most typical for operating system were formulated. Real factors of face outage and lost time to liquidate failure within the system of section transport were able to show major reasons of downtime for applied transport complexes. Duration of transport downtime for technical reasons were characterized as its operability, that is, capability to withstand overload and work for a long time without decreasing initial parameters.

Duration of unit downtime were characterized by its ratio which is relation of downtime h_d ratio for definite period to amount of time of actual transport operation h_f and downtime duration for the same period.

Taking into account abovementioned, transport downtime ratio η_d (another words, damage or failure) is expressed by the following equation :

$$\eta_d = \frac{h_d}{h_f + h_d} - \frac{1}{1 + \dfrac{h_f}{h_d}}.$$

In constantly changeable conditions of functioning underground transport durability of productive work of applied transport and technological systems to drive section development workings it was recommended to evaluate by the coefficient of their adaptation – successful operation of transport facilities in nontypical and close to critical maintenance conditions:

$$\eta_{ad} = \frac{h_f}{h_f + h_d} = 1 - \eta_d ,$$

where h_f – durability of transport actual work for a definite period, min / cm; h_d – transport downtime for the same period, min / cm.

Such approach for planning rates of driving section development workings including influence of adaptation coefficient of transport and technological systems is particularly urgent while continuous mining to the dip (to the rise) in condition of active soil swelling when driving air and assembly drifts is complicated by necessity of blasting swelled soil rocks, railway maintenance and working retimbering (Shyrin 2007). However, long period of field study and impossibility of physical modeling above mentioned processes are not able to predict immediately the behavior of transport and technological systems while mining stimulation.

Analysis of operating 88 development workings at mines of Western Donbas was carried out to determine factors limiting capacity of transport technological system. According to study program technological methods of continuous mining section development workings were classified according to the type of transport applied (Table 1).

Table 1. Flow chat of transport facilities applied within section development workings.

No	List of transport facilities	Field of application			Advantages	Disadvantages
		S, м^2	L, м	α, deg		
1	Electromotive (AM-8D)	8.5	–	2.86	There is no restriction along the length of transportation	Small transportation angles
2	Belt conveyor (1LU-80), surface ropeway (DKN)	10.3	1500	10	There is no restriction at big ranges of alternate profile	Small capacity at big transportation angles
3	Belt conveyor (1LT-80), electromotive (AM-8D)	10.3	–	2.86	There is no restriction along the length of transportation	Small transportation angles
4	Scraper conveyer (SP-63),belt conveyor (1LU-80), surface ropeway (DKN)	10.3	1500	10	There is no restriction at big ranges of alternate profile	Small capacity at big transportation angles
5	Scraper conveyor (SP-63), electromotive (AM-8D)	10,3	–	2.86	There is no restriction along the length of transportation	Small transportation angles
6	End haulage (LVD-34)	8.5	600	5	High capacity at big transportation angles	Difficult transportation within horizontal workings

It was stated that transport support of development headings within 36.4% driven workings is performed by electromotives AM8D, 39.8% headings are supported by surface cable ways (DKN) and 23.8% of haulage and load supply is performed with the help of one-ended cable installations. .

Comparative evaluation of technological effectiveness of abovementioned systems to drive development workings was carried out by the method of expert evidence. Taking into account specific factors it was found out (Rastsvetaev 2010) that at Western Donbas mines transport and technological systems of driving workings using continuous miner KSP-32 in combination with DKN are the most effective. To evaluate adaptive capability of this technological system basic and random factors characterizing conditions of interacting transport facilities with environment (rock massif) and elements of driven working (support, continuous miner) were formulated.

Basic factors are the following: working cross-section, length of transportation (working length), winning methods, gas abundance within transport working, seam thickness, fluctuation of transport facility track, working axis changeability, water content within transport working, rate of installation of temporary and permanent support, amount of transport units.

Rate of work performance connected with liquidation of emergency situations and rock pressure manifestation , temperature and country rock moisture, value and frequency of rock swelling within working, strain level of arch support, coefficient of cargo traffic irregularity, etc. are "random' factors.

To determine adaptive parameters of transport system, model of factor analysis which is suitable to many real situation of mining production was accepted (Okun 1974 & Harman 1972).

Ultimate goal of problem solving using factor analysis is to determine and reduce amount of factors greatly influencing the rate of driving development workings and their capacity, i.e. hypothesis estimation concerning minimum value of general factors which ideally reproduce available correlations.

In modern static computer programs various methods of factorisation of correlation matrix are used, but method of key factors (main axes) is more often used. In regard to this task minimum number of factors influencing adaptive transport capacity and rates of driving development workings was determined after computer-aided estimating matrix of factor loads.

In the course of calculation necessity to determine the rate of proper value of each factor took place. As it was mentioned in papers (Okun 1974 & Harman 1972), factoring ends up when the rate of proper value is slightly changed after sudden drop and diagram takes the form of horizontal line.

The next important estimation index of each factor meaningfulness is percent of referable dispersion of variables within correlation matrix. On the one hand, the task is to choose minimum number of factors which, from the other hand, could explain rather high percentage of entire dispersion of variables. Following condition should be taken into account to meet these two requirements, that is, total influence of basic factors on the whole process should be not less than 70-85% (Okun 1974; Harman 1972). After that it can be possible to show graphically factors influencing the process of driving workings before and after axis rotation.

According to results of analysis and synthesis of estimated data it was stated that factors connected with necessity to liquidate consequences of soil rock swelling greatly influence on the rates of driving section development workings and their capacity at mines of Western Donbas (Rastsvetaev 2010).

According to these results it's possible to come to conclusion that while soil rock swelling in technological process of driving section development workings instability of rail transport and ,as a consequence, reducing rates of driving and increasing time of developing extraction pillars can be seen. It is due to low adaption capacity of operating technological systems and traditionally applied rail transport while driving development workings in conditions of active soil rock swelling.

Analyzing factors limiting capacity of workings applying ground transport it was found out that creating technological system of secondary transport based on using alternative units with high adaptive capacity which will guarantee effective system operation in complicated mining and geological conditions is the way of problem solving.

Foreign experience of continuous mining shows that in condition of active soil rock swelling overhead monorails are used as a basic secondary transport. Such technological design is widely used at mines in Germany, Czech Republic, Poland and guarantee efficiency of material and equipment delivery. At some mines in the USA and RSA monorail transport is also used as basic one to transport minerals (Gilenko 1975).

At productive mines in Ukraine the field of overhead monorails application is limited. They are used only at stable operated mines of Central and Eastern Donbas. It is due to insufficient experience of this type of secondary transport maintenance, absence of study of monorail maintenance parameters and technical and economic substantiating fields of their effective application.

According to this study estimation of operating characteristics of overhead monorail at mines of Krasnoarmeyisk region in Donbas was performed to find out factors limiting field of their effective application. It is necessary to point out that entire monorail is based on general principles of moving large cargo units along monorail which consists of separate parts of 2.4…3.0 m length fixed by special locking joints. Monorail is packed by station and railway vehicles made up of cars and special platforms (Gilenko 1975).

Depending on functional role of workings and support type monorail parts can joint to the bars or legs of frame support or directly to working roof under anchor support . Monorail track is mainly in the center of air working. The place of monorail fixed to the bar of support is determined by the type of basic transport and its location relating to working cross-section within transport workings.

Despite high technical characteristic advertised by plants – manufacturers of monorail application of this type of secondary transport at mines in Krasnoarmeyisk region of Donbas causes a number of problems while its maintenance. Basic factors limiting field of overhead monorail effective application are rate deviation of cargo unit movement from design index and, as a consequence, time of delivering cargos to stoping and development faces.

Chronometration measurement of monorail maintenance parameters shows that the process of moving cargo units is characterized by frequent reducing movement rate of railway vehicles from 1.2 till 0.5 m / s. However, time of cargo delivery is increased as much as 1.3...1.6 times compared with estimated index.

According to expert evaluation of cargo transportation within section workings it was found out (Shyrin 2008) that the main reasons of irregularity of moving railway vehicles are the following: sign-variable track profile, deviation of sizes and weight of cargo units from normative standards, moisture and temperature change within working, as well as non-standard (emergency) situations taking place in the process of cargo delivery.

Estimation of functional state of section development workings equipped by overhead monorail showed that maximum curving of track profile is in zones of active stratification of roof rock. As a rule, these zones are situated above "bearing arches", i.e. frames of arch support where conjugated monorail components are attached (Figure 1).

Figure 1. Condition of development working support in zones of active roof rock displacement.

Instrument measurements showed (Rastsvetaev 2011) that in the process of "bearing arch" depression moved by roof rocks, deformation value of components of arch support of "bearing arch" 90...140 mm bigger than "conventional arch" located in the middle area of conjugated monorail parts.

Therefore, it is necessary to determine arch support deformation to predict in time zones of active rock pressure manifestation and carrying out measures directed to their prevention as well as increase of maintenance reliability of overhead monorail and development working capacity.

Results of modeling interaction of joint monorail parts with arch support components while passing railway vehicles showed that the process of moving large cargo units by monorail within workings using arch support should be considered as a single interactive system – "wall rocks-arch support-movable railway vehicle".

Mining study showed that behavior of this system unlike classic one "support-rock massif" will be determined not only by the state of roof rocks and support capacity but by the parameters of railway vehicles and, what is more important, by the place of fixing monorail on the support frame. The last one is stipulated by the fact that bearing element of this system is a bar of arch support. Static forces of rock pressure as well as forces produced by dynamic movement of monorail vehicle influence it. Depending on the character of rock pressure manifestation, place and method of fixing monorail on the bar of "bearing arch", support will be inadequately responded to dynamic loads while moving railway vehicle and transmit them to rock massif.

REFERENCES

Shyrin, L.N., Posunko, L.N. & Rastsveetaev, V.A. 2007. *Perspectives of developing adaptive systems of secondary transport at mines of Western Donbas*. School of underground mining: Materials of international scientific conference: Dnipropetrovs'k-Yalta: NMU: 374.

Technological design of underground transport within working area at coal mines (for flat seams with dip angle up to 18°). 1972. Moscow: Institute of Mining named by A.A. Skochinsky: 74-75.

Rastsvetaev, V.A., Posunko, L.N., Dyatlenko, M.G. & Shyrin A.L. 2010. *Complex estimation of transport and technological design of continuous mining development workings at mines of Western Donbas*. Materials of the V International scientific conference "Mining and ecological problems": 36-41.

Okun, Ya. 1974. *Factor analysis*. Moscow: Statistics: 200.

Harman, G.1972. *Contemporary factor analysis*. Moscow: Statistics: 488.

Gilenko, V.A., Kadyshev, V.V. & Kostyuchenko, S.I. 1975. *Monorail transport used while driving horizontal workings*. Moscow: Review.

Shyrin, L.N., Posunko, L.N. & Rastsvetaev, V.A. 2008. *Evaluating maintenance parameters of overhead monorail.* Geotechnical mechanics: Collaction of Scientific papers. Dnipropetrovs'k: Institute of Geotechnical Mechanics named by M.S. Polyakov of Academy of Science of Ukraine: Adition. 76: 91-96.

Rastsvetaev, V.A. 2011. *Peculiarity of forming additional loads on arch support of section workings with overhead monorail.* Scientific Bulletin of NMU, 4: 35-38.

Geomechanical Processes During Underground Mining – Pivnyak, Bondarenko, Kovalevs'ka & Illiashov (eds)
© 2012 Taylor & Francis Group, London, ISBN 978-0-415-66174-4

Influence of coal layers gasification on bearing rocks

V. Timoshuk, V. Tishkov, O. Inkin & E. Sherstiuk
National Mining University, Dnipropetrovs'k, Ukraine

ABSTRACT: Dependences of sandy-argillaceous rock permeability from volume deformations and thermal influence by results of laboratory research are determined. Research of bearing rock technogenic permeability at underground coal gasification with use of numerical geomechanical models is executed. The analysis of liquid gasification product composition is executed. Conclusions are drawn that power efficiency of UCG technology can be essentially raised by extraction and further use of condensed gasification products.

1 INTRODUCTION

The limited reserves of oil and natural gas in Ukraine leads to the fact that coal becoming the main natural source of energy. The increase in underground coal extraction is associated with the transition of mining operations to great depths, increasing the volume of waste dumps, air pollution and, consequently, environment stability violation (Arens 2001). Perspective way to rationalize the technology of coal extraction and processing is its underground gasification by thermochemical and mass transfer processes.

Domestic and foreign scientifically-practical experience of coal layers gasifying (Arens 2001) indicates vital importance of study bearing rock conditions which are exposed to influence of the big pressure and temperature differences in gasification process.

2 FORMULATING THE PROBLEM

All calculations of dangerous strain zones in the surrounding rocks of coal seam depend on the fire bottomhole (the gasifying circuit and gasified seam thickness). These data for underground coal gasification is quite difficult to determine analytically, so the location of these zones is calculated by creating mathematical numerical models of specific fields.

Their structural failure can lead to sharp increase in filtrational abilities and causing conditions for water overflow from overlying aquifers to the reactionary channel. Along with it, in works (Kreynin 1982 & Sadovenko 2001) the problem of underground coal gasification (UCG) power efficiency increasing and of its negative influence neutralisation by extraction of thermal energy and liquid gasification products being accumulated in roof rocks at mining coal layers was already considered. Thus, research of technogenic permeability character formed in underground gasifier vicinities and also the analysis of its influence on gasifying process and extraction of thermal and hydrocarbonic gasification products is important scientific and technical issue.

3 RESULTS

Study of changing permeability in covering strata during UCG process was carried out by executing of numerous tests of sample filtrational properties in triaxial stress conditions at various hydrostatic pressure values. Experimental specimens were presented by sandy differences from the sediments of the buchak-kiev suite composing overcoal waterbearing strata of Dneprovskiy basin. The range of loads in a test series was within 10…800 kPa, that corresponds to rock conditions in their shear zone. For studying temperature influence on bearing rock filtrational properties investigated samples were exposed to the temperature processing at 400 and 650 °C.

The hydraulic gradient was controlled by assign a hydraulic back pressure in the specimen and was equal 10 and 20 kPa that corresponded to real hydro-geological conditions of brown-coal deposits. Consolidation of sandy rock samples preceded carrying out of filtrational tests was lead at the loadings which correspond to geostatic pressure values, before obtainment of conditional deformation stabilisation, no more than 0.01% of volume change for the control time period. The water discharge control in free filtration conditions at the set pressure gradient was carried out up to the moment of its conditional stabilisation for the certain time period (a steady filtration mode). Filtration coefficient values were determined on the basis of measuring filtrate flow rate for each stage of the sample load according to the test program.

The results of laboratory tests presented on the fig. 1, illustrate existing dependences of filtration coefficients for medium and fine-grained sandy rocks on various degree of their consolidation or volume deformation size at presence or absence of sample preliminary temperature processing respectively.

The obtained data testify to existence of close linear dependence between filtration coefficient of tested sandy rocks and the value of volume deformation ($R^2 = 0.55...0.80$). In the range of effective loads 10…800 kPa and variations of volume deformations of rock samples from 0.0 to 0.12 the size of filtration coefficient changed in the range of 3.2…0.15 m / day. Dependence character of filtrational properties from volume deformation level keeps the kind for the tested sandy rocks in laboratory conditions. Considering character of specified dependence, and also effective loading level, it makes sense to speak about defining influence of active porosity and its changes at samples consolidation in triaxial stress state condition on change of sandy rock permeability.

Influence of high-temperature changes at temperature processing from 400 to 650 °C was revealed in growth of filtrational permeability of the tested samples in a range from 1.15 to 3.2 m / day in relation to their natural conditions.

The results obtained a basis for the next research stage which purpose was studying of filtrational fields in the deformed soil massifs. For this purpose the numerical geomechanical model of a rocky massif corresponding to a cross-section of Dneprovsky basin, and presented by the strata of sandy-argillaceous sediments containing a coal layer has been created.

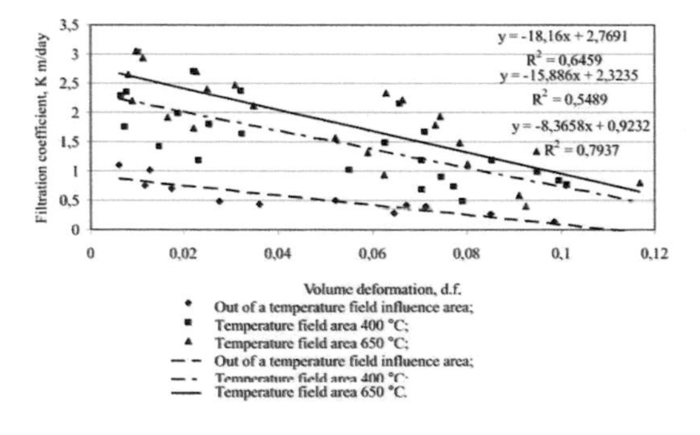

Figure 1. Results of filtrational tests of broken composition sandy rock specimens of the buchak-kiev suite of Palaeogene in triaxial stress state conditions.

The channel gasification width in the model was set on the basis of conditions for maximum gasifying fullness of coal seam, ensuring stable combustion heat of gas. Based on experience in gasification on Angren station "Podzemgaz" running on brown coal, lower combustion heat of gas is observed at a channel width of 15 m (Figure 2) (Kreynin 1982).

Considering the dependence of technogenic permeability of sandy-argillaceous sediments revealed in laboratory conditions on the level of deformations, representation of the character and zoning of a filtrational field is obtained on the basis of analysis of horizontal, vertical and volume deformation values for rocks in gasifying space vicinity.

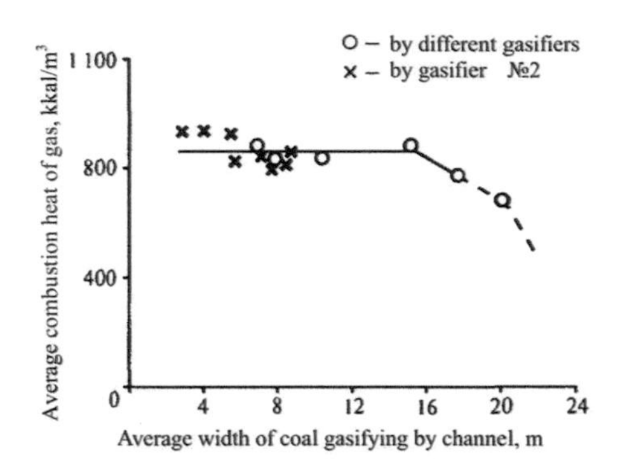

Figure 2. The influence of coal gasifying width along the channel on combustion heat of this gas.

Results of numerical modelling of rocky massif mode of deformation are obtained during problem solving in elastic-plastic statement. Design parameters of geomechanical model are specified according to fulfilled before research and results of studying mechanical properties on rock samples in three-dimensional stress state conditions.

For giving conditions at a coal layer capacity of 1.0 m and width of gasifying space about 9.0 m the field of technogenic permeability according to the reached level of deformations of containing rocks can be characterised by well expressed zoning by results of modelling is established. For a vertical deformation component relative tensile strains defining raised conductivity formation in a horizontal direction, concentrate in a roof and bottom of gasifying space. Horizontal tensile strains are fixed mainly on external contours of gasifying space and behind its limits, reaching values of the order 0.01…0.02 (Figure 3). Thus directly over gasifying space the zone of compressive strain is formed, defining thereby existence of the lowered values of permeability of containing rocks in a vertical direction.

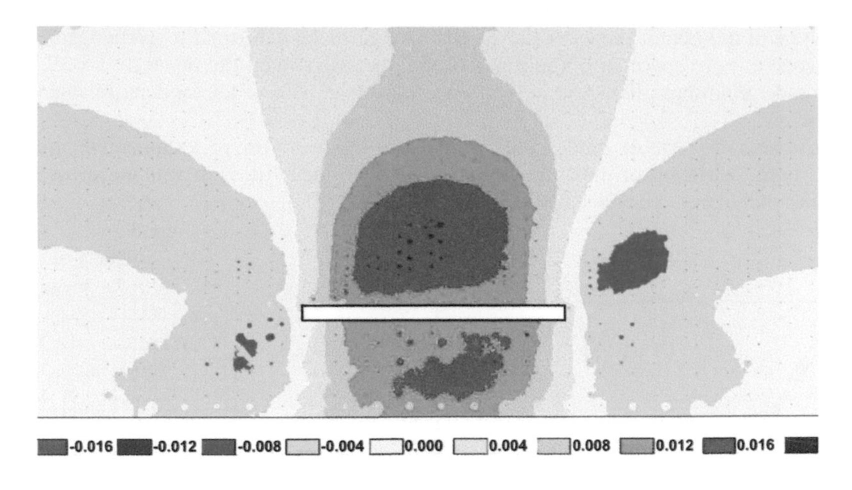

Figure 3. A field of relative horizontal deformations in gasifying space vicinity.

However, more integral representation of the character of technogenic permeability gives imposing of vertical and horizontal deformation fields – that is a field of volume deformations which in conditions of isotropic or quasi-isotropic on the properties of containing rocks is defining in formation of technogenic permeability. As follows the Figure 4, volume deformation distribution is characterised by prevalence of positive values, that in the presence of separate alteration zone will not result as a whole in essential growth of containing rocks permeability in a gasifying space vicinity.

The obtained representations about the character of technogenic permeability allow modelling of gazo-hydrodynamic processes in covering rocks at various stages of underground gasifier operation. Thus volume deformations field generated in a rocky roof will serve as a conductor for gas leaks which are formed in gasifying process of coal layers and being in superheated condition. In process of leaks advancing their pressure and temperature will decrease, while roof rocks will raise the temperature considerably. At inflowing of filtered gas in a zone with pressure below condensation pressure the part of gasification products will pass in a liquid condition. The condensate precipitated in the porous environment has not time to be filtered together with gas to a day surface that will lead to its accumulation in covering rocks. Change of temperature in gas leaks in covering rocks deformed by technogenic permeability is shown on the Figure 5 by results of calculations and physical simulating. The design procedure of the obtained results is described in details in work (Sadovenko 2007).

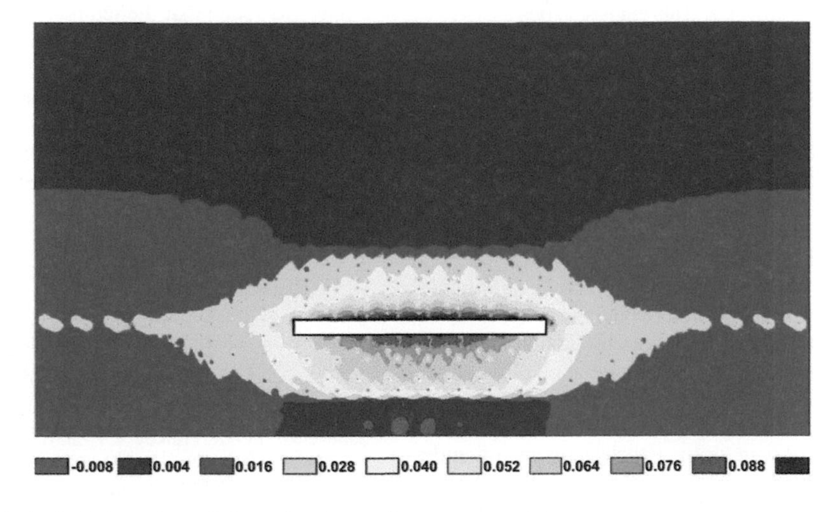

Figure 4. A field of relative volume deformations in a gasifying space vicinity.

The analysis of the obtained curves shows, that in roof rocks deformed by technogenic permeability, the saturation zone of pore space with liquid gasification products is on the distance 0.5…2.5 m from a rocky contour of underground gasifier. And in this zone, more than 96% of all condensate containing in UCG gas is deposited.

To identify area of probable condensate formation in the overlying rocks moisture content nomogram of natural gas used. Variation of moisture content of gas through filtration was constructed on the basis of known values of the pressure and temperature (Sadovenko 2007).

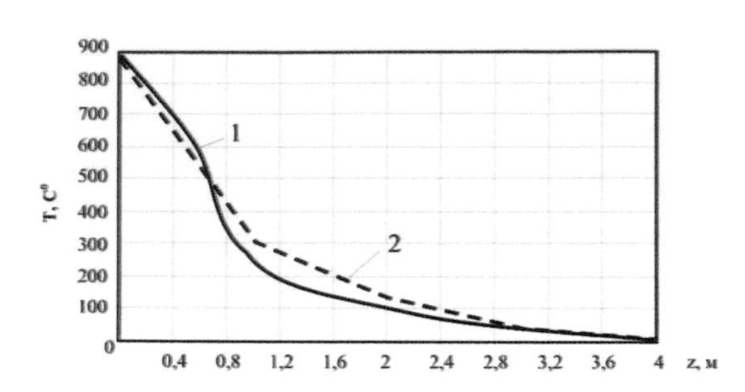

Figure 5. Change of temperature in gas leaks in covering rocks: 1 – calculated; 2 – obtained by physical simulating.

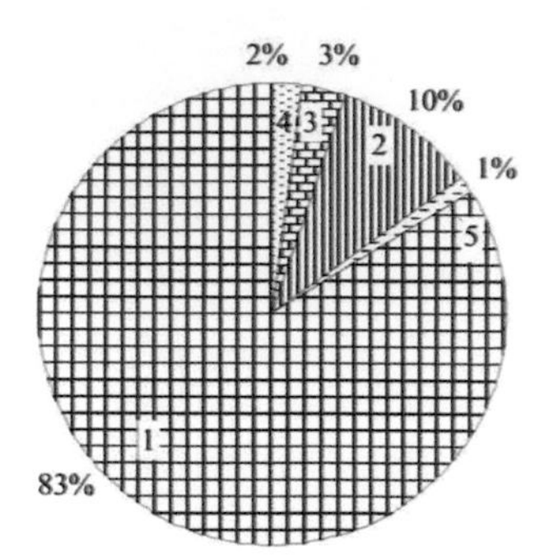

Figure 6. Composition of liquid gasification products being collected in roof rocks: 1 – neutral oils; 2 – phenols; 3 – the organic alkalies; 4 – asphaltenes; 5 – carboxylic acids.

Studying composition of liquid gasification products being accumulated in roof rocks was lead by the method of physical modelling with spectrophotometer UV2450 (Sadovenko 2007). The analysis of the obtained spectra has shown, that the precipitated condensate has low viscosity (less than $1.4 \cdot 10^{-2}$ Pa·c), and its density is $0.75…0.8$ g / cm^3. The basic components of a composition are the neutral oils representing the group of unsaturated, aromatic and paraffin hydrocarbons with additives of neutral oxygen compounds (Figure 6). The condensate differs by the low content of phenols (to 10%), and almost has no dust.

4 CONCLUSIONS

On the basis of experimental research and the analysis of gasification product distribution in roof rocks of coal layers the following conclusions are drawn:

– in gasifying process of coal layers the fields of technogenic permeability arising in covering rocks according to the reached level of rocks deformation it is characterised by well explicit zonality;

– saturation of a roof rocks by liquid gasification products occurs owing to precipitation of a condensate from a gas phase at abrupt change thermobaric conditions that forms an accumulation zone on distance 0.5…2.5 m from a rocky contour of underground gasifier;

– the liquid gasification products filling pore space of bearing rocks, are presented substantially as unsaturated, aromatic and paraffin hydrocarbons;

– power efficiency of UCG technology can be essentially raised by extraction and further use of condensed gasification products.

– close to linear correlation between filtration coefficient and volumetric strain value obtained on the results by seepage testing of sandy soil samples in triaxial instrument (filtration coefficient value was varied in the range 3.2...0.15 m / day at a value ε_{wol} from 0.0 to 0.12).

– it was found that the temperature effect on rock samples from 400 to 650 °C leads to an increase in their permeability from 1.15 to 3.2 m / day.

– according to the geomechanical modeling of rock mass for Dneprobass conditions the field of volume strain defined, which in almost isotropic enclosing rocks is dominant in the man-made permeability formation.

REFERENCES

Arens, V.Zg. 2001. *Physical and chemical geotechnology.* Tutorial, Moscow State Mining University: 656.

Kreynin, Ye.V. 1982. *Underground gasification of coal.* Moscow: Nedra: 394.

Sadovenko, I.O. & Tishkov V.V. 2001. *Assessing the possibility of taking away extra heat around the underground gas generator.* Dnipropetrovs'k: Naukovyi visnyk NGAU, 5: 133-134.

Sadovenko, I.O. & Inkin O.V. 2007. *Modelling of gasification product filtration in the overlying rocks.* Dnipropetrovs'k: Naukovyi visnyk NGU, 3: 11-15.

Sadovenko, I.O., Polyashov, O.S. & Inkin O.V. 2004. *Experimental studies of gasification product filtration mechanism.* Proc. Hyrnychodobuvna promyslovist Ukrainy I Polshchi: Aktualni Problemy I Perspektyvy. Dnipropetrovs'k: National Mining University: 598-603.

Lavrov, N.V., Kulakov, M.A., Kazachkova, S.Ts. and others. 1971. *About the underground gasification of brown coal deposit of Angren. Himiya tverdogo topliva.* Academy of Sciences USSR, 1: 73-79.

Geomechanical Processes During Underground Mining – Pivnyak, Bondarenko, Kovalevs'ka & Illiashov (eds)
© 2012 Taylor & Francis Group, London, ISBN 978-0-415-66174-4

Ecological aspects of the quantitative assessment of productive streams of coal mines

S. Salli & O. Mamajkin
National Mining University, Dnipropetrovs'k, Ukraine

ABSTRACT: Possibility of creationof "mines-dressing plants" aggregate system model is considered in the article. Complex index of economic and functional efficiency is used as an efficiency criterion.

1 INTRODUCTION

New economic, political and ecological situation which had developed in Ukraine, demands revision of priorities of development of the state since despite the declination in production among the European states it has the greatest integrated indicator of anthropogenic loads of surrounding environment practically on all territory. Officially recognized international status of our state concerning its ecological condition and level of environmental pollution is defined as a zone of "ecological disaster". Based upon it, there is a complex problem of a harmonious combination of acceleration of growth rates of market economic development with need of technical modernization of the mine enterprises which would provide a condition of self-supporting use of natural resources and environmental protection.

2 CONSIDERATION OF A PROBLEM

Starting point of preservation of the natural capital in system of formation of ecologically balanced economy in Ukraine is the determination of an ecological stock of assimilation potential of landscapes in the region. Calculations (Burkinsky 1999 & Veklich 2003), testify that in overwhelming majority of areas level economic loads of components of the landscape natural capital exceeds their assimilation potential (Table 1).

Moreover, since excess of level economic loadings is tracked at more than 72.6% of the territory of Ukraine, it is already a problem of a national scale. The objects of fuel and energy complex functioning under continuously changing conditions of environment and in many respects depending on their dynamics, also have on it corresponding influence with many difficult predictable and not always reversible consequences.

Table 1. Definition of an ecological reserve of assimilation potential components of the landscape natural capital of Ukraine.

Region	Assimilation potential	Load	Ecological stock
Donetsk	0.39	4.66	-4.27
Dnepropetrovsk	0.31	2.68	-2.37
Zaporozhzhe	0.29	1.40	-1.11
Kiev	0.59	1.85	-1.26
Ternopol	0.65	2.36	-1.71

The coal industry which is actively destroying all vital spheres of environment is most dangerous in this regard. Burning dumps, waste heaps, an intensive dust content and an air gas contamination, reservoirs brighteners and settlers, tailing pounds, pollution of surface and ground waters, dumping in a hydrographic network of the high mineral waters, dangerous geotectonic processes and invasion into the underground hydrosphere, provoking subsidence of a terrestrial surface, bogging of areas and regions, creation of artificially increased seismicity and so forth it is far not the full list of anthropogenic pressure on environment in mining regions. Even with closing of mines of a consequence of their former activity dozens of years will negatively affect the condition of environment and safety of life of the population of territories adjoining to them.

Miner's regions are practically zones of ecological catastrophe, and the enterprises of the coal industry are classified as ecologically dangerous. According to Goskomstat of Ukraine, in the coal-mining regions of Ukraine 158 coal mines operate, each producing 1000 t of coal from 150 to 800 t of the rock which forms waste heaps occupying huge areas, leading to intensive gas dust pollution of air and chemical poisoning of surface and ground waters, and also essentially changing hydrodynamic regime and level of underground waters. Development of coal fields negatively influences hydro-chemical mode of surface activity and underground water, increases pollution of air space, worsens fertility of lands (Burkinsky 1999 & Veklich 2003).

The total area of the earth which has been taken away under industrial platform of the coal-mining and coal remanufactures enterprises, makes about 22.5 thousand hectares. By data "Power strategy of Ukraine for the period till 2030", at carrying out mountain works from coal mines annually, by different estimates, it is allocated from 750 million m^3 to 2.7 billion m^3 of the methane which absolute majority is absorbed by the atmosphere. Among unorganized sources of emissions special place is occupied also by dumps of rock which can light up. Volume of the mine waters which are pumped out during coal mining, nearly 600 million m^3 a year whereas for economic and production needs of the enterprises of branch and for other consumers 250 million m^3 (40%) are used only make. Due to the extremely unsatisfactory purification of mine waters in the rivers over 1 million t of mineral salts is annually dissolved. It is counted up that for the prevention of negative consequences from activity of mines it is necessary to perform annually nature protection works for the sum of 230-240 million UAH (Bardas 2010 & Nedodayeva 2006).

All this testifies in favor of ecological certification of mines just because the coal enterprises don't make production, and allocate from the environment created without participation of the person, the finished product (coal) in the course of production occurs interaction industrial (the equipment, buildings and constructions) and natural factors and thus violation of balance of natural factors takes place. It is expressed in the following: allocation of methane, mine waters, delivery and warehousing of mine breed, violation of a terrestrial surface, including built up with the territory, under working of reservoirs, forests, etc.

Extent of this interaction depends on mine development in time both in space and from essence of the factor (Figure 1 and 2).

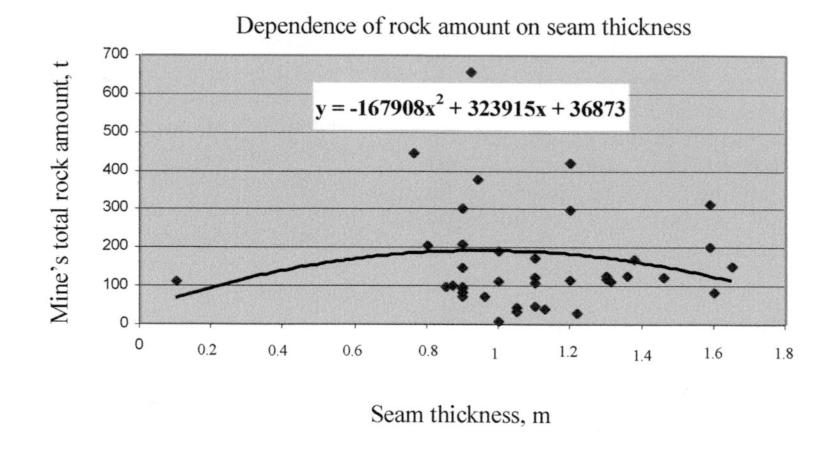

Figure 1. Volume of produced rock depending on layer capacity.

Separate natural factors make on mines beneficial or adverse effect on a production activity and economic results of work of mine; for example – change of capacity of a layer involves change of quantity by produced rock. At different times operation of mine is influenced by natural factors, for example increase in depth of mining, as a rule, leads to growth of gas volume and increase in released methane.

Finally work of mine is influenced by all factors in aggregate and consequently there is a need of the accounting of their joint action. Let's consider action and value of the specified factors from the point of view of participation in production costs of production and raw materials processing. The given factors aren't commensurable among themselves directly and consequently it is inexpedient to accept for definition of joint action a cost assessment.

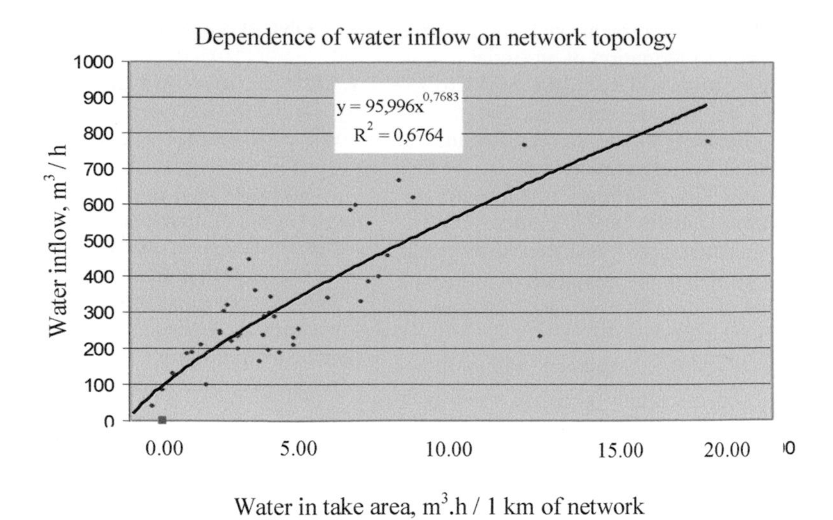

Figure 2. Dependence of water inflow on length of supported excavations.

Taking into account the stated above conditions it is possible to plot model of aggregate system "mines-concentrating factories", as criterion of productivity a complex indicator of economic KE and functional KF of an dispatch. And the indicator of KE should consider a condition of maximizing target parameters of streams of coal (D_i), breed (V_i), firedamp (M_i) and mine water (W_i), and an indicator of functional efficiency – to reflect the characteristic of an efficient condition of system (Salli, Bondarenko & Tereshchenko 2009).

Functionally aggregate system can be presented as follows

$$K_e = F \quad D_i, V_i, M_i, W_i$$

$$K_s = \{Z_{i1}, Z_{i2}, Z_{i3}, Z_{i4},$$

where K_{si} – characterizes stability of each diagnostic sign, and Z_{ij} – factors of stability of i-go of a sign.

Table 2. Rating of group of anthracite mines taking into account an ecological factor.

Mines	Extraction, thousand tons/year	Stock, mil. t	Prime cost of 1 ton, UAH	Exit of empty rock, thousand tons/year	Inflow of water, m^3 in hour	Values of complex criterion, λ_i
"Komsomolskaya"	1322	78.2	665.9	793	1350	0.60
"Partizanskaya"	257	10.8	917.9	77	290	0.85
"Schahterskaya Glubokaya"	449	123.5	776.4	224	190	0.71
"Progress"	466	74.9	665.1	186	340	0.72
"Tsentrosoyuz"	582	21.8	564.9	378	412	0.67
#81 "Kievskaya"	545	26.4	512.5	299	812	0.66
"Luganskaya"	222	6.8	931.8	118	1100	0.92
"n.a.Kosmonavtov"	491	29.4	611.9	294	600	0.74
"n.a. Frunze"	1663	35.7	554.3	997	637	0.58
1-2 "Rovenkovskaya"	234	18.4	889.2	117	400	0.79
"Zarya"	418	12.5	776.2	250	550	0.78
"n.a. Dzerzhynskogo"	251	18.2	894.1	130	350	0.82

Procedure of the multidimensional comparative analysis is reduced to formation of an initial matrix in the form of the standardized signs. Standardization is made by the standard methods, thus important stage of processing of the corrected values is their differentiation on importance by means of factors of hierarchy. The matrix of relationship between certain parameters is for this purpose formed. Finally there is a parameter λ and this parameter is closer to unit, the it is more deviation of values of diagnostic signs of mine from diagnostic signs of mine standard (Table 2).

The ecological factor λ_i changes within limits (0…1). It can't be essentially equal to zero as it would mean lack of natural factors, and only in some cases theoretically it can be equal to unit at equality of expenses for coal mining on natural and industrial factors. Smaller value of ecological factor is desirable that indirectly characterizes (in the generalized look) more favorable mining-and-geological conditions.

Operating mines if necessary can be distributed (are classified) on size of ecological factor. It can be demanded at the solution of various questions of management by branch.

It is important to note that from the point of view of integrated approach of assessment of the work of mines, the volume of production carries out rather important role, but not defining. Especially it is shown in respect of qualitative characteristics of coal and in the ratio given out and overworked (that remains in mine) breeds. Unfortunately, passing methane production in recent years on the majority of mines of Donbass it is not conducted. This circumstance can raise a mine rating on some positions.

3 CONCLUSIONS

1. Distribution of investments to support the capacity should be carryied out taking into account a given mine's rank by complex of its economic and ecological consequences.

2. Location of a mine in closed circle is necessary, but not an indispensable condition for its shutdown as at each enterprise there are reserves for increasing its economic and ecological rating.

3. Mines provided by balance reserves, can have a low rating as a consequence of insufficient investments, unique natural situation or irrational management of a mine's economy.

REFERENCES

Burkinsky, B.V., Stepanov, V.N., Burkinsky, B.V. & Kharichkov, S.K. 1999. *Management of natural resources: bases of economical and ecological theory.* Odessa: 350.

Veklich, O.O. 2003. *Economic mechanism of ecology managing RNBO Ukraine Ukrainian institute of environmental research and resources.* Kyiv: 88.

Bardas, A.V. 2010. *Foundations of ecological certification in Ukrainian coal extraction enterprices in condition of re-structured of sector.* National Mining University.

Nedodayeva, N. L. 2006. *Ecology-economical politics of management of natural resources in specific conditions of mining.* Donetsk: NAN of Ukraine. Institute of industrial economics: 356.

Salli, S.V., Bondarenko, Y.P. & Tereshchenko M.K. 2009. *Manage of technical – economic parameters of coal mines.* NMU. Gerda: 150.

Geomechanical Processes During Underground Mining – Pivnyak, Bondarenko, Kovalevs'ka & Illiashov (eds)
© 2012 Taylor & Francis Group, London, ISBN 978-0-415-66174-4

Development of mathematical foundations and technological support of the processes of complex hydro-pneumatic impact on coal seams

S. Grebyonkin, V. Pavlysh & O. Grebyonkina
Donetsk National Technical University, Donetsk, Ukraine

ABSTRACT: Physical principles and mathematical foundations of the process of integrated hydropneumatic impacts on coal seams are considered in order to reduce the intensity of the manifestations of the main hazards in underground coal mining.

1 RELEVANCE OF THE WORK

Application of methods and schemes of prior exposure on the coal seams, to fight with the manifestations of the main dangers, – is mandatory on mines and regulated by normative documents (DNAOP 1.1.30-1.XX-04 2004). To improve the efficiency of effects it is rational to apply methods of influence in the complex, using all the advantages of each method. Application of mathematical modeling allows to improve the procedure of design and calculation of parameters of technological schemes. In this regard, theme of the work is urgent.

Purpose of work – justification for principle and improvement of technology of preprocessing seams, using an integrated approach.

2 THE MAIN CONTENT OF THE WORK

In the works (Moskalenko 1971; Stern 1981; Pavlysh & Stern 2007) are reviewed three types of impact on the coal seams: hydraulic impact, pneumatic handling and degasification of bottomhole formation zone.

Among them to date the most widely implemented the hydraulic impact, which is used as a tool to control major hazards in underground coal mining. Through the efforts of many researchers is developed basic theories and technologies for hydro-impact. In the work (Pavlysh & Stern 2007) the theoretical foundations of the process have been further developed, that allowed to justify a modification of technology – the cascade method of hydrotreating. However in this area there are still problems as in the theoretical aspect and in the technological area. In particular, the task is to develop methods of engineering calculation of parameters in order to practices could, without the use of computational tools, conduct preliminary indicative calculations of parameters of technological schemes. In the area of improving the technology is the problem of implementation of the cascade technology for various conditions (for example- flat and abrupt, thick and thin layers, factors of disturbance of the seam's structure, especially occurrence, etc.).

As for the other two types of impact, their development is still in initial stage. However, at this stage, due to the results, obtained in (Stern 1981), have reason to set the task of the theoretical foundations and technology of complex hydropneumatic impact on coal seams, which includes three successive stages.

2.1 *Pneumatic treatment of unwetted seam*

At this stage removal of free and desorbing methane is provided. In addition, this method has the prospect in the aspect of changing of physical-chemical state of the seam and, probably, will reduce the tendency of the seam to spontaneous combustion.

2.2 *The hydraulic impact*

This type of exposure, through use of the developed technologies allows to produce a liquid saturation of the coal seam, that provides reduction of dust generation, helps to reduce gas emission and thus has a positive impact on working conditions in underground coal mining.

2.3 *Degassing of bottomhole zone of seam*

This type of exposure can reduce gas emission into the face.

Let us consider the main elements of the technological chain during the realization of a complex exposure.

3 PNEUMATIC PROCESSING

Recommended to apply on thin and medium-sized seams, developed by columnar system. The basic version of the technological scheme is shown on Fig. 1 that indicates recommended equipment and geometrical parameters.

The diameter of the well is usually taken 76 mm. Injection pressure P_H must exceed the gas pressure in the bed, but it is chosen minimum allowable (usually up to 20 kgf / sm²).

The most efficient and economical is a cyclic mode of pneumatic processing of coal seam by long wells, parallel lines of face. Even wells are injection wells, uneven – outflow.

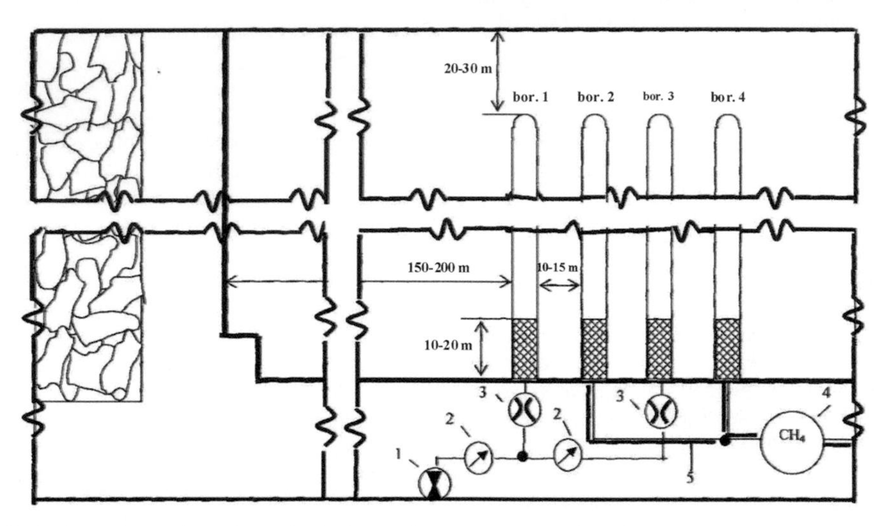

Figure 1. The basic version of the technological scheme of pneumatic processing of coal seam: 1 – compressor, 2 – pressure gauge; 3 – the counter of air flow; 4 – measuring of the concentration of methane (gas analyzer); 5 – mine gas pipeline.

Rate of injection:

$$q = 0.33 \cdot 10^{14} ml_F K \frac{P_N^2}{L_{M.S.}}, \text{ m}^3 / \text{min} \quad (1)$$

The duration of injection cycles is determined by the decrease in methane concentration in the outflow hole to 50-60%, the initial duration is determined by the formula:

$$t_N = 0.2 \cdot 10^{-15} \frac{L_{M.S.}^2 n_E P_G}{K P_H^2}, \text{ daily} \quad (2)$$

Total duration of exposure is determined by the intensity of removal of methane during the air injection and in typical conditions is 150 days. Reading of gas while pneumatic processing during this period by 35-40% higher than during degassing. According to this technology pneumatic processing of unwetted seam is conducted.

The same time pneumatic processing of coal seam allows to reduce its chemical activity almost to the safe limits. For seams with coal chemical activity index 0.06 ml / g hour (maximum value for bituminous coal), time for pneumatic processing , that is required to reduce this index to 0.015 ml / g hour (value, corresponding to the category of seams, that have low-risk of fire) is 130 days. In most cases pneumatic processing does not lead to significant heating of the coal seam: in all conditions the temperature increase is not dangerous in terms of spontaneous combustion.

The time interval between injection cycles corresponds to the restoration of methane concentration in the outflow hole to a maximum value.

Total time of pneumatic processing by a factor of gas-bearing reduction is determined by the moment, when the air injection ceases to significantly influence of the methane removal (when the duration of the discharge cycles, of the above condition, becomes practically equal to zero).

Reduction of gas bearing capacity of the array for all the exposure time is:

$$\delta X = \frac{a_{M0} - a_{M.out}}{1.1 a_{M0} - a_{M.out}} \cdot 100, \% \quad (3)$$

The time, that is required for reducing index of chemical activity of coal from the initial to the value a_0':

$$T_N = 2.5 \frac{a_{0out} - a_0'}{a_{out} a_0'}, \text{ days.} \quad (4)$$

3.1 *The hydraulic impact*

It is recommended as the second stage of the complex impact next to the pneumatic processing of unwetted seam.

The beginning of impact immediately following the end of pneumatic processing, thus it is important the fact that the wells, through which pneumatic processing was conducted, can be used as elements of the hydraulic impact technology.

Methodology, technology, mathematical models and calculation of parameters of this phase are presented in (Pavlysh & Stern 2007).

3.2 *Degassing of bottomhole zone*

This phase was proposed by MacNII (scientific-research institute) as the final stage of works on reducing dust and gas emission into mine workings. The method is directly applied as a means of reducing of gas emission into the face during the development of the seam.

Recommendations for use of technology are given in (Pavlysh & Stern 2007).

Theoretical and experimental investigations of the proposed complex impact could be the subject of further research.

4 CONCLUSIONS

Thus, we can assume that the complex hydropneumatic impact provides a purposeful change in the state of the coal seam, needed to increase the load on the working face, rate of conducting of mine workings and protection of labor.

New computational model for the investigation and calculation of parameters of the pneumatic processing process of coal seam is developed, recommended for use in combination with a hydraulic impact.

Pneumatic impact reduction of gas saturation of coal seam is in accordance with an exponential decreasing Theoretical foundations of the complex hydropneumatic impact on coal seams have been developed, which includes three stages: pneumatic treatment of unwetted seam, cascade hydrotreatment and degassing of bottomhole zone.

REFERENCES

DNAOP 1.1.30-1.XX-04. 2004. *Safe conduct of mining operations on seams prone to gas-dynamic phenomena (Revision 1)*. Kyiv: Fuel and Energy of Ukraine: 268.

Moskalenko, E.M. 1971. *Scientific basis of biochemical and physico-chemical methods of combating with methane in the coal mines*: Dis. dr. ... technical science. Moscow: Sciences: 508.

Stern, Y.M. 1981. *Improvement of technology of fluid injection into a coal seam in order to improve the efficiency of its processing to combat with methane and dust in mines*: Dis. candidate technical science: 05.05.04. Moscow: 211.

Pavlysh, V., Stern, Y.M. *Fundamentals of the theory and technology parameters processes of hydropneumatic impact on the coal seams*. Monograph. Donetsk: "VIC": 409.

Geomechanical Processes During Underground Mining – Pivnyak, Bondarenko, Kovalevs'ka & Illiashov (eds)
© 2012 Taylor & Francis Group, London, ISBN 978-0-415-66174-4

Analysis of test methods of determining antidust respirator quality

S. Cheberyachko, O. Yavors'ka & T. Morozova
National Mining University, Dnipropetrovs'k, Ukraine

ABSTRACT: Comparison of calculation procedures of vagueness domestic and harmonized with European DSTU is conducted. Efficiency of tests for the different amount of samples is analyzed. Results are given, that it is necessary to increase the amount of samples for the increasing of results quality.

1 INTRODUCTION

Antidust respirators are one of the most important means to protect workers against such diseases as pneumonia and dust bronchitis. Labor productivity depends on their quality as they are additional source of load to people breath system. Therefore, it is very important to determine real indices of protection efficiency and breathing resistance which are the basic criteria of evaluating means of individual protection of respiratory organs (MIPRO) because calculation accuracy of dust load and people productivity depends on them. Standard methods given in GOST 12.4.041-89 were used to determine them. However, regulations harmonized with European ones which represent demands and list of methods and technique concerning test of MIPRO have come into operation in Ukraine since 2004. The aim of specialized laboratories is to determine measurement errors. It is not possible to get the permission of regulatory authority to carry out the tests without its solving. The task is complicated by the absence of methods concerning error evaluation in European regulations in contrast to domestic ones where after described procedures of determining measurement results it is shown how to check their accuracy. So, the task is to determine the indices validity of MIPRO according to harmonized DSTU.

2 BASIC PART

Protective efficiency of antidust respirators is determined by coefficient of penetration K_n (Petryanov 1984):

$$K_n = \frac{C}{C_0}, \tag{1}$$

where C – concentration of harmful substance under respirator, $\mathrm{mg/m^3}$; C_0 – concentration of harmful substance in atmosphere, $\mathrm{mg/m^3}$.

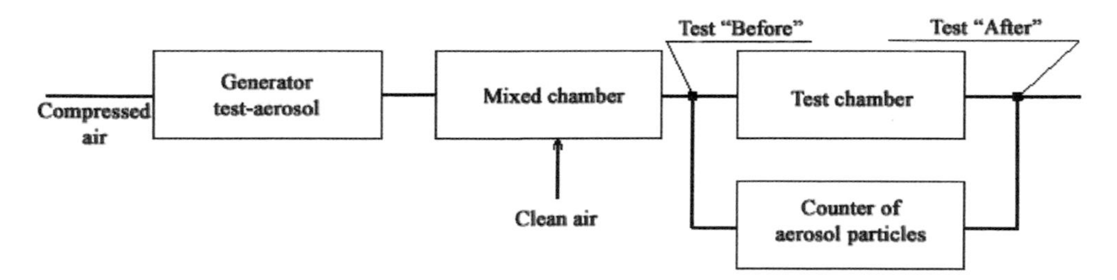

Figure 1. Principle scheme of respirator tests by test-aerosol.

Figure 1 shows principle scheme of respirator tests to determine its protection efficiency. Tests are carried out in the following way: compressed air is delivered to generator with special liquid, test-aerosol is delivered from generator to mixing chamber where it is diluted by the clean air up to the necessary concentration and finally it goes to the chamber with respirator. Input and output concentration of test-aerosol is determined by counter of aerosol particles and coefficient of penetration is determined by formula (1).

The second important task of MIPRO is to determine breath resistance according to the pressure difference fixed by micromanometer. It happens because certain quantity of air current passes through the MIPRO (Figure 2). It can be determined by:

$$R = (n_i - n_0)K_1 , \qquad (2)$$

where n_i – resistance registration performed by micromanometer, mm. Aq; n_0 – internal resistance of micromanometer mm. Aq; K_1 – temperature and air pressure correction coefficient (Methods… 2001).

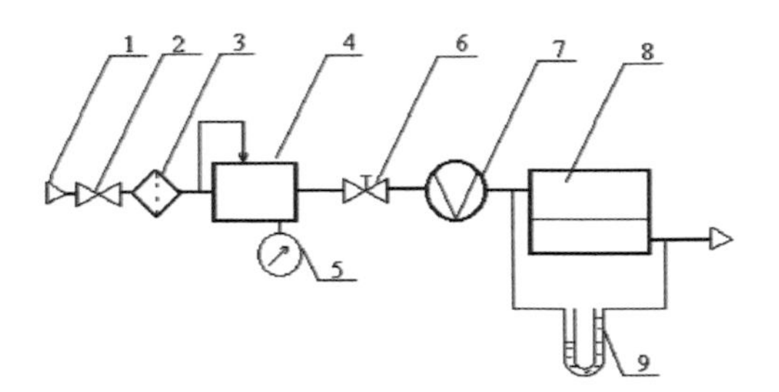

Figure 2. Principle scheme to determine resistance of MIPRO against constant air current: 1 – entrance from thoroughfare, 2 – uncontrolled valve, 3 – air cleaning filter, 4 – reducing gear- stabilizer -, 5 – manometer, 6 – controlled valve, 7 – diaphragm, 8 – test chamber, 9 – micromanometer.

However, completely new approach concerning quality estimation of MIPRO is the basis in European standards research methods are principally the same. The advantage is preparing respirators to the tests which is absent in domestic regulations whereas it is provided in the countries of European Community (EN 133, EN 136, EN 140, EN 143, EN 149). The idea of the test is the following: the first series of samples (in most cases series consist of three respirators or filters) are first subjected to thermal effect (daily time delay is 24 hours at a temperature of +70 °C and with the same duration at a temperature of -30 °C).The second series of samples pass through “modeling wearing schedule”. It means that double air current with artificial air humidification is conducted through filter half mask fixed on the “Sheffild” head model. The third series of samples are subjected to mechanical effect (imitation of low frequency vibration) and undergo inflammation test. Furthermore, according to GOST EN 149-2003 protective items are tested not only with air flow at 30 l / min but with the flow rate at 95 l / min.

Values of protective efficiency and breath resistance after procedures provided for “preparation” will differ between series of samples that makes difficult to determine true results due to their possible sufficient differences. Taking into consideration that series consist of three MIPRO it is rather hard to determine the error of final result according to each procedure provided for “preparation». Calculation of measurement error can be performed according to the GOST 8.207-76 “Direct measurements with numerous observations. Methods of processing observation results”. Error of measurement results consists of evaluating deviation (random error) and systematic error which depends on measuring tools. To determine random error it is required:

– to calculate average value of n-number of the item by formula:

$$\overline{R} = \frac{1}{n}\sum_{i=1}^{n} R_i , \qquad (3)$$

where n – the number of items to be tested; R_i – induce value of i-item determined by tests (Novikov & Kotsuba 2001; Volodarsky & Kosheva 2008);

– to determine standard deviation S by formula:

$$S = \sqrt{\frac{1}{n-1}\sum_{i=1}^{n}\left(R_i - \overline{R}\right)^2} . \qquad (4)$$

– to check on abnormal results, i.e. to determine correlation U_{max} or U_{min} by formulas:

$$U_{max} = \left(R_{max} - \overline{R}\right)/ S \qquad (5)$$

or $$U_{min} = \left(\overline{R} - R_{min}\right)/ S , \qquad (6)$$

where R_{max}, R_{min} – maximum and minimum values of obtained induce according to selection determined by tests, values U_{max} and U_{min} are compared with threshold value β, at the level of significance $\alpha = 0.05$ (Methods…2001): if $U_{max} \geq \beta$ or $U_{min} \geq \beta$, then the result of observation is R_{max} or R_{min}, correspondingly abnormal and should be excluded from selection;otherwise the result is considered to be normal and is not excluded after exclusion of normal value; average value $\overline{R}$ – determined again and S – calculated;

– to determine relative standard deviation of measurement results by formulas:

$$S' = S/\sqrt{n} \qquad (7)$$

or $$S' = S'/\sqrt{n-1} , \qquad (8)$$

– to find out confidence interval ε of random error

by formula:

$$\varepsilon = tS', \tag{9}$$

where t – Student coefficient which depends on confidence probability P and number of observations n.

Calculation results are written in the form of:

$$R = \overline{R} \pm \varepsilon, \%. \tag{10}$$

Dependence of general systematic error on measuring tools is determined by formula:

$$\theta = k\sqrt{\sum_{i-1}^{m} \theta_i^2}, \tag{11}$$

where k – coefficient determined according to confidence probability; θ_i – limit of systematic error of i – measuring tool:

$$\theta_i = \frac{100\Delta_i}{X_i},$$

where Δ_i – error of i – measuring tool; X_i – value of measured one on i – measuring tool.

To calculate errors of penetration coefficient according to the test-aerosol and breath resistance several dozens of antidust respirators SHB-1 "Lepestok-200" were chosen at SIE "Standart" (Dnipropetrovs'k). The first step was to carry out a number of tests to determine basic indices of several series of respirators according to the GOST 12.4.028-76 (respirators SHB-1 "Lepestok"). The first series consist of three samples. Next series have 3 respirators more to compare with previous ones. After that average values of measurement results for each group of MIPRO and their standard deviations were calculated by formulas (4-8). Figure 3 and 4 show the results.

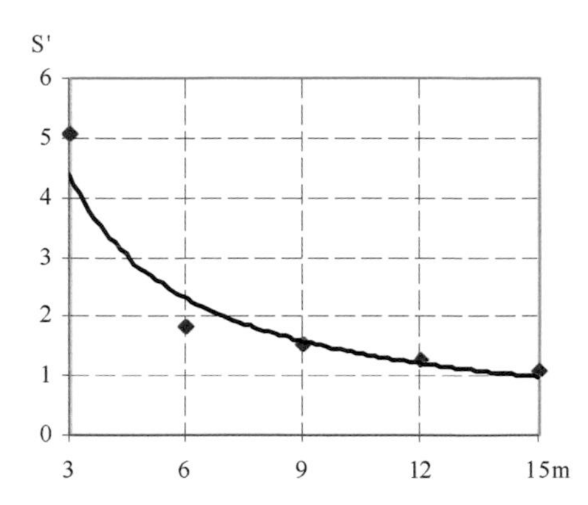

Figure 3. Dependence of relative standard deviation of breath resistance on the number of observations.

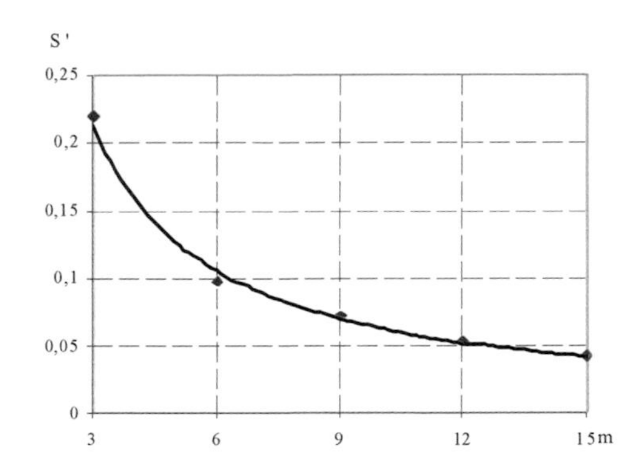

Figure 4. Dependence of relative standard deviation of penetration coefficient according to test-aerosol "oil fog" on the number of observations.

Figure 3 and 4 shows that with the increase of the number of observations standard deviation reduces, and its value stays constant if the number of respirators is not more than ten. So, minimum number of test samples should be not less than 10 to guarantee necessary accuracy of measurement. The next step is to carry out test of MIPRO by harmonized methods given in GOST EN 149-2003.Their peculiarity is so-called procedure of respiratory preparation described above. To determine optimal number of observations several series of samples which were different in quantity of respirators (next series have 3 respirators more to compare with the previous ones) were taken. The first group of MIPRO was subjected to temperature effect and the second one passes through "wearing schedule modeling" before determining coefficient of penetration according to test-aerosol "paraffin oil" and breath resistance. Relative standard deviation for each group of samples (Figure 5 and 6) was calculated by formula (8).

Analyzing diagrams we can come to the following conclusion: relative standard deviation stays constant when the number of samples is not more than 10; preliminary preparation improves homogeneity of quality induces of MIPRO, particularly while obtaining coefficient of penetration according to test-aerosol "paraffin oil" where deviation after delivering is two times more for the same number of samples than for the samples passed through preliminary preparation.

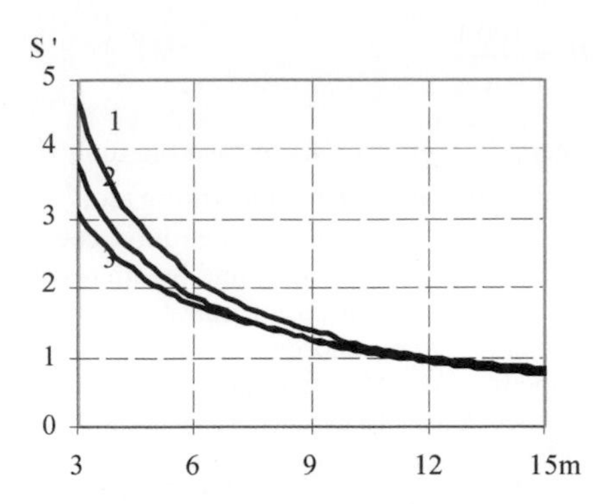

Figure 5. Dependence of relative standard deviation of breath resistance on a number of observations: 1 – samples without preliminary preparation, 2 – samples subjected to temperature effect, 3 – sampled passed through "wearing schedule".

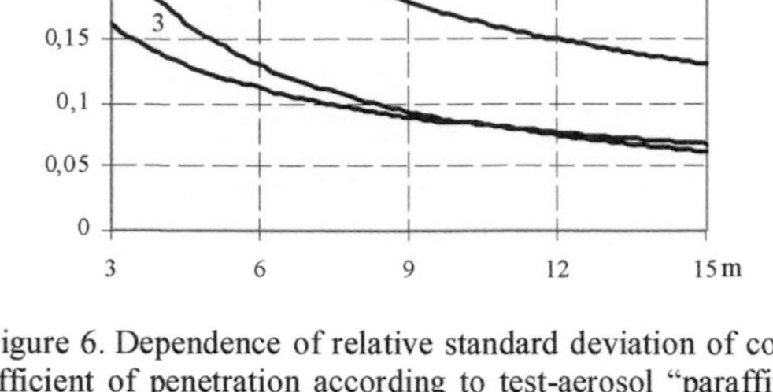

Figure 6. Dependence of relative standard deviation of coefficient of penetration according to test-aerosol "paraffin oil" on a number of observations: 1 – samples without preliminary preparation, 2 – samples subjected to temperature effect, 3 – samples passed though "wearing schedule".

3 CONCLUSIONS

Tests show high probability of error while determining quality of three samples as harmonized standards suggest. Total number of .samples (more than 12 while determining each induce) is sufficient for high accuracy of measurement taking into account the procedure of preparation. However, preliminary preparation considerably influences quality induce of MIPRO, particularly coefficient of penetration according to test-aerosol. It sufficiently increases boundary range of occasional error and makes difficult to determine the type of respirator protection. To carry out the whole complex of tests and guarantee high accuracy, it is required to spend much time for preparation and study. It will lead for increase of test cost. Registration of operating conditions of MIPRO and preliminary preparation of samples which reflect their future sphere of application are the ways of this problem solving.

REFERENCES

Petryanov, I., Koscheev V., Basmanov P. & jthers. 1984. *Lepestok (Light respirators)*. Moscow: Science: 218.

Methods of measuring resistance of constant air current of filter boxes to gas masks and respirators, filter respirators, and filter materials. 2001. RND 37.001-2001. Dnipropetrovs'k: NMAU: 12.

Novikov, V & Kotsuba, A. 2001. *Basics of metrology and metrological activity*. Part 2. Manuscript. Kyiv: Nopaprint: 210.

Volodarsky, Ye. & Kosheva, L. 2008. *Statistical data processing*. Manuscript. Kyiv: NAS: 308.

«Respirators SHB-1 «Lepestok» Technical conditions». GOST 12.4.028-76.

Geomechanical Processes During Underground Mining – Pivnyak, Bondarenko, Kovalevs'ka & Illiashov (eds)
© 2012 Taylor & Francis Group, London, ISBN 978-0-415-66174-4

Investigation of the geomechanical processes while mining thick ore deposits by room systems with backfill of worked-out area

E. Chistyakov
State enterprise "Scientific-research mining ore institute"

V. Ruskih
National Mining University, Dnipropetrovs'k, Ukraine

S. Zubko
JSC " Zaporizhzhya iron ore," Dniprorudne, Ukraine

ABSTRACT: Mining of iron mills in Ukraine reach a depth of 1000 m or more. This leads to a significant increase in the stress-strain state of rock, and as a consequence, the deterioration of mining conditions. The article shows the basis of new technical solutions to mine steep ore deposits under conditions of low stability of the hanging-wall host rock. The results of investigations of polarization-optical models and simulation of equivalent materials are given. On mine, "Expluatacionnaya" JSC "Zaporizhzhya iron ore" an experimental section of a mine field has been selected, where on the floor in a 740-825 m tested the results of research.

1 INTRODUCTION

Iron ore industry in Ukraine holds one of the leading places in the world by production volume. However, providing the industry with high quality raw steel production continues to be an actual problem. The growth of consumption of the mineral resource base requires expansion, providing the necessary quality of salable ore, maintaining the profitability of mining and ore-processing plants.

For a long time mining has reached a depth of 1000 m and deeper. As a consequence, stress-strain state of rock at such depths has increased as well, that led to the deterioration of mining conditions of deposits. To save the profitability and increase the efficiency of underground mining can be reached mainly due to the intensification of production.

The efficiency of extraction and safety of mining operations depend on optimal size of the main structural elements and sequence of the chamber.

2 FORMULATING THE PROBLEM

One of the mining and metallurgical industry of Ukraine, which faced the task is represented, is JSC "Zaporizhzhya iron ore", based at the Uzhno-Belozyorsk deposits of rich iron ore.

Zaporizhzhya iron ore complex develops steep deposit with thickness from 20 to 180 m with help of chamber system and backfill of goaf. The main production is carried out with help of chambers from 15 to 30 m in width and 100...200 m in height in the twin floor 640...840 m (Figure 1).

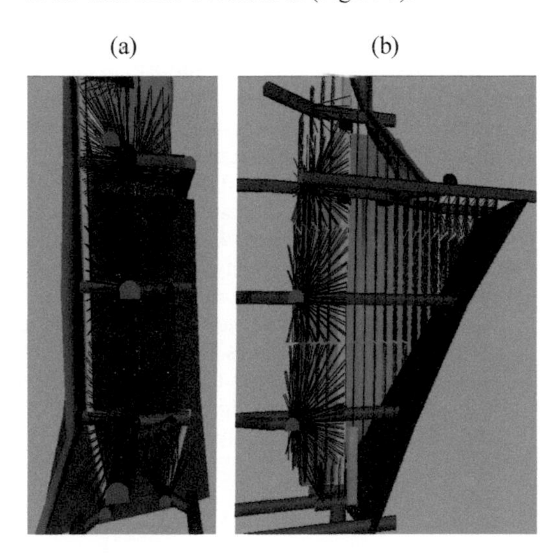

Figure 1. Imitational model of drilling the chamber at 690-840 m level: (a) along the strike of the ore deposits, and (b) across the strike of the ore deposits.

Surveying measurements in the drifts, and hanging-wall gain, with the development of stoping, value of the subsidence and horizontal displacements have been determined. In the gains at depths of 605, 640, 715, 740 m, and also in the drifts at 740 and 840 m observation stations have been established. Ongoing observations allowed to establish the nature of displacement of the rock mass around the existing and laid down chambers. At area of axes 7-11 floors of 740-840 m 14 chambers have been worked out and 9 chambers were laid down within three years. In this case, from year to year in the diagram (Figure 2) it is clearly observed the dynamics of displacement of the massif from 0 to 90 mm. This should be considered when designing deeper levels. Figure 2 shows the results of observations in the hanging wall drift 740 m. The diagrams show both the areas of individual chambers influence and the consolidated dislocation zones of undermined strata.

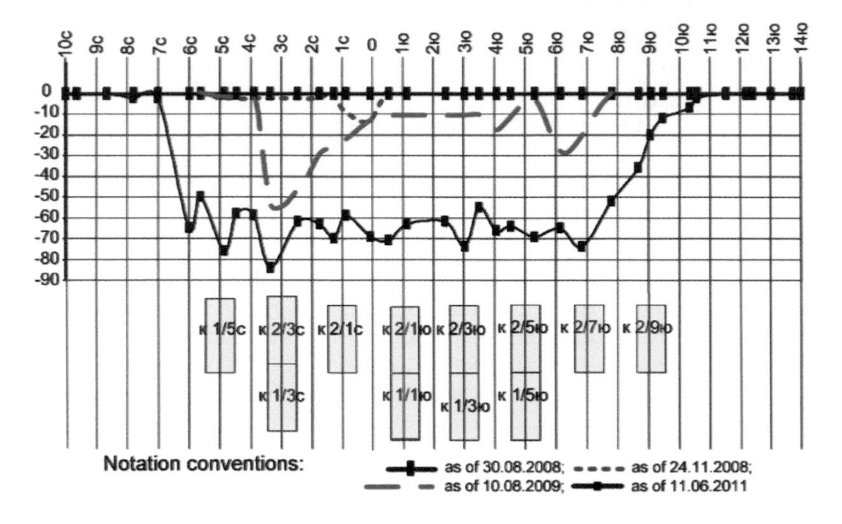

Figure 2. The results of measurements of subsidence along the hanging wall drift profile, depth 740 m.

While developing levels 640-740-840 m on a room-pillar, plant, in some parts of deposit, aced with the problems of stability of the country rocks in the hanging-wall with room parameters: height – 100-130 m, width – 30 m, length – 40-50 m (Figure 3). The large size of rooms led to a decrease in the stability of the roof and hanging wall side, which later was reflected in some reduction of quality indicators, due to increased pollution of mined ore. Made in the southern part of the mining of the ore body changes in technology and BSB parameters, namely, leaving pillars of ore, separating itself from the camera hanging wall rocks, downward blasting, the transition to work after two pillars did not fully resolve the problem.

3 MATERIALS UNDER ANALYSIS

Order to determine rational technology to mine hanging wall rooms in floors 740...840 m in order to prevent dumping on the basis of detailed geological prospecting and analysis of results observations of mine, have been identified settlement and acceptable treatment options designed room on the dependence of (1-3) and studies performed on physical models.

$$L_k = \pi^2 \sqrt{1 + \sin 2i_k \frac{R}{K_{ph} \cdot \gamma H}} \tag{1}$$

$$L_n = \frac{\pi^2}{\pi - 2\sin\beta} \sqrt{\frac{R}{K_{ph} \cdot \gamma H}} \tag{2}$$

$$L_b = \pi^2 \sqrt[3]{3f} \sqrt{\frac{R}{K_{ph} \cdot \gamma H}} \tag{3}$$

where L_n – inclined or vertical outcrop along the strike of the deposit, m; L_k – horizontal outcrop along the strike of the deposit, m; L_b – vertical exposure across the strike of the deposit, m; i_k – dip angle roof of room along the strike, degrees; β – dip angle of hanging-wall rocks, degree; γH – Pressure of above lying rocks t / m^2;

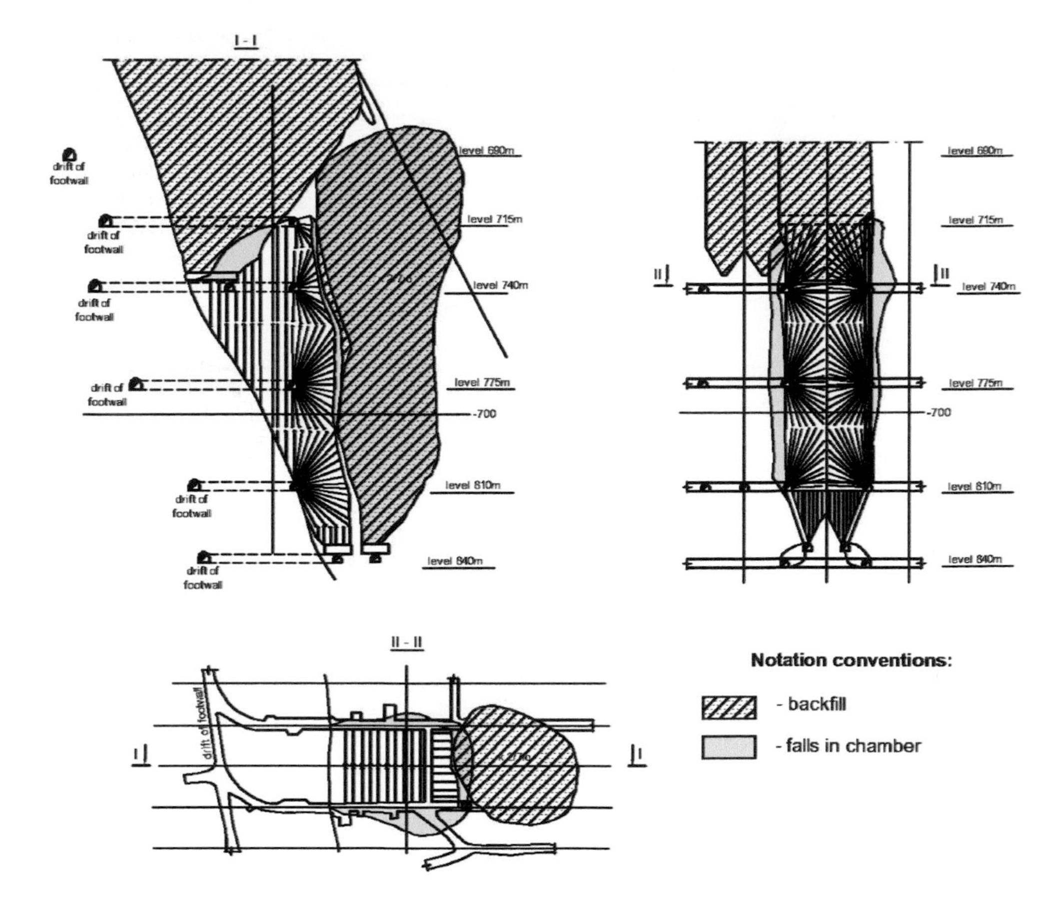

Figure 3. Areas of dumped in mined rooms.

3.1 *Polarization-optical modeling*

Studies were performed on models of the polarization-active material – igdantin in gravitational forces field according to classical methods.

When the polarization-optical modeling of the volume problem was solved in a plane stress state (Chystyakov 1992). Reproduced in the model plane cut at a certain depth is a picture of stress distribution in the vicinity of an infinite number of rooms, the debugging on a "room-pillar".

As a result, the model to mined rooms 2/1 on the levels 605 and 740 m 1/2 with the levels 715... 840 m, taking into account the backfill mining related rooms with a levels 605 and 715...740...840 m, as well as rooms overlying levels set the maximum concentration factors stresses in the zone of influence of rooms. Curves of maximum stress concentration factors around the room levels 605 1/2s and 1/1 with 740...840 m 715 levels when they are working out a joint are presented in Figure 4.

3.2 *Simulation on equivalent materials*

If the polarization-optical modeling of the form stress (biaxial or triaxial) is not important, then test for stability in static models of equivalent materials the role of the stress state is very important. This necessitates the study of the stress state in addition to the array in the plane stressed state, to investigate the stability of the system components to carry out development under conditions of plane strain, providing triaxial stress state.

Studies on models of equivalent materials in compliance with the terms of full-scale similarity model provided data on the stability of the elements of systems development at different playable on models developed shapes and sizes of spaces. The nature of the deformation also depends on the sequence of mining chambers. Therefore, when modeling on equivalent material was taken into account the sequence of mining.

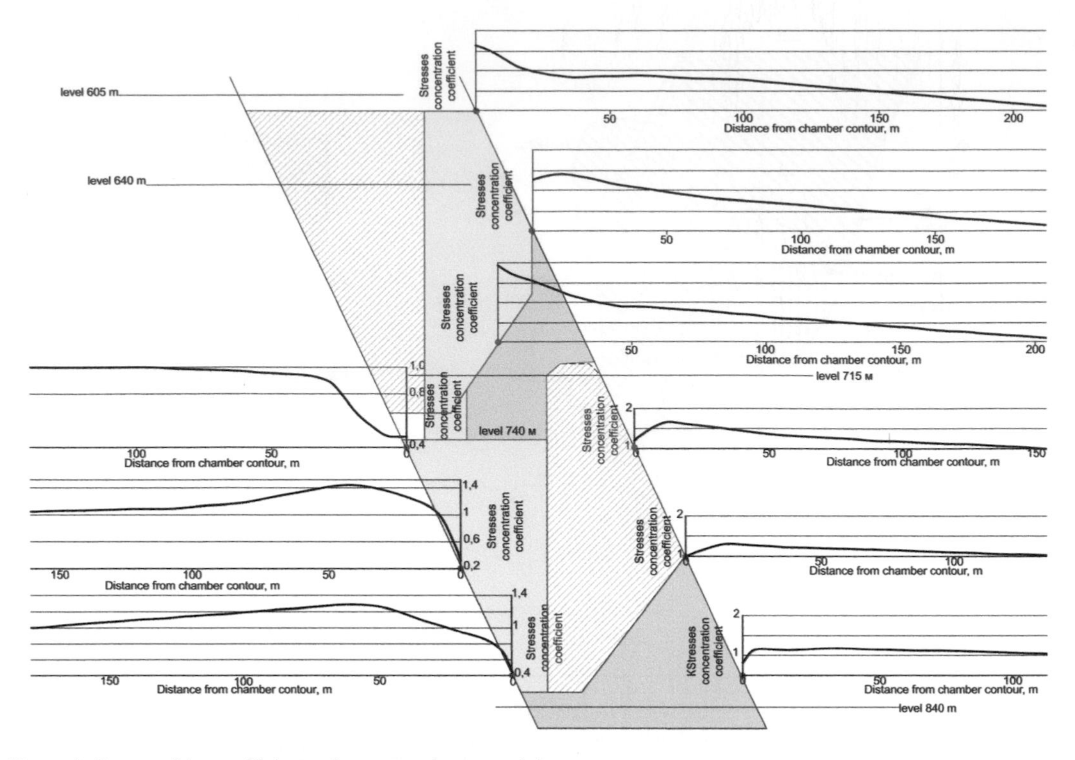

Figure 4. Curves of the coefficients of concentration around the room.

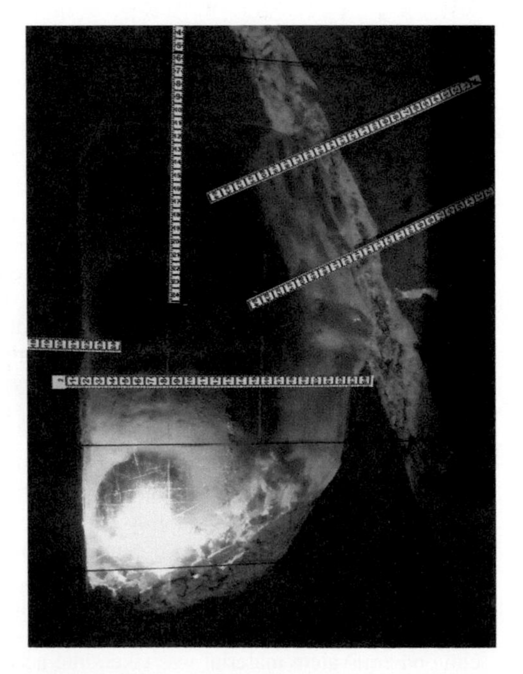

Figure 5. Collapse of overlying rock when modeling mining of room level 940...1040 m according to the "room-pillar".

Simulation performed on equivalent materials in plane strain on the stand with a transparent front wall. Figure 5 shows an example of modeling the chamber floor 940...1040 m according to the "chamber-pillar" mining deposits on the stage with an exposure of the hanging wall. As a result of simulation data obtained on the size of exposure chamber in which the possible collapse of the host rocks.

3.3 *Analysis of the study results*

This research of mining of the steep ore deposits in low stability of the country rocks in the hanging wall of the polarization-optical models and models of equivalent materials showed the following:

– the intensity of the stresses in the pillars in the room equal to the width and the pillars 15 and 15 m, 30 and 30 m is the same, the stability of pillars at equal strength is determined by the shape of ore;

– in areas of transition to the testing of rooms 15 m wide stress in the 15 meter pillars adjacent to the rooms 30 m wide by 15...20% higher than when working off a solid camera and pillars between the room of equal width;

– when working out on a “room-2 pillars” zone of mutual influence of cells in contact, the stresses in the cross section of the rear sight are distributed curvilinear with a maximum at the walls and a minimum at the center; at working on a “camera-pillar” in the rear sight tension are distributed evenly, not exceeding the maximum intensity of the stresses in the walls of the chambers, the debugging on a “room-2 pillars”;

– in the ceiling pillars (the bottoms of the overlying cells), regardless of the form of cameras, the stress concentration at the hanging wall are increases by 2.5-3.0 times;

– in rooms with a sloping bottom in the end wall of the inside / side stress increases with the approach of breaking in to the inside / side. In this case areas of concentration are in the heel arch and in the bottom of the interface closer until merger after breaking of the last layer.

According to the analysis and synthesis of laboratory research recommended procedure for breaking a room with breaking the hanging wall of the triangle as a last resort.

4 TECHNOLOGICAL SECTION

After the submitted studies for testing rooms from the hanging wall in the floor of 740-840 was selected an experimental plot where the mine field of development adopted by the room system, followed by purification of the space filled with a hardening tab at the bottom of the mountains. 825 m, a width of 15 m and room height from 35 to 80 m depending on the contours of the ore deposits. Breaking the cells sublevel, the vertical layers in the same plane, or in advance of the upper sub stage, is made on a pre-cutting a slot in split vertically located across the room. Drilling of the ore produced an array of cameras ascending and descending fans wells drilled from the vectors and the sublevel drifts. The bottom of the room set up in ore array flat. It is formed by the explosion of the fans of wells drilled from a slit to the expansion of the unit vector cutting slits located across the chamber (Figure 6). Haulage of ore from the camera produced by Scoop trams TORO-400E (D).

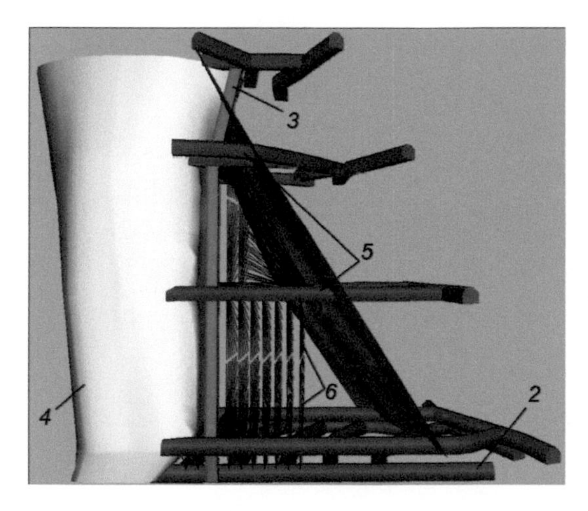

Figure 6. Model of mining the ore from an array of hanging-walls: 1 – haulage unit vector; 2 – slash unit vector; 3 – cutting of raise; 4 – backfill; 5 – drilling unit vector; 6 – fans of boreholes.

5 CONCLUSIONS

Given the accumulated experience of operation of the South White Lake Field, and the results of the investigations the following conclusions:

– when sewage excavation work in the first chamber in the footwall and in the central part of the deposits as being due to the protective effect of backfill in more favorable conditions on the distribution and magnitude of stress, when the concentration ratio does not exceed 1.7;

– sewage treatment notch camera in the hanging wall of the ore to carry out the abandonment of the triangle to prevent collapses of unstable rocks and dumped the hanging wall. Refinement of the triangle made after excavation and laying sewage chamber, possibly with Scoop trams;

– practice of cameras in the hanging wall at the site location of the low stability of the oil shale production on a “two-pillar chamber”.

Based on these results a number of cameras in low stability of the country rocks overlying the South White Lake Field have been successfully developed, which confirms the validity of studies.

REFERENCES

Chystyakov, E.P. 1992. *The solution volume of geomechanical problems on fragmented physical models*. Proc. The development of metallic and nonmetallic deposits in Ukraine. Kryvy Rig: 49-61.

Geomechanical Processes During Underground Mining – Pivnyak, Bondarenko, Kovalevs'ka & Illiashov (eds)
© 2012 Taylor & Francis Group, London, ISBN 978-0-415-66174-4

Spontaneous combustion and coal dust explosibity related mapping of South African coalfields

B. Genc & T. Suping
School of Mining Engineering, University of the Witwatersrand, Johannesburg, South Africa

A. Cook
Latona Consulting (Pty) Ltd, Johannesburg, South Africa

ABSTRACT: Spontaneous combustion and coal dust explosibility are two of the paramount challenges faced by the South African coal mining industry. Lotana Consulting (Pty) Ltd. is a consulting firm that obtained a number of coal samples from different coal mines in South Africa to determine the probability of spontaneous combustions and the risk of coal dust explosibility in the majority of the South African coal mines over the last three years. The database that Lotana Consulting (Pty) Ltd. produced consists of information collected through several tests of coal samples from various mines. These tests were performed in order to determine both the Wits-Ehac index and explosibility index (K_{ex}). The Wits-Ehac index measures the spontaneous combustion liabilities of the coal samples while (K_{ex}) measures the explosibility of the coal dust. Using this database, maps of the major coalfields have been generated showing the liability of both spontaneous combustion and coal dust explosibility for the coal mines. This paper plots where the high-risk areas are in terms of spontaneous combustion and coal dust explosibility, and it enables the South African coal mining industry to take important steps to improve safety in coal mines.

1 INTRODUCTION

Spontaneous combustion and coal dust explosibility are two of the paramount challenges faced by the South African coal mining industry. Lotana Consulting (Pty) Ltd. is a consulting firm that obtained a number of coal samples from different coal mines in South Africa to determine the probability of spontaneous combustions and the risk of coal dust explosibility in the majority of the South African coal mines. More than 80 coal samples have been obtained over a three year period from various coal mines in South Africa to determine the liability of spontaneous combustion and the coal dust explosibility index.

The spontaneous combustion tests were carried out at the University of the Witwatersrand to determine both the Wits-Ehac index and the crossing point temperature, which are combined to obtain the coal liability. The coal dust explosibility index (K_{ex}) was carried out at the CSIR Kloppersbos Explosion facility in a 40 litre pressure vessel to determine the rate of pressure rise.

This has provided a database of results, collated and maintained by Latona Consulting, to review and evaluate South African coal seams, and using this database, maps of the major coalfields have been generated showing the liability of both spontaneous combustion and coal dust explosibility for the coal mines.

2 EXPLOSIBILITY PREDICTIVE INDEX

Explosibility of the coal dust is determined experimentally as an explosibility index (K_{ex}). An apparatus called a 40-litre explosion vessel is used to determine the K_{ex} value of the coal dust. According to Gouws and Knoetze (1995), the K_{ex} value is calculated from the plotted curves of gas pressure versus time graphs. It is important to understand that coal dust is considered non-explosive when the K_{ex} value is less than 70 bar / s, whilst coal dust that is highly likely to propagate explosion have a K_{ex} value of more than 90 bar / s. Figure 1 shows the 40-litre explosion vessel used during determination of the K_{ex} value of the coal dust (Gouws & Knoetze 1995).

The process of predicting the K_{ex} value is based on the relationship between explosibility and energy released by the combustion of coal volatiles. The calorific value (CV) of the volatiles in coal is determined by finding the difference between the energy released by combustion of the fixed portion of carbon of coal and the calorific value of coal. There is a need to assume that inorganic components of coal do not contribute to the volatile content of coal. It is also important to assume that the amount of energy liberated during the combustion of a given quantity of fixed carbon is the same as that would be liberated from a similar quantity of graphite. The relationship between the total calorific content and fixed amount of carbon is expressed as follows:

CV of total volatile material = $(100 \times$ CV of the coal $- 32.8 \times$ FC$) / 100$,

where CV – calorific value, MJ / kg; FC – fixed carbon, %; 32.8 – calorific value of graphite, MJ / kg (Gouws & Knoetze 1995).

Figure 1. A 40 litre explosion vessel used to determine the explosibity index. (Gouws & Knoetze 1995).

It was found that K_{ex} values were predicted best by the calorific value of volatiles using the following relationship; $K_{ex} = 15.81$ (CV of volatiles) – 6.35. The equation was valid for K_{ex} values of between 90 and 120 bar / s. It was found that the calorific value of volatiles has a correlation coefficient of 0.87 and the complete relationship to find the K_{ex} value is as follow $K_{ex} = 0.1581(100 \times$ CV of coal $- 32.8 \times$ FC$) - 6.35$. (Gouws & Knoetze 1995).

3 SPONTANEOUS COMBUSTION INDICES

The propensity of coal spontaneous combustion can be determined using various laboratory techniques, namely; the ignition temperature tests technique (crossing-point temperature tests (X.P.T), the differential thermal analysis (D.T.A)), and, the adiabatic calorimetry technique.

In order to determine the crossing point temperature measurement, it is necessary to compare the relationship of how the temperature increases against time for coal and an inert material. Crossing-point temperatures tests consist of samples of coal and an inert material both placed in similar sample holders and placed in an oil bath. Thereafter oil is heated at a constant rate so that the crossing-point temperature can be measured. The crossing-point temperature is defined as the temperature at which the coal sample equals that of an inert material sample. (Gouws & Knoetze 1995).

When using differential thermal analysis, the difference between the temperatures of the coal sample and an inert material sample is measured and plotted against the temperature of the inert material sample. It is important to understand that in differential thermal analysis, three stages are realized. Initially, the temperature of an inert material sample is higher than the temperature of the coal sample (Stage I) and this is based on the cooling effect of the evaporation of moisture content in the coal. Next, evaporation of the moisture content, the coal sample starts to heat up at a faster rate than the heating rate of the inert material (Stage II) and this is based on the tendency of coal to self-heating and attempting to reach the temperature of the surrounding temperature (i.e. Oil bath temperature). Finally the high exothermicity is reached at a point where the line crosses the zero base line and is referred to as the crossing point temperature (Gouws & Knoetze 1995).

Uludag et al, (2001) also confirmed the three stages and explained that stage I begins with minimal differential and gradually increases towards the crossing-point temperature where the differential is zero. Stage II is a continuation from the crossing point temperature to the point called kick-point. Furthermore, Stage II is referred to as one of the best indicators of spontaneous combustion. Stage III is when the coal begins to burn beyond the kick–point. Figure 2 shows a typical differential thermogram.

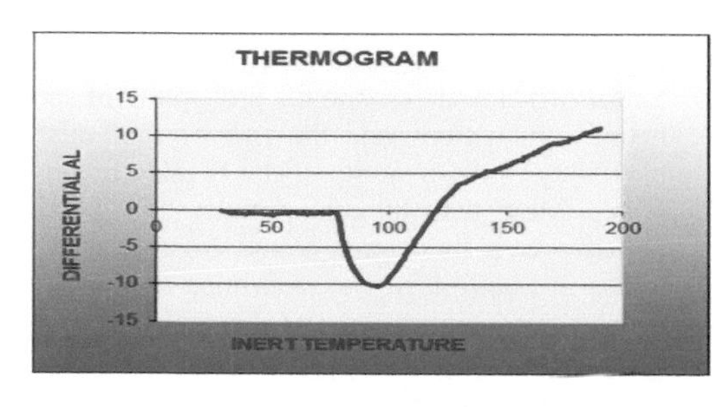

Figure 2. Typical differential analysis thermogram.

It is important to understand that the kick–point temperature means the temperature at which coal starts to burn. Stage II is a good indicator of spontaneous combustion liability and that the liability of spontaneous combustion increases with the increasing slope of Stage II. However the most reliable indicator of spontaneous combustion is the Wits-EHAC Index. The Wits-EHAC Index is defined as:

Wits-EHAC Index = 0.5 × (Stage II slope / Crossing-point temperature) × 1000 (Gouws 1987).

According to Gouws (1987), the characteristics of the curves plotted using the obtained results (i.e. ignition temperature tests) are used to determine the propensity of coal to self-heat according to the Wits-EHAC liability index. It is important to understand that when an index value of coal is more than five, there is a high propensity to spontaneous combustion and when an index value is less than three, there is a low propensity to spontaneous combustion. An index value of between three and five indicates that the coal sample has a relatively medium risk to spontaneously combust (Gouws & Knoetze 1995).

The testing apparatus used for the Wits-EHAC Index consists of an oil bath, six coal and inert material cell assemblies, a circulator, a heater, a flow meter used for airflow monitoring, an air supply compressor and a computer. The oil bath is a 40 L stainless-steel tank. The oil used has a low viscosity and high flash point and break down temperature. Generally, the oil used has a flash point of 210 °C and a maximum usable temperature of 320 °C. Figure 3 shows an apparatus setup of the Wits-EHAC testing apparatus.

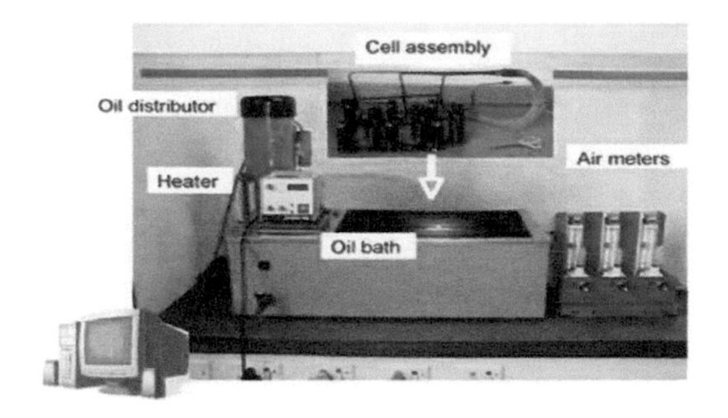

Figure 3. Wits–EHAC apparatus setup.

The cell assembly consists of six cells whereby three of them are used for coal samples whilst the remaining three are filled with inert material such as calcined alumina. The cell assemblies have to be oil tight. It is important to understand that each cap used for closing the cells has a platinum resistance thermocouple (PRT) inserted through the middle. Fumes produced during heating of coal samples in the cells is released through the chimney built into the caps. Air supplied is directed to the bottom of the cells through a copper tube spiral. This spiral is long enough to allow air to reach the oil temperature before entering the cell. It is important to understand that the airflow rate is controlled using the flow meter located between the compressor and sample holder which are connected with a plastic tubing. Figure 4 shows the inserted platinum resistance thermocouple cell caps and the chimney.

When preparing the coal samples, it is important to ensure that the coal remains as fresh as possible once taken from the mine. This is attained by storing coal in a sealed container until the preparation time of the cell assembly and testing arrives. Coal preparation is done using a glove box that is filled with nitrogen. The coal samples required size of (212 μm) is obtained through crushing and sieving. Thereafter, coal is weighed on an electronic scale. A 20 g coal sample is used for each coal cell. When testing commences, the test environment must be initially at 30 °C temperature. This is achieved by sealing the chimneys on the caps until the temperature stabilizes. The test commences by switching on the heater, air compressor and then initiating the data capturing program on a personal computer (PC).

The Adiabatic calorimeter testing apparatus was initiated based on an idea of overcoming the problems related to the ignition temperature test apparatus and the simplification of the testing procedure. There are three testing options when using the calorimeter apparatus, namely; the incubation test, the minimum self-heating temperature test and the crossing-point temperature test.

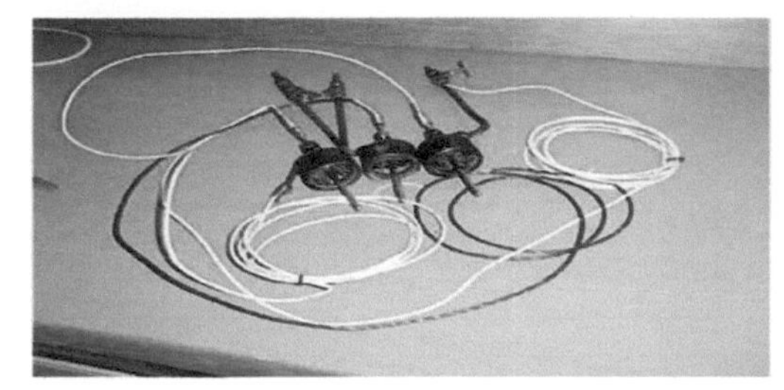

Figure 4. The cell caps with the inserted platinum resistance thermocouple (PRT) and the chimney.

When using the incubation test, the apparatus and coal samples are heated to a predetermined commencing temperature for a chosen stabilization period. Thereafter, air is infiltrated through the samples and then the oven and the intake gas are heated to the temperature of the sample. Oxidation of the sample takes place. It is important to understand that for the incubation test, the propensity of coal to self-heat is predicted by using the initial rate of heating of the coal sample and the total temperature rise factors The minimum self-heating temperature test is simply an incubation test with increments of starting temperatures. The test is started at a specific temperature then heated to a higher temperature if there is no self-heating of coal samples. The incubation test is repeated until the coal samples self-heat. The starting temperature at which self-heating occurs is taken as the sample`s minimum self-heating temperature. It is important to understand that coal with a low minimum self-heating temperature is more prone to spontaneous combustion than a coal with a higher minimum self-heating temperature (Gouws & Eroglu 1993).

The crossing temperature test involves heating of the coal sample using the same rate as that of the ignition temperature test and the observation of thermal behaviour with respect to datum temperature, but the experiment was initially unsuccessful when using the calorimeter apparatus. Instead a band-heater controlled by a variac was used to heat the coal sample. It was necessary to set the variac so that it can result in the heating of the inert material at the required rate. In order for the oven to match the rise in temperature of the sample, it was required to use the incubation mode of the calorimeter. The test is repeated using the coal sample. The data obtained during heating of the inert material is used as a datum against that of the coal heating so as to examine the characteristics (Gouws & Eroglu 1993).

The technique used for this experiment is called the rising temperature technique (or the ramping temperature technique) and possess similar characteristics as the differential thermal analysis test. It is important to notice that both tests measures the heating rate of the coal with respect to the rate at which inert material are heated in similar circumstances.

4 RESULTS

The results entail the explosibility of coal in bar / s, crossing–point temperature in degrees Celsius and the Wits–EHAC index. The tests results are shown from 2009, 2010 and 2011 respectively. The name of the coal mines have been disguised.

Table 1. Spontaneous Combustion and Explosibility Test Results (2009).

Name of mine	K_{ex} (bar / s)	Wits – EHAC Index	Crossing-Point Temperature (°C)	Spontaneous Combustion Liability
ND	75	4.24	127.6	Medium
S1	11	5.72	120.2	High
Bo	139	4.79	115.3	Medium
Mo	10	5.09	123.3	High
M1	27	5.01	108.7	High
Po	116	5.27	136.4	High
Op	118	5.41	119.3	High
Sp	135	5.31	111.9	High
Sp	112	4.88	120.8	Medium
Sp	235	5.46	119.9	High
Bw	45	5.35	111.4	High
Pa	134	4.14	133.4	Medium
DR	131	5.31	121.9	High
Ss	63	4.53	130.9	Medium
M2		5.62	119.2	High
M3		5.02	116.4	High
M4		5.18	121	High
M5		5.33	118.4	High

Table 1 shows the 2009 tests results for spontaneous combustion and explosibility. It is evident from the 2009 spontaneous combustion liability and explosibility results that all of the tested coal samples from various South African collieries ranged from medium to high propensity of coal to spontaneously combust and most of the collieries have higher propensity. The minimum calculated Wits-EHAC index was about 4.1 whilst the maximum index found was just below 5.8. The explosibility index of those collieries is ranging from non-explosive coal dust at K_{ex} value of 10 (bar / s) to the one that is highly likely to propagate an explosion at a K_{ex} value of 235 bar / s.

Table 2 shows 2010 tests results for spontaneous combustion and explosibility. The similar range (i.e. medium to high) is still found from the 2010 spontaneous combustion liability results and most of the collieries were still having high propensity. The minimum spontaneous combustion index found was just above 4.6 and the maximum found was on 5.64. The explosibility index results of 2010 also followed a similar trend as the 2009 results but the minimum K_{ex} value was 12 bar / s whilst the maximum K_{ex} value was 142 bar / s.

Table 2. Spontaneous Combustion and Explosibility Test Results (2010).

Name of mine	K_{ex} (bar / s)	Wits – EHAC Index	Crossing-Point Temperature (°C)	Spontaneous Combustion Liability
Gr 4	142	5.64	98.6	High
Gr 5	96	5.44	102.6	High
Xs 5	131	5.31	105.2	High
Ma	106	5.14	109.3	High
Ta	106	5.47	110.8	High
SW	131	5.26	117.3	High
S1	12	5.58	104.8	High
D 16	139	4.91	118.8	Medium
D 15	128	5.51	102.9	High
Si		5.32	113.7	High
M 1		4.64	117.8	Medium
NC		5.02	119.3	High
nK		5.38	108	High
Op		5.49	108.2	High
Op		5.23	120.9	High
Op		5.1	110.8	High
Op		5.52	113.6	High
Op		5.27	117	High
Wy		5.24	114.7	High
DE		5.6	121.2	High
VaA		5.58	95.2	High
KE		4.86	120.2	Medium
KW		4.71	113.9	Medium
Xs2		4.91	116.7	Medium
SW		4.92	121.9	Medium
VaG		4.86	108.5	Medium
Mo		5.33	124.7	High
Xs2N		4.91	116.7	Medium
Xs5		5.31	105.2	High

Table 3 shows the tests results of spontaneous combustion and explosibility. The number of tests done on spontaneous combustion out figured the explosibility tests during the year 2011 because of some operational problems with the explosibility testing equipment. A big difference was noticed because most of the coal samples tested from various collieries had a medium level of propensity to spontaneous combustions. This could have been improved by reduced critical factors that influence spontaneous combustion. The minimum calculated Wits-EHAC index was 3.1, which is still above the low range identified by Gouws (1987), while the maximum Wits-EHAC index was approximately 5.9. It was also found that the minimum K_{ex} value was 9 bar / s whilst the maximum K_{ex} value was 217 bar / s.

The spontaneous combustion liability bar chart on Figure 5 shows the behaviour of the selected South African collieries with regards to spontaneous combustion. The Wits-EHAC index margins are indicated using colours. The blue part shows the coal samples that have a value of more than 3 but less than 5 which are known to possess a medium risk to spontaneous combustion. The red part indicates coal samples that have a value of more than 5 and are known to have a high risk to the propensity of spontaneous combustion. It can be seen from Figure 5 that almost half of the tested collieries possess a medium risk to propensity of spontaneous combustion, whilst the other half possess a high risk to propensity of spontaneous combustion.

The explosibility index bar chart on Figure 6 shows the behaviour of the tested selected South African collieries. The explosibility index margins which indicate the behaviour of coal are indicated using the colours. The blue part (bottom) shows coal samples which have a K_{ex} value of less than 70 bar / s and is considered non explosive. It can be seen from Figure 6 that only a third of the tested collieries have K_{ex} values of less than 70 bar / s. The red colour indicates coal samples with a K_{ex} value of more than 90 bar / s and the white part indicates coal samples with a K_{ex} value between 70 and 90 bar / s. Figure 6 shows that most of the collieries have a K_{ex} value of more than 90 bar / s which indicate that their coal samples are highly likely to propagate an explosion.

Table 3. Spontaneous Combustion and Explosibility Test Results (2011).

Name of mine	K_{ex} (bar / s)	Wits – EHAC Index	Crossing-Point Temperature (°C)	Spontaneous Combustion Liability
Ta	142	4.73	124.3	Medium
S1	10	4.41	129.4	Medium
Kr	199	4.81	114.2	Medium
Gr4	217	4.74	118.7	Medium
Gr5		4.9	112.9	Medium
Kr	159	4.81	114.2	Medium
Kr	122	4.67	126.2	Medium
Dr		5.01	118.7	High
M1	28	5.36	126.8	High
M1	28	5.33	123.4	High
Sp		4.56	121.2	Medium
Sp		4.79	117.3	Medium
Tu		4.73	129.3	Medium
TshG		3.76	137.4	Medium
TshV		3.65	144.8	Medium
TshM		3.63	145.8	Medium
KG		4.3	130.3	Medium
KD		4.21	133.2	Medium
SW		4.62	126.9	Medium
M1	28	5.91	127.8	High
GGV2		5.12	123.2	High
CCV4		4.66	128.1	Medium
M2		5.18	125.9	High
M3		4.21	134.6	Medium
COA1		3.14	161.9	Medium
COA2		3.46	150.5	Medium
COA3		3.58	154.3	Medium
COA4		3.86	131.3	Medium
KW	9	4.71	113.9	Medium
S1	10			
D	85			
B4	67			
B3	116			
B2	47			
B1	156			
B	132			
A10	28			
A8	60			

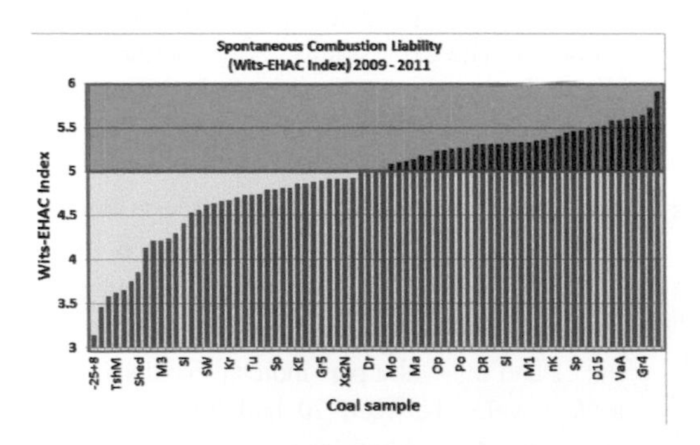

Figure 5. Spontaneous Combustion Liability bar chart.

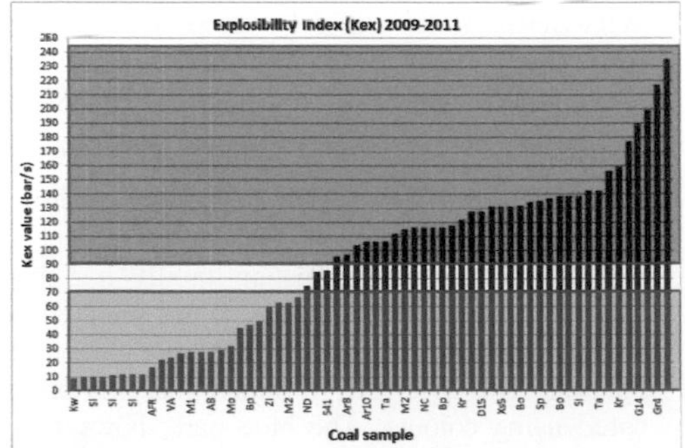

Figure 6. Explosibility Index bar chart.

5 SPONTANEOUS COMBUSTION AND EXPLOSIBILITY RISK MAPS

To address spontaneous combustion and coal dust explosibility problems, using the database, maps of the major coalfields have been generated, showing the liability of both spontaneous combustion and coal dust explosibility for the coal mines. The explosibility index map on Figure 7 shows the risk ratings of the selected South African collieries. The map is generated by adapting the mine boundaries of the South African collieries from Bartholomew and Associates, and is based on the explosibility test results of the coal samples. The map shows that most of the collieries located in the Witbank and Highveld coalfields have high index ratings, particularly the Kriel area. It can also be noticed that some of the mines located in the Ermelo area (Waterberg coalfields) have high index ratings. These high risk areas have an explosibility index range of 150 bar / s to 250 bar / s. Although most of the collieries in the Witbank and Highveld coalfields are rated high risk, a low risk rating can also be seen on the south of Ogies. The anthracite coalfields have a low risk rating for propagating explosions.

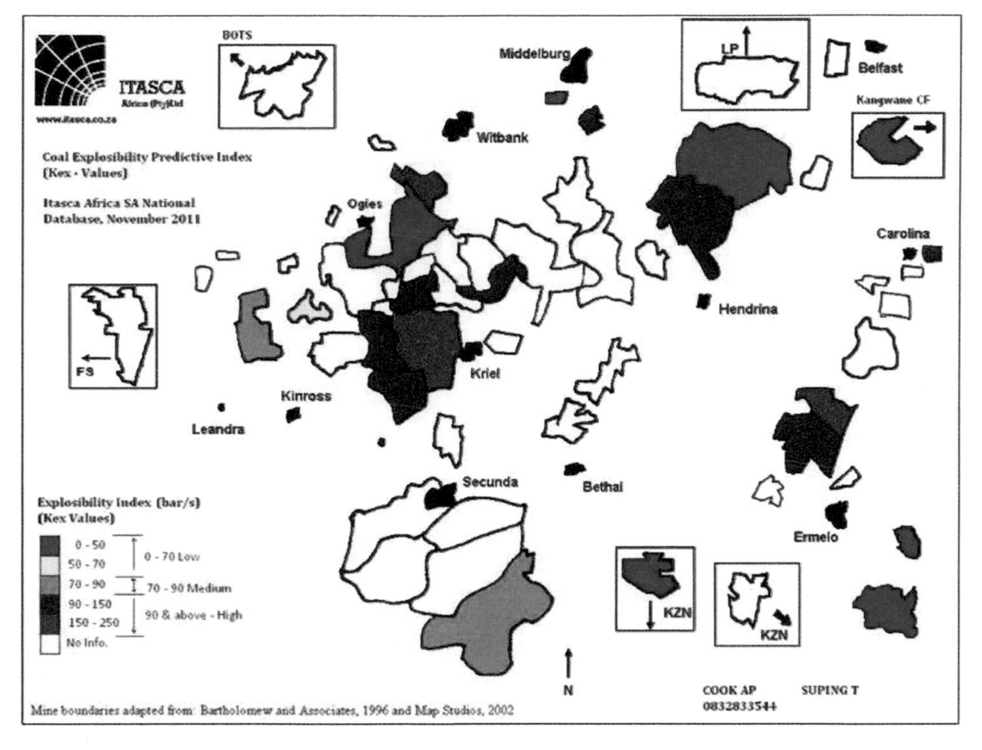

Figure 7: The Coal Dust Explosibility Index Map.

The spontaneous combustion liability map on Figure 8 shows the risk ratings of the selected South African collieries in terms of the Wits-Ehac index. The map shows that most of the collieries located in the Witbank and Highveld coalfields have a high risk to induce spontaneous combustion. The high risk areas also include the north eastern part of Ogies. The southern parts of Waterberg coalfields in the Ermelo area are also rated high in terms of risk. Although most of the selected collieries possess high risk ratings, medium risk ratings can also be seen. The map also shows that the northern parts of Ermelo have medium risk to propensity of spontaneous combustion.

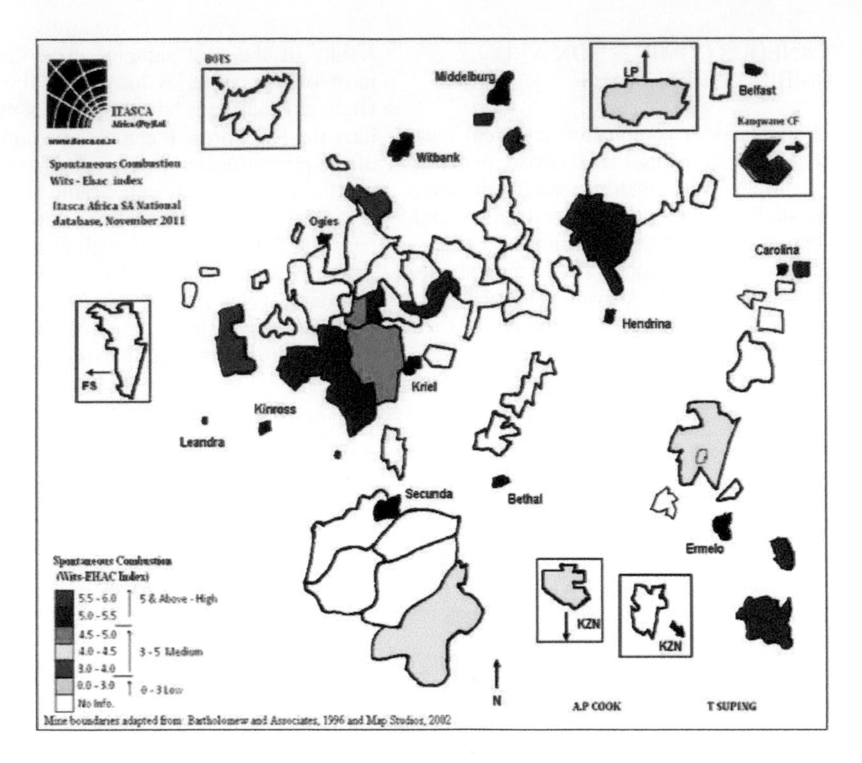

Figure 8. Spontaneous Combustion Liability Map.

6 CONCLUSIONS

To address spontaneous combustion and coal dust explosibility problems, using the database, maps of the major coalfields have been generated, showing the liability of both spontaneous combustion and coal dust explosibility for the coal mines. The state of spontaneous combustion and explosibility of the selected South African collieries was analysed and classified through the series of laboratory tests. It is evident that South African collieries are predominantly challenged by spontaneous combustion and explosibility. Most of our collieries have high risk ratings of coal dust explosibility and the propensity of spontaneous combustion of our collieries ranges from medium to high. There is an important notice made during analysis of the spontaneous combustion test results which showed that there are no low results found and this shows the significance of monitoring spontaneous combustion in our collieries. Based on the tests results and the generated maps it was found that there are higher spontaneous combustion risks and explosibility risk in the Witbank and Highveld coalfields. Relatively low spontaneous combustion risk and explosibility risk were only found in the Kwa-Zulu Natal province where particularly our anthracite coal is embedded.

REFERENCES

Gouws, M.J. 1987. *Crossing point characteristics and Differential Thermal Analysis of South African Coals.* M.Sc. Dissertation, Department of Mining Engineering, University of the Witwatersrand, Johannesburg.

Gouws, M.J., Wade, L. & Phillips, H.R. 1987. *An Apparatus to Establish the Spontaneous Combustion Propensity of South African Coals.* Symposium on Safety in Coal Mines, CSIR, Pretoria: 7.1-7.2

Gouws, M.J. & Eroglu, H.N. 1993. *A Spontaneous Combustion Liability Index.* Istanbul: the 13th Mining Congress of Turkey: 59-68

Gouws, M.J. & Knoetze, T.P. 1995. *Coal Self-heating and Explosibility.* The Journal of The South African Institute of Mining and Metallurgy: 39-41.

Uludag, S., Phillips, H.R. & Eroglu, N.H. 2001. *Assessing Spontaneous Combustion Risk In South African Coal Mines Using A GIS Tool.* Proceedings of the 17th International Mining Conference and Exhibition in Turkey.

Uludag, S. 2007. *The Spontaneous Combustion Index and its application: Past, present, and future, in Stracher, G.B., ed. Geology of Coal Fires: Case Studies from Around the World*: Geology Society of America Reviews in Engineering Geology: Vol. XVIII: 15-22.

Geomechanical Processes During Underground Mining – Pivnyak, Bondarenko, Kovalevs'ka & Illiashov (eds)
 ISBN 978-0-415-66174-4

Monitoring of quality of mineral raw material by method of conductivity

O. Svetkina
National Mining University, Dnipropetrovs'k, Ukraine

ABSTRACT: It is shown that the high sensitiveness of conductivity to the changes of the superficial state of hard materials allows to watch quality of mineral raw material, and also phase transformations of it on the initial stages of process of enriching.

An important meaning during the carrying out of the process of enriching has a maintainance or increasing of metallurgical value of ore , which is determined by the presence of useful and harmful admixtures, for example, for iron ore useful admixtures are manganese, chrome, nickel, titan, vanadium, cobalt, harmful – sulphur, phosphorus, copper, arsenic, zinc, lead, tin.

To metallurgical properties of ores, determining the quality of the produced concentrate, such, as recovery and basicity. Under recovery we understand speed of taking away of oxygen from a mineral by gas-repairer in the process of melting, under basicity the attitude of oxides of calcium and magnesium toward a silica and alumina in ore. One of parameters, allowing to estimate quality of the received concentrates is a of change conductivity which depends on admixtures, on one hand, and on other – on absorption oxygen on-the-surface of the enriched ore.

Measuring of conductivity of materials of semiconductor type (foremost – oxides of transitional metals) received wide spreading in 50-60th. These researches were directly related to the new approach to the explanation of mechanism of catalysis. The new approach is related to the use of theoretical bases of physics of solid in an totality with the notions of chemical connection, that was formulated in the theory of chemosorption border layer, on one hand, and in the electronic theory of chemosorption and catalysis on semiconductors – on other.

Minerals, applied in the process of enriching in most cases, have semiconductor character, in this connection possibility of application of the electronic approach to the problems of enriching of minerals, control and of the quality of minerals management, appears which supposes direct correlation between chemosorption and electronic properties of the grounded up mineral. From here appeared the necessity of experimental determination of electric descriptions of the activated ores with the set purpose: to find correlations between mechanical activity and electronic structure of these objects.

Difficulties, related to verification of electronic theory, are aggravated by doubts in reliabity of the experimental material used for this purpose. It is specified in works, that in the conditions at which the most of the data on the conductivity of minerals were got, – i.e. by the method of direct-current on the powders – electric descriptions of not so much object are measured, but contacts between grains of powder, and that on the basis of such information it is impossible for sure to differentiate the charged and unloaded forms of adsorption.

Obviously, exactly the absence of sufficient clarity in the question about , what information is given (in the case of polycrystalline objects) by the method of direct-current generated among experimenters that mistrust to electric information, which in 70th sharply reduced interest to measurings of conductivity of minerals. The use of alternating current for by-passing of intercrystalline contacts brings in new complications, related with dependence on frequency. Now it is already obvious, that at research of the superficial phenomena there are both tasks which are better described in terms of area theory (collective co-operations) and problems, better interpreted from the point of view local co-operation between a molecule on-the-surface and superficial center.

The purpose of this work is to develop the highly sensitive method of control of quality of concentrates on metallurgical properties of ores, based on measuring of conductivity in an appendix to research the surface of polycrystalline minerals. Researches of the state of surface and the processes taking place on it, and also descriptions of the size of contact of potential reflecting individual properties of object got here, can serve as basis of such highly sensitive method.

Conductivity of minerals submits exponential dependence on a temperature:

$$\sigma = \sigma_0 \, exp\left(-E_\sigma / kT\right), \quad (1)$$

where σ – specific conductivity, $Om^{-1} \cdot cm^{-1}$, reciprocal of specific electric resistance of $\rho = 1/\sigma$; σ_0 – pre-exponential multiplier, poorly temperature-dependent; E_σ – energy of activating of conductivity, eV.

For single-crystals the measured value σ and the value E_σ calculated on equation (1) directly characterize all totality of the phenomena of transfer of carrent in a volume and on-the-surface solid and that is why the changes of these parameters reflect the changes of the electronic state in volume and on-the-surface and can serve for strict comparisons with the theoretical models of these processes. In case with polycrystalline materials of the ground up minerals (powders from them and pills pressed or pressed and then sintered) conductivity will depend on quality of contacts between grains of powder (crystallines).

Researches of electric, dielectric and some other properties of ores, in a highlydispersed (polycrystalline) form, after the process of despersivity, showed that the behaviour of such systems can be described, using a two-phase (two-layer) model: areas with semiconductor conductivity (actually material of object), with intermittent relatively thin dielectric layers (intergrain contacts). Chart of this model presented on Figure 1.

Such two phase model is suitable not only to the case of the pressed powders (with simple mechanical contacts between grains) but also to powders, sintered at a high temperature (higher 600-800 °C), when thin bridges (bridges, necks, "structure of swiss cheese") appear between grains. The approximation analysis of properties of the system on the basis of this model shows, that measurings on a direct current and on low frequencies (about кГц) give information mainly about properties of contacts between grains of powder, and measurings on high-frequencies, shunting contact resistances, expose properties of the studied matter itself.

(a)

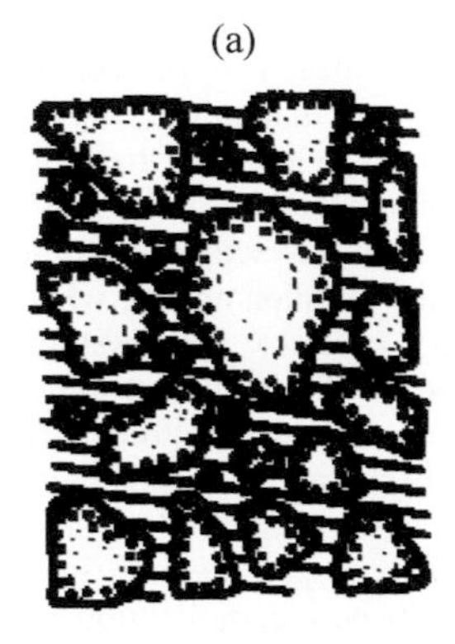

(b)

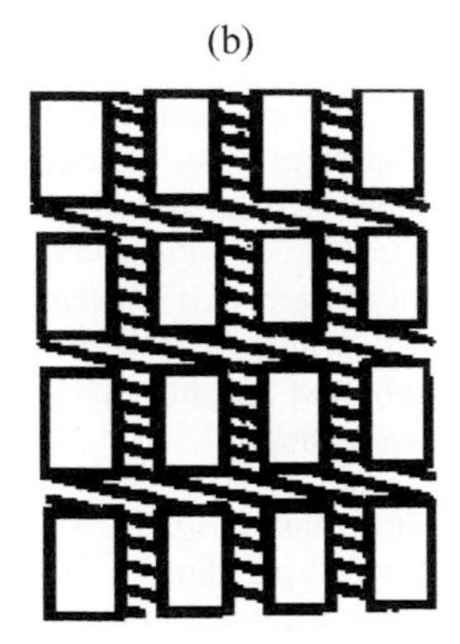

Figure 1. Polycrystalline object (a) diagrammatic picture of conglomerate of particles (pains is the area of partial charge (b) formal model.

However the high-frequency measurings are utilized not often, because of difficulties in interpretation of these results, due to large heterogeneity of structure of the real polydispersive systems, not covered by a simple twolayer model, and deposit (especially on SVCH) of polarization losses; in addition, a sensitiveness and exactness of measurings of conductivity go down on high-frequencies. Consequently, the method of alternating current does not solve the problem of measuring of conductivity of polycrystalline objects, especially for the systems, when we are interested not only in superficial layers of grains which on high-frequencies just are "turned" off but also volume ones.

According to the theory of Petric, the conductivity of such sample, consisting of grains, parted by contact barriers (which got practically all the tension attached to the sample), depends, certainly, on a form and height of barriers, but at the same time it remains proportional to the conductivity of grain itself.

According to Petric specific conductivity of polycrystalline object is expressed by the equation

$$\sigma = \frac{qM\rho}{kT} e^{-q\varphi / kT}, \quad (2)$$

where M – coefficient, determined by a form and width of barrier; ρ – density of basic transmitters in a crystal; φ – height of barrier (in relation to the edge of valency area); q – charge of electron; k – constant of Bol'cman; T – absolute temperature.

According to Sleter, the conductivity of the dispersible system is also proportional to the multiplier of $exp[-H/kT]$, thus for the height of barrier of H expression is given

$$H = E_0 + \left(N_0 q^2 / 4\varepsilon\varepsilon_0\right) \cdot X^2, \quad (3)$$

where E_0 – energy of electron on the border of barrier layer; N_0 – concentration of admixtures in a volume; X – thickness of barrier layer; q – charge of electron, ε and ε_0 – dielectric constant of environment and vacuum accordingly.

It goes from a formula (3), that the height of barrier must change proportionally to the concentrations of admixture in a volume. Such dependence is observed, according to the calculations of Sleter, until this concentration is less than some threshold value, for example, for PbS with the width of banned area a $\Delta E = 0.37$ eV such value N_0 is made by 10^{18} cm^{-3}, that correspond to the maximal thickness of barrier layer of $X = 382$ Å.

Thus, a theoretical analysis shows that conductivity of the polycrystalline system is though determined by contact barriers, but through properties of these barriers depends also on the electronic state of volume of crystallites and at the observance of certain border conditions changes proportionally to the change of this state.

Consequently, such measurings for polycrystalline objects give information mainly about the presuperficial layer of crystallites. The thickness of this layer, or length of the debaev screening (L), hesitates from 10^{-6} cm for the enriched or invertible layers in semiconductor minerals with high conductivity to 10^{-4} cm (and even to 10^{-3} cm) for the impoverished layers in highom (wide zone) ores. Thus, measured prisuperficial conductivity contains, according to equations (1)-(3), information also about the volume of these crystallites. Moreover, exactly due to the peculiarities of structure in the dispersible state, when we observe extraordinarily high correlation of surface to a volume and to the numerous superficial contacts, measurings of conductivity on a direct current, is the integral reflection of the electronic state of contacts. Such measurings appear very sensible, as compared to measurings on single-crystals and on an alternating current, by the mean of research of processes on the surface.

All of restoration properties of concentrates depend on adsorption of oxygen on the surface of minerals, as exactly speed of desorption will be responsible for the fall of temperature of restoration from one side, and from other – for maintenance of admixture elements on the surface the enriched minerals. In the process of enriching of ores there is formation of new active surface on the different stages and as a result there is a process of adsorption of both oxygen and preventing admixtures.

Indeed, at grinding down of minerals there is a process of adsorption in a greater or less rate in different gridding devices. In a Table 1 kinetic information is presented on adsorption of oxygen of the oxidized quartzites at grinding in the different grindings down and environments (alcohol).

Table 1. Content of combined O_2 in the power of the oxidized quartzites.

Grinder	Number of the tested sample		1	2	3	4	5	6	7
MVV	Time of grinding	Star.	0.5	1	2	3	4	5	5*
	O_2, %	0.67	1.60	2.00	2.30	2.30	2.50	2.10	2.00
GMV	Time of grinding	Star.	5	10	15	20	30	40	
	O_2, %	0.67	0.97	1.27	1.30	1.30	1.35	1.30	

* it is grinding down in an alcohol.

For control of quality of mineral we offered the method of termovacuum curves of conductivity (TVE-curves). This method at present is one of the most informing and universal methods of realization of "monitoring of the stages of enriching of ores on conductivity". In this method of observing the state of surface of minerals is carry out depending on the changeable temperature of warming up – both preliminary and during measurings. Measurings as a rule are carried out in a vacuum, as here we have the best display of the dependence of conductivity on desorption, and at the temperatures high enough – and on beginning restoration. However in principle the method can be widespread and on the conditions of warming up in any atmosphere – by restoration, oxidation, neutral. Essentially the described method is the combination of topochemical method of "monitoring" depending on a temperature and method of diagrams "conductivity – composition" and can be, consequently, utilized for the construction of three-dimensional "surface" of changes of electric properties of the complicated system in coordinates "composition – a temperature".

A main objective of the method of TVE-curves is an analysis of superficial phase composition and tracing of its changes depending on different factors, affecting process of enriching of minerals (temperature and atmosphere of treatment, superficial topochemical reactions and so on.). Measurings of conductivity are carried out for ordinary polycrystalline objects on a direct current and thus give information about the composition of presuperficial layer of particles in $10^{-6} - 10^{-4}$ cm thick.

It is necessary to mark that conditions close to the described method, were utilized in a number of works of other authors. However in none of such works conclusion was formulated that the similar use of the electric measurings can be put in basis of original method of phase analysis in the process of enriching of ores. Reason of it is apparently, in that traditional opinion, that the electric parameters of polycrystalline object are not suitable as individual characteristic of matter because of their too high sensitiveness to biography of this sample and to the state of its surface. In addition, measurings not in vacuum conditions more frequent give at different temperature the graphs which don't differ from each other in quality. The basic feature of the described method consists in that not simply size of conductivity is fixed (or its temperature coefficient in the conditions of the pseudostationary state of object), but process of irreversible changes σ (or energy of activating of conductivity of E_{σ}) at warming up of object in a vacuum. Thus the mentionel source of instability of electric information (high sensitiveness of conductivity to the change of the state of solid) becomes an advantage, as it appears that considerable changes of conductivity usually observed at the step increasing at the temperature of warming (and reflecting the specific for this matter process of changing of its state) overcover casual (biographic) variation of electric values at each of these temperatures. Therefore got graphs – termovacuum curves of conductivity, or TVE-curves – have a certain form for this chemical matter regardless of biography of concrete sample, and consequently, can be utilized as standards in a phase analysis.

We will consider more concrete, how characteristic of the got TVE-curves is conditioned. The specific type of these graphs is created by dependence of conductivity on the number of factors (electronic, adsorption, topochemical) the relative deposit of which is determined by chemical nature of object and is different in different temperature areas. Basic from these factors are as follows.

A) The typical for semiconductors convertible growth of conductivity with a temperature by the law of Arrhenius, characterized by the large range of values of E_{σ} for different matters – from ~ 0 to a few eV (equation (1)).

B) Desorption of gases and steams (oxygen, water), which is the most intensive at warming up in vacuum till 200-300 °C. And changes of conductivity caused by this are not only different in interval of value, but also (for minerals of n – type and p – conductors) are contrary in sign.

C) Restoration in a vacuum (removal of the lattice oxygen) with formation of greatly defect or quagimetallic structures, that is most marked in area of 300-400 °C and for different matters greatly depends on energy of grate.

D) Other topochenical and other processes in a grate (grinding down, decomposition of remaining hydroxides and carbonates, structuring, crystallization and so on) – if temperature of preliminary processing it is not high enough.

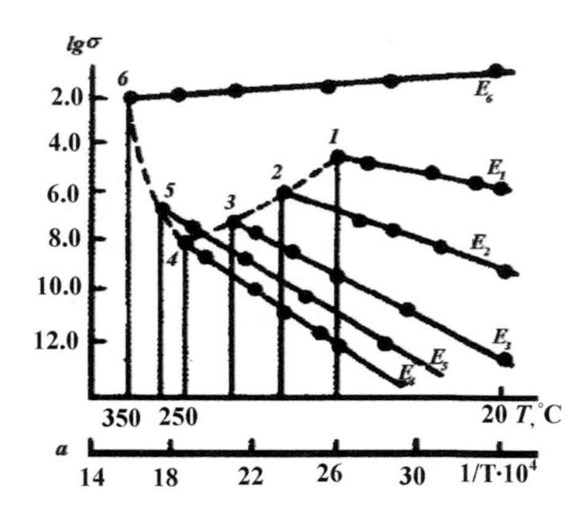

Figure 2. Arrhenius deprendences conductivity of the sample of oxidized quartzite.

The method of formation of TVE-curves is that the pill of the matter is warmed up in vacuum at temperatures from 100 till 400 °C (these temperatures are marked in Figure 2 T_{vak}). At this in every 50 °C the sample is kept till the constant value of conductivity and we determine the temperature dependence of specific conductivity б in interval from this T_{vak} till 20 °C (after warming up at 100 °C – interval from 100 °C till 20 °C, then after warming up at 150 °C till 20 °C and so on). The values of conductivity at each this top temperature T_{vak} are marked by figures from

1 till 6. As a result we have the series of arrenisove curves, each with its inclination giving the value of energy of activation of conductivity from E_1 till E_2. It can be seen than in chosen example the values of conductivity depending on top T_{vak} go through the minimum and the values of the energy of activation – the value of inclination – through the maximum in the given case at 250 °C.

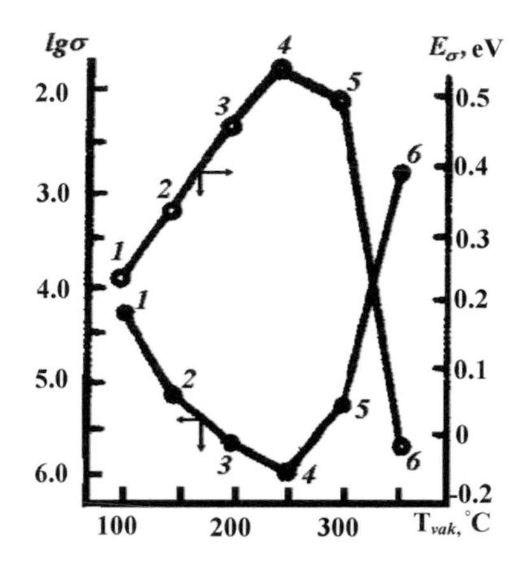

Figure 3. TVE- curves formed for $lg\sigma$ и E_σ on the data Figure 2.

According to the results of tests the built curves are presented on Figure 3. The method of TVE-curves opens wide possibilities for the different variants of "monitoring" of process of enriching. We will bring a few illustrations of it. The exposure of remaining (nonenriched) phase, not noticed already sciagraphy, was carried out on the example of hydrooxide phase, remaining in ore at not enough high temperature of thermolysis. Finding out of new superficial mixture in the process of grinding of the oxidized quartzites appeared possible on the most initial stage of formation of this mixture, when a sciagraphy method was powerless.

We will add that conclusions done on the basis of TVE-curves conform with by information of other used methods – roentgenophose analysis of contact difference of potentials, magnetic receptivity and EPR. It is necessary to suppose that in a perspective, at the accumulation of the reliable standard graphs, given on a Figure 3, for the large number of matters, the method of TVE-curves can be widely utillized in combination with roentgenophose and other analyses as standard or as an expressmethod of control of process of enriching.

Geomechanical Processes During Underground Mining – Pivnyak, Bondarenko, Kovalevs'ka & Illiashov (eds)
 ISBN 978-0-415-66174-4

Imitating modeling stability of mine workings

O. Vladyko, M. Kononenko & O. Khomenko
National Mining University, Dnipropetrovs'k, Ukraine

ABSTRACT: The analysis of the research to improve the stability of underground mine workings with a random distribution of stability parameters of the containing massif. Possibility of application of the "Simulink" program is opened for imitating modeling of stability of mine workings. The imitating model and algorithm of calculation for definition probability of destruction of excavation is developed. Dependence of change in the probability of destruction of mining production from the depth of laying and moisture of a rock massif are established. The obtained results allow to predict the stability of mine workings with the change of conditions carrying out mine workings.

1 INTRODUCTION

During the construction and exploitation of underground mine workings is important their reliability, that takes into account the random deviations of calculated values from their average values. At calculation of stability of underground mine workings deterministic values of physico-mechanical properties of the containing rock massif are replaced by random. Numerous scientific researches devoted to the solution of this task, the main results of which belong to Rzhanitsyn A.R., Bolotin V.V., Shashenko A.N. and others (Shashenko 1988). It is established that with increasing depth of mining during the operation of solid mineral deposits increases the stresses on the contour of mine workings. This leads to decrease in rates of preparatory and cut works and causing more accidents.

2 IMITATING MODELLING

For the effective design of mine workings is often resorted to the use of mathematical modeling as natural experiments are not always available. A mathematical model is an abstract description of the object with mathematical equations. One type of modeling is the imitating model in which the logical-mathematical model of studied system is algorithm of the system realized on computer using software products (Sovetov & Yakovlev 1999). As statistical modeling understand machine reproduction of functioning of probabilistic models, or researches of the determined processes defined in the form of mathematical models of logic elements by means of statistical tests (Monte-Carlo method). The peculiarity of statistical modeling is the random assignment information with the known distribution law, and as a consequence probabilistic evaluation of the characteristics of the investigated processes. Statistical modeling is an effective method of Research poorly organized system with a simple logic operation.

In most cases, modern modeling tools allow you to provide a high level of adequacy of the model. One such tool is a software product "Simulink" – an interactive tool for modeling, simulating and analyzing dynamic systems. It allows you to build graphical block diagrams, simulate dynamic systems to investigate their performance and improve the scientific and research projects. "Simulink" is fully integrated with the software product "Matlab", providing access to a wide range of tools for analysis and design of various industrial processes. As an example it is offered to consider imitating model of probabilistic calculation of destruction of mine working with use of software product of "Simulink".

3 LGORITHM OF IMITATING MODELLING

The authors propose the following sequence imitating modeling:

– analysis of the physico-mechanical properties of rocks containing underground mine workings;

– determination of the law of distribution strength of rocks around the workings;

– modeling of the characteristic of safety Δ with useing software product of "Simulink";

– determination of probability of destruction mine working contour using the Gause integral probability.

The analysis of acting drilling and blasting passports which are used at carrying out mine workings in the operating conditions of the South Belozersky deposit, allowed to establish probability of destruction of rock massif on the contour of mine workings, which corresponds to the quite products worked methods, bases on calculating factor of safety (strength reserve) and characteristics of safety (Rzhanitsyn 1982):

$$\Delta = \frac{n-1}{\sqrt{n^2 v_{com}^2 + v_\theta^2}}, \quad (1)$$

where n – factor of safety; v_{com} and v_θ - corresponding coefficients of variation of strength and stress.

The value of factor of safety

$$n = \frac{\overline{\sigma}_{com}}{\sigma_\theta}, \quad (2)$$

where $\overline{\sigma}_{com}$ and σ_θ – massif strength and stress on the contour of mine working.

Stress on the contour of mine working taken as constant

$$\sigma_\theta = 2\gamma H, \quad (3)$$

where γ and H – massif strength and stress on the contour of mine working.

The coefficient of variation of strength v_{com}, %

$$v_{com} = \frac{s}{\overline{R}} \cdot 100, \quad (4)$$

where s - standard deviation of individual test results of the average strength $\overline{R}$.

$$s = \sqrt{\frac{\Sigma\left(R - \overline{R}\right)}{k-1}}. \quad (5)$$

The average strength

$$\overline{R} = \Sigma \frac{R}{k}, \quad (6)$$

where R - ultimate strength a single sample; k - number of tested samples.

Variation coefficient of stress v_θ, taken as variability of rock $v_\theta = v_{com}$.

The probability of destruction contour of mine working, expressed in parts of its perimeter determined by the Gause integral probability

$$V = 0{,}5 - \frac{1}{\sqrt{2\pi}} \int_0^{\Delta} e^{-\frac{t^2}{2}} dt. \quad (7)$$

4 EXAMPLE OF CALCULATION

The South Belozersky deposit of rich iron ore, which operates on the basis of Closed Joint Stock Company "Zaporozhsky iron-ore plant" (CJSC "ZIOP"), represented by steep-falling deposits with an average dip angle 68°, capacity 10-120 m. Hanging block represented by ferruginous quartzite, recumbent block – by shales quartz-chlorite-sericite.

Aim of imitating modeling is to determine the probability of destruction of mine workings wich laid down at depths from 300 to 1200 m. Modelling was performed using Simulink-model block diagram is shown in Figure 1.

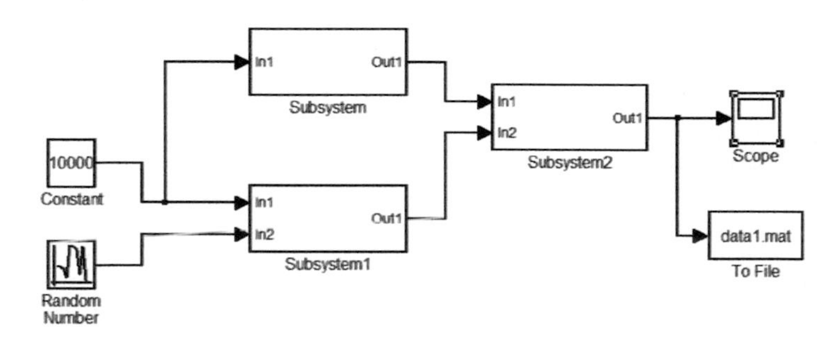

Figure 1. The general structure of Simulink-model.

Figure 1 presents the basic notation and variables used in the mathematical model of the mine working destruction: Subsystem – the voltage on the circuit making, Subsystem 1 – strength of the mine working, Subsystem 3 – security index Δ, the block Constant – imitate the average strength of the rock, block Random Number imitate the distribution of rock strength around mine working. The results of modeling are written to the results file with the using block To File, and also recorded into the block Scope. An example of the modeling results presented in Figure 2.

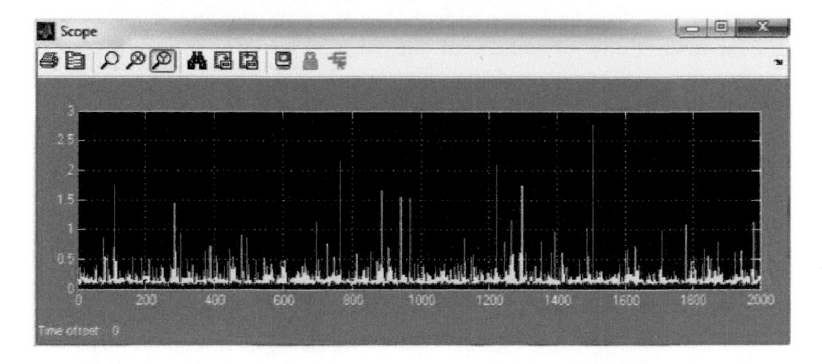

Figure 2. Results of imitating modeling in block Scope.

5 RESULTS OF MODELLING

The obtained results imitating modeling of stability of mine workings established that with increasing depth of the production increases the probability of its destruction, depending on the strength of the surrounding rock massif. This is manifested in the form of flaking, peeling and collapse of rock in the mine workings in the mines, "The operational" and "Heading" CJSC "ZIOP". The nature change in the probability of destruction mine working on the depth of laying is shown in Figure 3.

From the dependencies in Figure 3, we can establish that the value of the probability of destruction carried out in quartzites mine working almost did not increase with the depth laying. In the hematite-martite ore with strength 10-12 and in quartz-chlorine-sulfure shales, the value of mine working destruction probability with increasing depth to 1000-1200 m growth from 3 till 40%. For the hematite-martite ore with strength 6-8, actually the mine working stability is observed to a depth of 600 m.

Further studies of the probability of mine working stability with increasing humidity of the surrounding rock massif, allowed to establish graph of dependence probability of mine working destruction from the rock's humidity, which are presented in Figure 4.

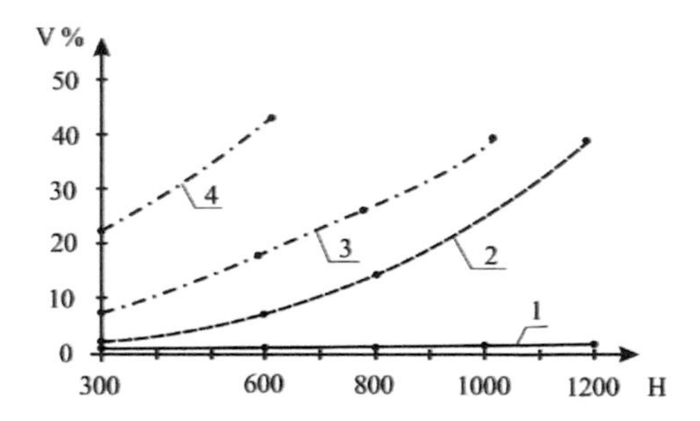

Figure 3. Graph of dependence probability of mine working destruction from the depth of its laying: 1 – hematite-martite quartzites with strength 14-15; 2 – hematite-martite ore with strength 10-12; 3 – quartz-chlorine-sulfure shales with strength 7-9; 4 – hematite-martite ore strength 6-8.

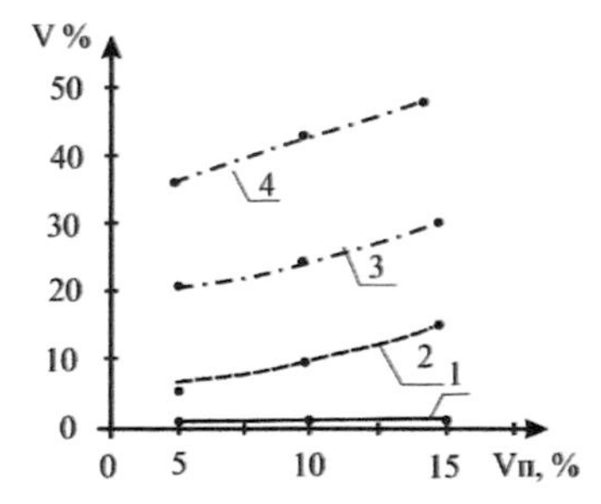

Figure 4. Graph of dependence probability of mine working destruction from the rock's humidity: 1 – hematite-martite quartzites with strength 14-15; 2 – hematite-martite ore with strength 10-12; 3 – quartz-chlorine-sulfure shales with strength 7-9; 4 – hematite-martite ore strength 6-8.

In Figure 4, we can determine that probability of mine working destruction from the humidity of the surrounding rock massif is increased by 1-15%. Thus in hematite-martite quartzites with strength 14-15 probability of mine working destruction will be near 1-2%, in the hematite-martite ores with strength 10-12 – 5-15%, in the quartz-chlorine-sulfure shales with strength 7-9 – 21-29% and in the hematite-martite ores with strength 6-8 – 35-47%.

6 CONCLUSIONS

1. Completed analysis of the current researches of the underground mine workings stability with the use of the random distribution of values of the strength of containing massif. Established that with increasing depth of the mine workings increases strength on the contour. Simulation model, which was developing, let to predict the stability of unfixed mine workings at different depths of laying.

2. On the test calculations example for the conditions of CJSC "ZIOP" establish dependence probability of mine working destruction from the depth of its laying and from the rock's humidity of containing massif. In the 300-900 m depth intervals the probability of the mine working destruction increases from 5 till 30%. This matches with real data of deposit's operation.

3. Further extraction of iron ore at the plant is associated with increased depth of mining to 900-1200 m. Performed analysis of stability mine working at these depths allows us to establish the probability destruction of mine working, which is in the range of the depth increases to 40%.

REFERENCES

Shashenko, A.N. 1988. *Stability of underground workings in a heterogeneous rock massif.* Dr. Tech.Sc. thesis. Dnipropetrovs'k: 495.

Sovetov, B. Ya.& Yakovlev, S.A. 1999. *Systems modeling.* Moscow: Higher school: 271.

Rzhanitsyn, A.R. 1982. *Building mechanics*: Tutorial. Moscow: Higher school: 400.

Geomechanical Processes During Underground Mining – Pivnyak, Bondarenko, Kovalevs'ka & Illiashov (eds)
© 2012 Taylor & Francis Group, London, ISBN 978-0-415-66174-4

Irregularity of deposition of hard coal in thin seams in Poland according to the criterion of the thickness variation seams

A. Krowiak
Central Mining Institute, Katowice, Poland

ABSTRACT: This paper presents the result of analysis of deposition irregularities in the thin hard coal seams which are in the geological resources in the active Polish mines. Thin seams have been classified as the ones with thickness from range 0.6 to 1.6 m. By irregularity in covering of the hard coal in the meaning of this article, it is the changeability of thickness of seams. In peculiarity it was analysed the stores according to criterion of ranges of average thickness seams as well as the ranges of depth of covering, including the relative positive and negative deviations from average of thickness of seams. Clustering methods in analysis was applied as well as tools of automatic neuronal nets (algorithm Kohonena) with programme STATISTICA v.9.0.

1 INTRODUCTION

It has been abandoned exploitation of thin seams in the Polish coal mining industry, since many years. Resources in these seams as an outside industrial retraining, having still relative affluence of thick seams. In a situation where in some mines already resources in the thick seams are getting lower or getting access to them requires a very large investment in deepening of shafts and the construction of new levels of greater and greater depths, it is worth considering whether or not to return to the exploitation thin seams treating it as completion of exploitation of thick seams. Positive experiences in exploitation thin seams are known in Ukraine, where most seams are thin ones. They consider to be a profitable exploitation seams over thick 0.65 m. This ain of this article is to show, of course only on selected thematic section, which geological resources of coal in thin seams we dispose, and what mining-geological conditions are for them characteristic one.

By irregularities in bedding of hard coal in the sense of this article is variation of thickness of seams. In particular, the average thickness of the seams was analyzed and the maximum relative positive and negative deviations from this thickness.

Irregularity of the deposition of seam is a large trouble during their exploitation. Thinning of the thickness of the exploited seams in long wall technique and other techniques may need to cut allowance ceiling and floor rock, which affects the yield of commercial coal. Significant thickening if the seam can cause loss of coal by leaving his allowance in the resources. Significant irregularities in bedding of seams can have influence on choice parcels to exploitation and selection of technical equipments.

There were analyzed geological resources of hard coal in the seams lying on in active mines in Poland (except mine "Bogdanka"). It was assumed criterion that the thin seams will fall seams an average thickness located in the range from 0.6 to 1.6 m.

The basis for an analysis were the source data prepared by the Geological Service of individual mine, on the basis of the documentation made according to the rules described in literature (Nieć 1982; Materiały źródłowe 2010). Dispersed resources in total volume of 540 794 thousands tons, the abundance of several to tens of thousands of tons, for which there were no detailed geological analysis, because they not qualify for potential mining were excluded from the analysis.

2 METHOD OF ANALYSIS

Following intervals of relative deviation of maximum positive thickness from value of average were accepted: from 0 to 10%; from 10.01 to 20%; from 20.01 to 30%; above 30%.; for the negative deviations the following ranges were accepted: from 0 to 10%; from 10.01 to 20%; from 20.01 to 30%; from 30.01 to 40.0% as well as above 40%. Deviation was calculated as a quotient of the relative difference in deviation from the mean maximum and minimum thickness of the seams to the average thickness of the seams.

Analyses were conducted for two thematic sections: the division on ranges of average of thickness seams as well as the division on intervals of average depth of bedding. In the first section of analysis the range of the average thickness of seams were accepted: from 0.6 to 0.8 m; from 0.81 to 1.0 m; from 1.01 to 1.2 m; from 1.21 to 1.4 m as well as above 1.4 m.

For the second cross-sectional analysis, the following ranges of average dept of bedding of seams were accepted: from 0 to 200m, from 201 to 400 m, from 401 to 600 m, from 601 to 800 m, from 801 to 1000 m and above 1000 m.

Source data for analysis consist of several thousands of records; each of those ones describes in detail the parameters of the field or seam. For purposes of analysis it was created a set consisting of selected information of source data records: the average thickness of the seam, the maximum thickness of the seam, the minimum thickness of the seam, the average depth of bedding, the amount of resources.

For each of the records it was calculated minimum relative deviation (relative to the minimum thickness of the seam) and the maximum relative deviation (relative to the maximum thickness of the seam). Each a relative deviation is assigned of the natural numbers from 1 to 4, depending on where the range of variation of the relative deviation of the data is located.

Then it was created a working subset containing 20 combinations of relative deviations. For this set it was prescribed sequence records that meet the criteria of cross-sections of thematic analysis, the average thickness of the seam range or an average depth of bedding range. Each subset of the working was subjected to a computer treatment with the use clustering methods and of neural automatic networks (Cichosz 2000; Tadeusiewicz 1993). Application of the cluster methods allows the separation of subsets characterized by similarity of variables relative to each other (Everitt, Landau & Leese 2001).

The logic of clustering methods can be, descriptively, presented in the following manner, based on the analyzed examples. Each of the records from the working set can be represented in a multi dimensional Cartesian system in the form of vectors. Top coordinates of each of the vectors are described in different compartments natural numbers corresponding to the maximum and minimum deviation of the relative thickness of the seam. These variables will be called describing variables. Clustering method groups the received beam vectors in Cartesian space in subgroups, such whose tops are most similar to each other. Making a summary of resources assigned to these groups of vectors, we obtain the quantity resources that meet the formed criteria. To solve this problem, we can use different methods of taxonomy (Bukietyński 1969; Krowiak 2005a; Krowiak 2005b & Krowiak 2004. Application of the Kohonen algorithm and automatic neural network allows multiple reduction of computation labour intensity of compared to other methods (Tarczyński 2011).

A practical tool for the implementation of this analysis was a computer program STATISTICA v. 9.0 (Dokumentacja programu STATISTICA 2010). The results of the analysis are presented in the tables contained in the article text.

3 RESULTS OF ANALYSIS

Table 1a contains division of the size of all resources deposited in thin seams in subgroups according to thickness ranges of relative deviation and the Table 1b contains the percentages of these subgroups in the total resources deposited in thin seams. Graphical illustration of the data contained in the Table 1a is shown in Figure 1. Table 2a is presenting the allocation of the resources into subgroups according to the average thickness of the seams ranges and Table 2b is presenting the percentages of these subgroups in the total resources deposited in thin seams.

Table 1 (a). Division of resources based on the thickness variation ranges of maximum and minimum average thickness of seams (size of resources in thousands Tons).

Ranges of relative negative deviation from the average thickness	Ranges of relative positive deviation from the average thickness				SUM
	0-10.0%	10.01-20.0%	20.01-30.0%	above 30 %	
0 – (-) 10.0%	263 628	96 802	167 506	0	527 936
(-) 10.01 – (-) 20.0%	154 552	209 533	987 159	0	1 351 244
(-) 20.01 – (-) 30.0%	52 155	191 880	2 950 659	-	3 194 694
(-) 30.01 – (-) 40.0%	27 198	358 607	2 843 535	78	3 229 418
above 40%	76 549	144 298	1 818 738	4 492	2 044 077
SUM	574 082	1 001 120	8 767 597	4 570	10 347 369

Source: own elaboration.

Table 1 (a). Division of resources based on the thickness variation ranges of maximum and minimum average thickness of seams (percentage of the total resources in thin seams).

Ranges of relative negative deviation from the average thickness	Ranges of relative positive deviation from the average thickness				SUM
	0-10.0%	10.01-20.0%	20.01-30.0%	above 30%	
0 – (-) 10.0%	0.99%	0.82%	4.79%		6.61%
(-) 10.01 – (-) 20.0%	1.19%	1.58%	36.35%		39.12%
(-) 20.01 – (-) 30.0%	1.19%	2.64%	28.52%		32.36%
(-) 30.01 – (-) 40.0%	1.12%	4.18%	8.47%		13.78%
above 40%	1.04%	0.45%	6.60%	0.04%	8.13%
SUM	5.55%	9.68%	84.73%	0.04%	100.00%

Source: own elaboration.

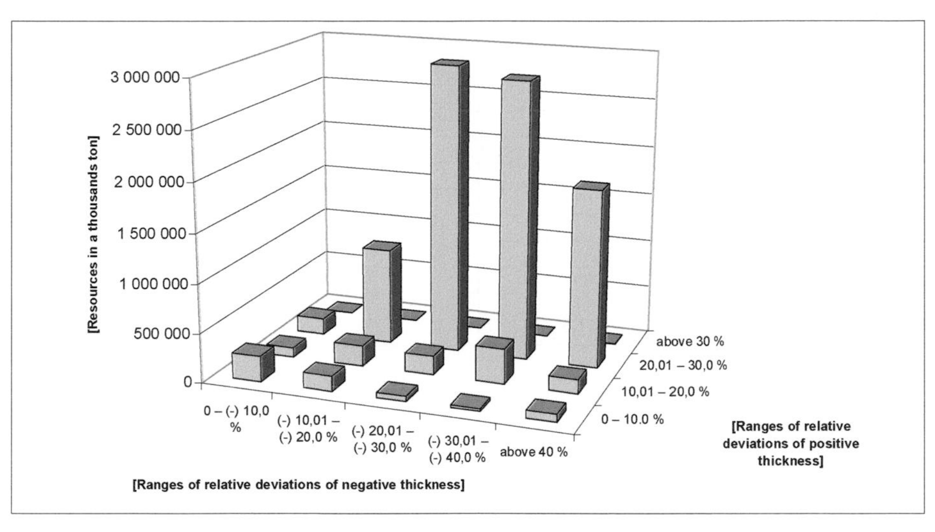

Source: own elaboration.

Figure 1. Quantities of resources depending on the ranges of positive and negative deviations from the mean thickness of the seams.

Table 2 (a). Division of resources according to criterion of ranges of deviations maximum and minimum thickness from average thickness with division on ranges of average thickness (absolute value).

Ranges of relative negative deviation from the average thickness	Ranges of relative positive deviation from the average thickness				SUM
	0-10.0%	10.01-20.0%	20.01-30.0%	above 30%	
Amount of resources [thousands ton]					
Average seam thickness range 0.6 – 0.8 m					
0 – (-) 10.0%	85 132	28 392	67 805		181 329
(-) 10.01 – (-) 20.0%	16 828	56 844	292 518		366 190
(-) 20.01 – (-) 30.0%	924	49	56 811		57 784
(-) 30.01 – (-) 40.0%			33 800		33 800
above 40%			44 400		44 400
SUM	102 884	85 285	495 334		683 503
Average seam thickness range 0.81 – 1.0 m					
0 – (-) 10,0 %	45 990	36 545	85 962		168 497
(-) 10.01 – (-) 20.0%	75 032	45 850	639 657		760 539
(-) 20.01 – (-) 30.0%	695	66 292	2 007 516		2 074 503
(-) 30.01 – (-) 40.0%	1 735	14 428	1 007 059	78	1 023 300
above 40%			21 544		21 544
SUM	123 452	163 115	3 761 738	78	4 048 383

Continuation of the Table 2 (a).

Ranges of relative negative deviation from the average thickness	Ranges of relative positive deviation from the average thickness				SUM
	0-10.0%	10.01-20.0%	20.01-30.0%	above 30%	
Average seam thickness range 1.01 – 1.2 m					
0 – (-) 10,0 %	68 529	10 696	12 760		91 985
(-) 10.01 – (-) 20.0%	13 475	34 672	32 676		80 823
(-) 20.01 – (-) 30.0%	21 719	103 975	865 748		991 442
(-) 30.01 – (-) 40.0%	14 571	87 034	1 530 728		1 632 333
above 40%	5 304	37 175	509 384		551 863
SUM	123 598	273 552	2 951 296		3 348 446
Average seam thickness range 1.21 – 1.4 m					
0 – (-) 10,0 %	41 188	17 334	401		58 923
(-) 10.01 – (-) 20.0%	30 180	50 637	11 087		91 904
(-) 20.01 – (-) 30.0%	9 433	19 149	19 661		48 243
(-) 30.01 – (-) 40.0%	10 072	254 241	251 570		515 883
above 40%	25 404	91 616	593 557		710 577
SUM	116 277	432 977	876 276		1 425 530
Average seam thickness range above 1.4 m					
0 – (-) 10,0 %	22 789	3 835	578		27 202
(-) 10.01 – (-) 20.0%	19 037	21 530	11 221		51 788
(-) 20.01 – (-) 30.0%	19 384	2 415	923		22 722
(-) 30.01 – (-) 40.0%	820	2 904	20 378		24 102
above 40%	45 841	15 507	649 853	4 492	715 693
SUM	107 871	46 191	682 953	4 492	841 507
TOTAL	574 082	1 001 120	8 767 597	4 570	10 347 369

Source: own elaboration

Figure 2 shows the size of resources depending on the relative deviations of positive and negative to the average thickness of seams for the average thickness of seams ranges from 0.81 to 1.0 m. The diameter of the "bubble" on the chart is proportional to the size of resources that meet the criteria for the coordinates of its measure. Similarly, in Figure 3 are given the resources for the size of an average seam thickness range from 1.01 to 1.2 m, and in Figure 4 for an average seam thickness range from 1.01 to 1.4 m.

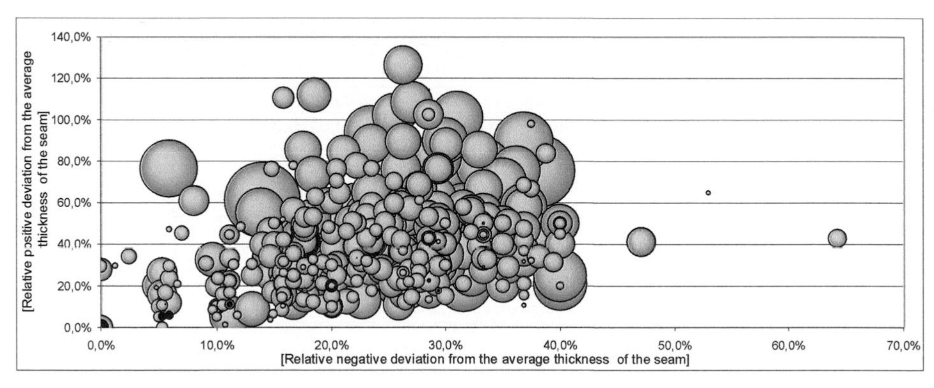

Source: own elaboration.

Figure 2. The size of the resources depending on the relative deviations of positive and negative to the average thickness the seam – for an average seam thickness range from 0.81 to 1.0 m

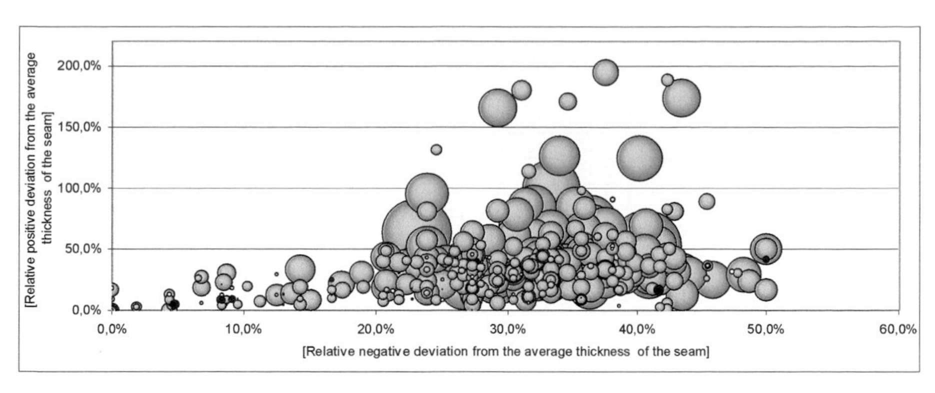

Source: own elaboration.

Figure 3. The size of the resources depending on the relative deviations of positive and negative to the average thickness the seam – for an average seam thickness range from 1.01 to 1.2 m.

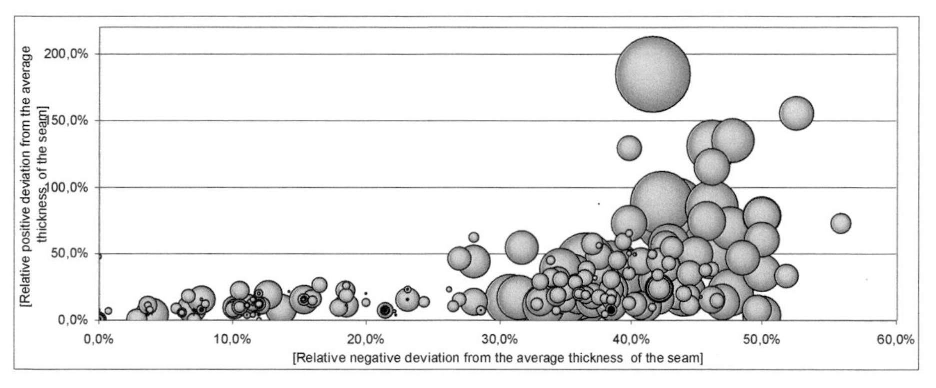

Source: own elaboration.

Figure 4. The size of the resources depending on the relative deviations of positive and negative to the average thickness the seam – for an average seam thickness range from 1.21 to 1.4 m.

Table 2 (b). Division of resources according to criterion of ranges of deviations maximum and minimum thickness from average thickness with division on ranges of average thickness (percentages).

Ranges of relative negative deviation from the average thickness	Ranges of relative positive deviation from the average thickness				SUM
	0-10.0%	10.01-20.0%	20.01-30.0%	above 30%	
Percentages					
Average seam thickness range 0.6 – 0.8 m					
0 – (-) 10.0%	12.46%	4.15%	9.92%		26.53%
(-) 10.01 – (-) 20.0%	2.46%	8.32%	42.80%		53.58%
(-) 20.01 – (-) 30.0%	0.14%	0.01%	8.31%		8.45%
(-) 30.01 – (-) 40.0%			4.95%		4.95%
above 40%			6.50%		6.50%
SUM	15.05%	12.48%	72.47%		100.00%
Average seam thickness range 0.81 – 1.0 m					
0 – (-) 10.0%	1.14%	0,90%	2.12%		4.16%
(-) 10.01 – (-) 20.0%	1.85%	1.13%	15.80%		18.79%
(-) 20.01 – (-) 30.0%	0.02%	1.64%	49.59%		51.24%
(-) 30.01 – (-) 40.0%	0.04%	0.36%	24.88%	0.002%	25.28%
above 40%			0.53%		0.53%
SUM	3.05%	4.03%	92.92%	0.002%	100.00%

Continuation of the Table 2 (b).

Ranges of relative negative deviation from the average thickness	Ranges of relative positive deviation from the average thickness				SUM
	0-10.0%	10.01-20.0%	20.01-30.0%	above 30%	
Average seam thickness range 1.01 – 1.2 m					
0 – (-) 10.0%	2.05%	0.32%	0.38%		2.75%
(-) 10.01 – (-) 20.0%	0.40%	1.04%	0.98%		2.41%
(-) 20.01 – (-) 30.0%	0.65%	3.11%	25.86%		29.61%
(-) 30.01 – (-) 40.0%	0.44%	2.60%	45.71%		48.75%
above 40%	0.16%	1.11%	15.21%		16.48%
SUM	3.69%	8.17%	88.14%		100.00%
Average seam thickness range 1.21 – 1.4 m					
0 – (-) 10.0%	2.89%	1.22%	0.03%		4.13%
(-) 10.01 – (-) 20.0%	2.12%	3.55%	0.78%		6.45%
(-) 20.01 – (-) 30.0%	0.66%	1.34%	1.38%		3.38%
(-) 30.01 – (-) 40.0%	0.71%	17.83%	17.65%		36.19%
above 40%	1.78%	6.43%	41.64%		49.85%
SUM	8.16%	30.37%	61.47%		100.00%
Average seam thickness range above 1.4 m					
0 – (-) 10.0%	2.71%	0.46%	0.07%		3.23%
(-) 10.01 – (-) 20.0%	2.26%	2.56%	1.33%		6.15%
(-) 20.01 – (-) 30.0%	2.30%	0.29%	0.11%		2.70%
(-) 30.01 – (-) 40.0%	0.10%	0.35%	2.42%		2.86%
above 40%	5.45%	1.84%	77.22%	0.53%	85.05%
SUM	12.82%	5.49%	81.16%	0.53%	100.00%
TOTAL for the whole resources	5.55%	9.68%	84.73%	0.04%	100.00%

Source: own elaboration.

Table 3 is presenting the allocation of the resources in the subgroups intervals according to the average depth of bedding of seams and in Table 3a we can observe the percentages of these subgroups in the whole of these resources deposited in thin seams.

Figure 5 shows the size of resources depending on the relative positive and negative deviations to the average thickness of seams for depth of betting in the range from 201 to 400 m. The diameter of the "bubble" on the chart is proportional to the size of resources that meet the criteria for the coordinates of it's the measure.

Similarly, in Figure 6 are given the size of resources for depth of bedding in the range from 401 to 600 m and in Figure 7 for range from 601 to 800 m.

Table 3 (a). Division of the resources according to criterion of ranges of deviations maximum and minimum thickness from average thickness with division on ranges of average depth of bedding (absolute values).

Ranges of relative negative deviation from the average thickness	Ranges of relative positive deviation from the average thickness				SUM
	0-10.0%	10.01-20.0%	20.01-30.0%	above 30%	
Amount of resources [thousands ton]					
Average depth of bedding 0 – 200 m					
0 – (-) 10.0%	29 788	7 986	11 179		48 953
(-) 10.01 – (-) 20.0%	11 574	12 200	42 079		65 853
(-) 20.01 – (-) 30.0%		7 102	93 486		100 588
(-) 30.01 – (-) 40.0%		6 546	47 540		54 086
above 40%	9 087	4 635	51 748		65 470
SUM	50 449	38 469	246 032	0	334 950
Average depth of bedding 201 – 400 m					
0 – (-) 10.0%	62 307	29 970	18 209		110 486
(-) 10.01 – (-) 20.0%	32 249	30 269	157 357		219 875
(-) 20.01 – (-) 30.0%	7 031	34 395	459 029		500 455
(-) 30.01 – (-) 40.0%	5 374	28 383	365 165		398 922
above 40%		37 720	429 231		466 951
SUM	106 961	160 737	1 428 991		1 696 689

Continuation of the Table 3 (a).

Ranges of relative negative deviation from the average thickness	Ranges of relative positive deviation from the average thickness				SUM
	0-10.0%	10.01-20.0%	20.01-30.0%	above 30%	
Average depth of bedding 401 – 600 m					
0 – (-) 10.0%	56 326	11 915	27 866		96 107
(-) 10.01 – (-) 20.0%	32 597	51 743	311 795		396 135
(-) 20.01 – (-) 30.0%	32 318	16 674	705 081		754 073
(-) 30.01 – (-) 40.0%	7 031	19 866	796 162		823 059
above 40%	15 674	51 894	423 273		490 841
SUM	143 946	152 092	2 264 177		2 560 215
Average depth of bedding 601 – 800 m					
0 – (-) 10.0%	79 033	19 817	91 100		189 950
(-) 10.01 – (-) 20.0%	43 004	72 464	252 632		368 100
(-) 20.01 – (-) 30.0%	8 333	63 915	1 176 912		1 249 160
(-) 30.01 – (-) 40.0%	11 981	138 179	865 445	78	1 015 683
above 40%	20 267	30 806	329 166	4 492	384 731
SUM	162 618	325 181	2 715 255	4 570	3 207 624
Average depth of bedding 801 – 1000 m					
0 – (-) 10.0%	28 957	25 873	19 152		73 982
(-) 10.01 – (-) 20.0%	35 128	32 667	138 467		206 262
(-) 20.01 – (-) 30.0%	4 473	34 428	263 495		302 396
(-) 30.01 – (-) 40.0%	1 948	72 818	675 920		750 686
above 40%	31 521	16 820	382 981		431 322
SUM	102 027	182 606	1 480 015		1 764 648
Average depth of bedding above 1000 m					
0 – (-) 10.0%	7 217	1 241			8 458
(-) 10.01 – (-) 20.0%		10 190	84 829		95 019
(-) 20.01 – (-) 30.0%		35 366	252 656		288 022
(-) 30.01 – (-) 40.0%	864	92 815	93 303		186 982
above 40%		2 423	202 339		204 762
SUM	8 081	142 035	633 127		783 243
TOTAL	574 082	1 001 120	8 767 597	4 570	10 347 369

Source: own elaboration.

Table 3 (b). Division of the resources according to criterion of ranges of deviations maximum and minimum thickness from average thickness with division on ranges of average depth of bedding (percentages).

Ranges of relative negative deviation from the average thickness	Ranges of relative positive deviation from the average thickness				SUM
	0-10.0%	10.01-20.0%	20.01-30.0%	above 30%	
Percentages					
Average depth of bedding 0 – 200 m					
0 – (-) 10.0%	8.89%	2.38%	3.34%		14.62%
(-) 10.01 – (-) 20.0%	3.46%	3.64%	12.56%		19.66%
(-) 20.01 – (-) 30.0%	0.00%	2.12%	27.91%		30.03%
(-) 30.01 – (-) 40.0%	0.00%	1.95%	14.19%		16.15%
above 40%	2,.71%	1.38%	15.45%		19.55%
SUM	15.06%	11.48%	73.45%		100.00%
Average depth of bedding 201 – 400 m					
0 – (-) 10.0%	3.67%	1.77%	1.07%		6.51%
(-) 10.01 – (-) 20.0%	1.90%	1.78%	9.27%		12.96%
(-) 20.01 – (-) 30.0%	0.41%	2.03%	27.05%		29.50%
(-) 30.01 – (-) 40.0%	0.32%	1.67%	21.52%		23.51%
above 40%	0.00%	2.22%	25.30%		27.52%
SUM	6.30%	9.47%	84.22%		100.00%

Continuation of the Table 3 (b).

Ranges of relative negative deviation from the average thickness	Ranges of relative positive deviation from the average thickness				SUM
	0-10.0%	10.01-20.0%	20.01-30.0%	above 30%	
Average depth of bedding 401 – 600 m					
0 – (-) 10.0%	2.20%	0.47%	1.09%		3.75%
(-) 10.01 – (-) 20.0%	1.27%	2.02%	12.18%		15.47%
(-) 20.01 – (-) 30.0%	1.26%	0.65%	27.54%		29.45%
(-) 30.01 – (-) 40.0%	0.27%	0.78%	31.10%		32.15%
above 40%	0.61%	2.03%	16.53%		19.17%
SUM	5.62%	5.94%	88.44%		100.00%
Average depth of bedding 601 – 800 m					
0 – (-) 10.0%	2.46%	0.62%	2.84%		5.92%
(-) 10.01 – (-) 20.0%	1.34%	2.26%	7.88%		11.48%
(-) 20.01 – (-) 30.0%	0.26%	1.99%	36.69%		38.94%
(-) 30.01 – (-) 40.0%	0.37%	4.31%	26.98%	0.002%	31.66%
above 40%	0.63%	0.96%	10.26%	0.14%	11.99%
SUM	5.07%	10.14%	84.65%	0.14%	100.00%
Average depth of bedding 801 – 1000 m					
0 – (-) 10.0%	1.64%	1.47%	1.09%		4.19%
(-) 10.01 – (-) 20.0%	1.99%	1.85%	7.85%		11.69%
(-) 20.01 – (-) 30.0%	0.25%	1.95%	14.93%		17.14%
(-) 30.01 – (-) 40.0%	0.11%	4.13%	38.30%		42.54%
above 40%	1.79%	0.95%	21.70%		24.44%
SUM	5.78%	10.35%	83.87%		100.00%
Average depth of bedding above 1000 m					
0 – (-) 10.0%	0.92%	0.16%			1.08%
(-) 10.01 – (-) 20.0%		1.30%	10.83%		12.13%
(-) 20.01 – (-) 30.0%		4.52%	32.26%		36.77%
(-) 30.01 – (-) 40.0%	0.11%	11.85%	11.91%		23.87%
above 40%		0.31%	25.83%		26.14%
SUM	1.03%	18.13%	80.83%		100.00%
TOTAL for the whole resources	5.55%	9.68%	84,.73%	0.04%	100.00%

Source: own elaboration.

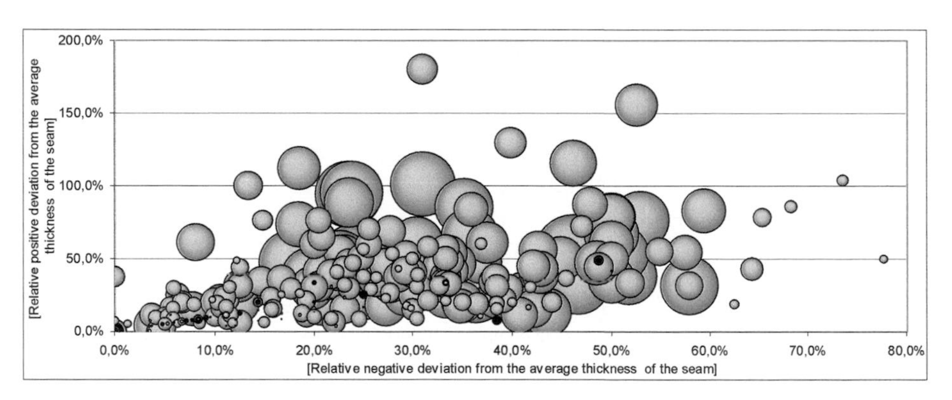

Source: own elaboration.

Figure 5. The size of the resources depending on the relative deviations of positive and negative to the average thickness the seam – for the range depth of bedding from 201 to 400 m.

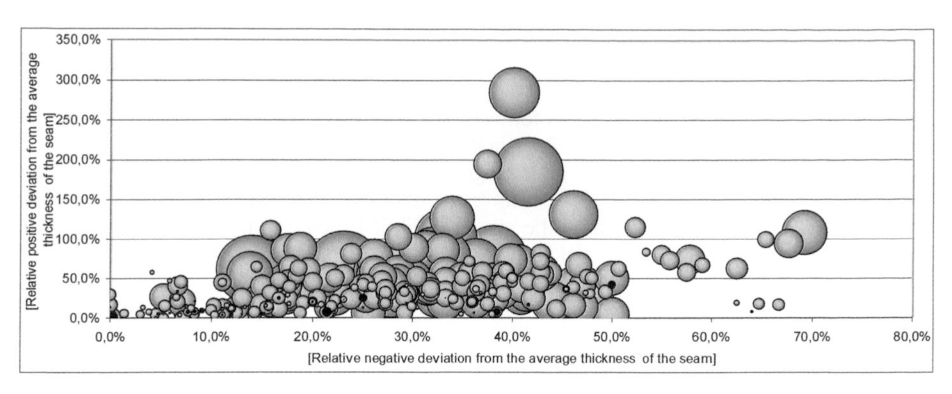

Source: own elaboration.

Figure 6. The size of the resources depending on the relative deviations of positive and negative to the average thickness the seam – for the range depth of bedding from 401 to 600 m.

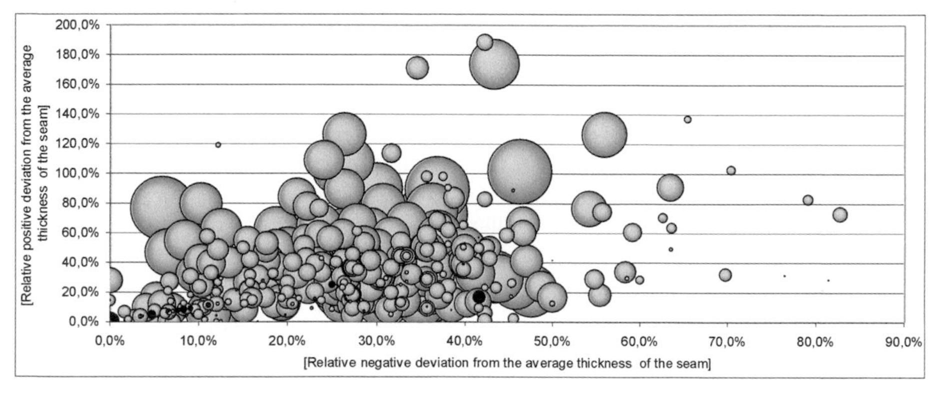

Source: own elaboration.

Figure 7. The size of the resources depending on the relative deviations of positive and negative to the average thickness the seam – for the range depth of bedding from 601 to 800 m.

4 CONCLUSIONS FROM THE ANALYSIS

Geological resources of active mines in Poland (except mine "Bogdanka"), in thin seams covers 10 888 163 thousands ton of hard coal, from which in article was subjected the resources of 10 347 369 thousands tonnes.

It covers in the ranges of average of thickness of seams as follows: in the range from 0.6 to 0.8 m – 683 503 thousand ton i.e. the 6.61% whole of stories; from 0.81 to 1.0 m – 4 048 383 thousand ton i.e. the 39.14% whole of stories; from 1.01 to 1.2 m – 3 348 446 thousand ton i.e. the 32.33% whole of stories; from 1.21 to 1.4 m – 1 425 520 thousand ton i.e. the 13.78% whole of stories; in range above 1.4 m – 841 507 thousand ton i.e. the 8.14% whole of stories.

It covers in the ranges of average depth of bedding as follows: in the range from 0 to 200 m – 334 950 thousand ton i.e. the 3.24% whole of stories; in the range from 201 to 400 m – 1 696 689 thousand ton i.e. the 16,40% whole of stories; in the range from 400 to 600 m – 2 560 215 thousand ton i.e. the 24.74% whole of stories; in the range from 601 to 800 m – 3 207 624 thousand tonnes i.e. the 31.00% whole of stories; in the range from 801 to 1000 m – 1 764 648 thousand ton i.e. the 17.05% whole of stories; in the range above 1000 m – 783 243 thousand tonnes i.e. the 7.57% whole of stories.

It covers in the ranges of relatively positive deviations of thicknesses from of average value as follows: in the range 0 to 10% – 574 082 thousand ton i.e. the 5.55% whole of stories; in the range from 10,01 to 20.0% – 1 001 120 thousand ton i.e. the 9.68% whole of stories; in the range from 20.01 to 30.0% – 8 767 597 thousand ton i.e. the 84.73%

whole of stories; in the range over 30% – 4 570 thousand ton i.e. the 0.04% whole of stories.

It covers in the range of relatively negative deviations of thickness from of average value as follows: in the range from 0 do (-) 10.0% – 527 936 thousand ton i.e. the 6.61% whole of stories; in the range from (-) 10.01 to (-) 20.0% – 1 351 244 thousand ton i.e. the 39.12% whole of stories; in the range from (-) 20.01 to (-) 30.0% – 3 194 694 thousand ton i.e. the 32.36% whole of stories; in the range from (-) 30.01 to (-) 40.0% – 2 044 077 thousand ton i.e. the 13.78% whole of stories;

In the average seam thickness range from 0.6 to 0.8 m size of resources in the various ranges positive relative deviations are, respectively: in the range from 0 do 10% – 102 884 thousand ton i.e. the 0.99% whole of stories; in the range from 10.01 to 20.0% – 85 285 thousand ton i.e. the 0.82% whole of stories; in the range from 20.01 to 30.0% – 495 334 thousand ton i.e. the 4.79% whole of stories.

In the average seam thickness range from 0.81 to 1.0 m size of resources in the various ranges positive relative deviations are, respectively: in the range from 0 to 10% – 123 452 thousand ton i.e. the 1.19% whole of stories; in the range from 10.01 to 20.0% – 163 115 thousand ton i.e. the 1.58% whole of stories; in the range from 20.01 to 30.0% – 3 761 738 thousand ton i.e. 36.35% whole of stories; in the range above 30% – 78 thousand ton i.e. the 0.001% whole of stories.

In the average seam thickness range from 1.0 to 1.2 m size of resources in the various ranges positive relative deviations are, respectively: in the range from 0 to 10% – 123 598 thousand ton i.e. the 1.19% whole of stories; in the range from 10.01 to 20.0% – 273 552 thousand ton i.e. the 2.64% whole of stories; in the range from 20.01 to 30.0% – 2 951 296 thousand ton i.e. the 28.52% whole of stories.

In the average seam thickness range from 1.21 to 1.4 m size of resources in the various ranges positive relative deviations are, respectively: in the range from 0 to 10% – 116 277 thousand ton i.e. the 1.12% whole of stories; in the range from 10.01 to 20.0% – 432 977 thousand ton i.e. the 4.18% whole of stories; in the range from 20.01 to 30.0% – 876 276 thousand ton i.e. 8.47% whole of stories.

In the average seam thickness range above 1.4 m size of resources in the various ranges positive relative deviations are, respectively: in the range from 0 to 10% – 107 871 thousand ton i.e. the 1.04% whole of stories; in the range from 10.01 to 20.0% – 46 191 thousand ton i.e. the 0.45% whole of stories; in the range from 20.01 to 30.0% – 682 953 thousand ton i.e. the 6.60% whole of stories; in the range above 30.0% – 4 492 thousand ton i.e. the 0.04% whole of stories.

In the average seam thickness range from 0.6 to 0.8 m size of resources in the various ranges negative relative deviations are, respectively: in the range from 0 to (-) 10.0% – 181 329 thousand ton i.e. the 1.75% whole of stories: in the range from (-) 10.01% to (-) 20.0% – 366 190 thousand ton i.e. the 3.54% whole of stories; in the range from (-) 20.01% to (-) 30.0% – 57 784 thousand ton i.e. 0.56% whole of stories; in the range from (-) 30.01% to (-) 40.0% – 33 800 thousand ton i.e. the 0.33% whole of stories; in the range above (-) 40% – 44 400 thousand tone i.e. the 0.43% whole of stories.

In the average seam thickness range from 0.81 to 1.0 m size of resources in the various ranges negative relative deviations are, respectively: in the range from 0 to (-) 10.0% – 168 thousand ton i.e. the 1.63% whole of stories: in the range from (-) 10.01% to (-) 20.0 % – 760 539 thousand ton i.e. the 7.35% whole of stories; in the range from (-) 20.01% to (-) 30.0% – 2 074 503 thousand ton i.e. 20.05% whole of stories; in the range from (-) 30.01% to (-) 40.0% – 1 023 300 thousand ton i.e. the 9.89% whole of stories; in the range above (-) 40% – 21 544 thousand tone i.e. the 0.21% whole of stories.

In the average seam thickness range from 1.01 to 1.2 m size of resources in the various ranges negative relative deviations are, respectively: in the range from 0 to (-) 10.0% – 91 985 thousand ton i.e. the 0.89% whole of stories: in the range from (-) 10.01% to (-) 20.0% – 80 823 thousand ton i.e. the 0.78% whole of stories; in the range from (-) 20.01% to (-) 30.0% – 991 442 thousand ton i.e. 9.58% whole of stories; in the range from (-) 30.01% to (-) 40.0 % – 1 632 333 thousand ton i.e. the 15.78% whole of stories; in the range above (-) 40% – 551 863 thousand tone i.e. the 5.33% whole of stories.

In the average seam thickness range from 1.21 to 1.4 m size of resources in the various ranges negative relative deviations are, respectively: in the range from 0 to (-) 10.0% – 58 923 thousand ton i.e. the 0.57% whole of stories: in the range from (-) 10.01% to (-) 20.0% – 91 904 thousand ton i.e. the 0.89% whole of stories; in the range from (-) 20.01% to (-) 30.0% – 48 243 thousand ton i.e. 0.47% whole of stories; in the range from (-) 30.01% to (-) 40.0% – 515 883 thousand ton i.e. the 4.99% whole of stories; in the range above (-) 40% – 710 577 thousand tone i.e. the 6.87% whole of stories.

In the average seam thickness range above 1.4 m size of resources in the various ranges negative relative deviations are, respectively: in the range from 0 to (-) 10.0% – 27 202 thousand ton i.e. the 0.26% whole of stories: in the range from (-) 10.01% to (-)

20.0% – 51 788 thousand ton i.e. the 0.50% whole of stories; in the range from (-) 20.01% to (-) 30.0% – 22 722 thousand ton i.e. 0.22% whole of stories; in the range from (-) 30.01% to (-) 40.0% – 24 102 thousand ton i.e. the 0.23 % whole of stories; in the range above (-) 40% – 715 693 thousand tone i.e. the 6.92% whole of stories.

Maximum amount of resources of 2 004 516 thousand tonnes representing the 18.65% whole of stories there are located in the seams meeting the following criteria: the average thickness of seams in the range from 0.81 to 1.0 m; relative positive deviation in the range from 20.01 to 30.0%; relative negative deviation in the range in the range of (-) 20.01 to (-) 30.0%.

Second in order of size of the maximum resource of 1 530 728 thousand tonnes representing the 14.79% whole of stories is located in the seams meeting following criteria: the average thickness of seams in the range from 1.01 to 1.2 m; positive relative deviation in the range from 20.01 to 30.0%; negative relative deviation in the range from (-) 30.01 to (-) 40.0%.

5 CONCLUSIONS

The method of analysis using multiple criteria simultaneously, presented in the article, allows to perform analysis of different thematic cross sections. Published results of the analysis should, thus, be treated as one of many possible to perform thematic cross section. Presented results of analysis can be applied, at least, as an interesting addition to geological documentation already hold by the mines and in the strategic management of this industry. It could be useful as one of the elements supporting the decision-making process concerning the further development of mines in the strategic management. Subject of the analysis were the geological resources of hard coal bedding in thin seams. From the perspective of practitioners, it is important how many and which resources can be classified as industrial resources in terms of economic viability of their exploitation. However, this requires additional word study, which sooner or later, should be taken. Assuming, however, for the time being theoretically that approximately 5% of them qualify for economically reasonable use, without incurring huge expenditures on the construction of new levels and mining excavations, then we have, right now, the resources ca. 100 million tons.

Source data for the performed analyses were collected from the project the Polish – Ukrainian "Technical and technological capabilities and economic merits of exploitation of thin seams of the hard coal" (The decision of the Minister of Science and Higher Education number 654/N-UKRAINA/2010/0).

REFERENCES

Bukietyński, W., Hellwig, Z., Królik, U. & Smoluk, A. 1969. *Uwagi o dyskryminacji zbiorów skończonych*. Prace Naukowe WSE Wrocław.

Cichosz, P. 2000. *Systemy uczące się*. WNT Warszawa.

Dokumentacja programu STATISTICA. V. 9.0, 2010.

Everitt, B.S., Landau, S. & Leese, M. 2001. *Cluster analysis*. Londyn, Arnold, New York, Oxford University Press.

Krowiak, A. 2005 (a). *Wyznaczanie podobieństwa grupy kopalni węgla kamiennego metoda podzbiorów izotropowych w oparciu o zmienne ceny standaryzowanej oraz koszty jednostkowego na produkcji węgla*", 6. Materiały konferencyjne "Szkoła Eksploatacji Podziemnej-2005". Kraków: Wydawca IGSMiE: 459-475.

Krowiak, A. 2005 (b). *Poszukiwanie podobieństwa relacji wielkości opisujących kopalnie węgla kamiennego w oparciu o zmienne unormowane*. Publikacje naukowe AGH Kraków, Szkoła Ekonomiki i Zarządzania w Górnictwie Krynica: 379-390.

Krowiak, A. 2004. *Grupowanie jako metoda określania podobieństwa relacji ekonomicznych w zbiorze kopalń węgla kamiennego*. Kraków: AGH Górnictwo i Geoinżynieria, 4/2: 135-139.

Materiały źródłowe z kopalń węgla kamiennego. 2010.

Nieć, M. 1982. *Geologia kopalniana*. Warszawa: Wydawnictwa Geologiczne.

Tadeusiewicz, R. 1993. *Sieci neuronowe*. Warszawa: AOW RW.

Tarczyński, G. 2011. *Algorytm Kohonena w analizie danych ekonomicznych*". Wrocław: Wydawnictwo Uniwersytetu Ekonomicznego we Wrocławiu.

Geomechanical Processes During Underground Mining – Pivnyak, Bondarenko, Kovalevs'ka & Illiashov (eds)
© 2012 Taylor & Francis Group, London, ISBN 978-0-415-66174-4

Experimental investigation of aeroelastic and hydroelastic instability parameters of a marine pipeline

Y. Kyrychenko, V. Samusia & V. Kyrychenko
National Mining University, Dnipropetrovs'k, Ukraine

O. Goman
Dnipropetrovs'k state university, Dnipropetrovs'k, Ukraine

ABSTRACT: The article contains the results of experimental research of unstable aeroclasticity type of wind resonance and galloping for the elements of pipe-line of deep-water hydraulic handling. The experiments have being carried in aerodynamic pipe on a scale models while using the strain measure balance. The laws exposed and some qualitative results can be useful while investigating the self-excited oscillating of the constructions under the influence of wind load.

1 INTRODUCTION

Each known problem solving of elastic construction dynamics under the influence of wind loads supposes there is indispensable for the calculations information about aerodynamic forces which can be obtained experimentally only. This is the main trouble of all aeroelastic problems as to obtain reliable information concerning the working aerodynamic forces needs a great number of experiments and finally it is connected with sizeable expenditures than subsequent motion equation solution. In this connection the practical significance of similar investigations exceeds the bounds of oscillating bodies behavior in a wind flow owing to technical actuality of analogous researches for liquid flowing rod construction, for example, submarine pulp feed-lines. That's why it is advisable to considerate from the scientific point of view more common problem of self-excited oscillation beginning mechanism for elastic constructions being flown by air or liquid.

While designing the construction to be interacted with the flow it is necessary to pay much attention to its element oscillation calculations to find such geometric and dynamic characteristics which can exclude possible self-excited oscillation excitation.

Each section of pipe-line is a set of pipe-lines. Intake pipe-line is the main one and it is surrounded with some auxiliary pipe-lines which have less diameter.

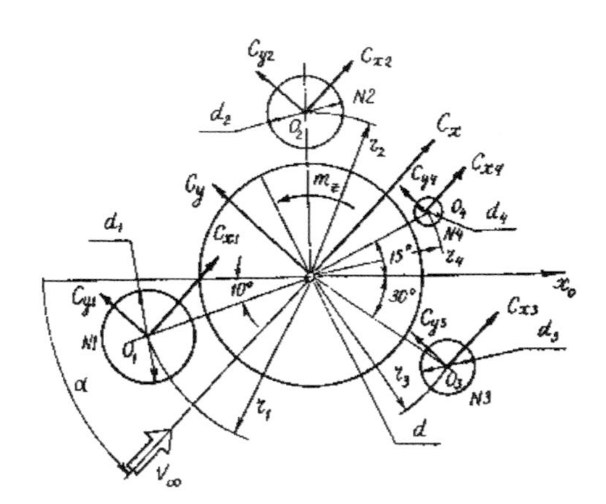

Figure 1. Set #1 Scheme.

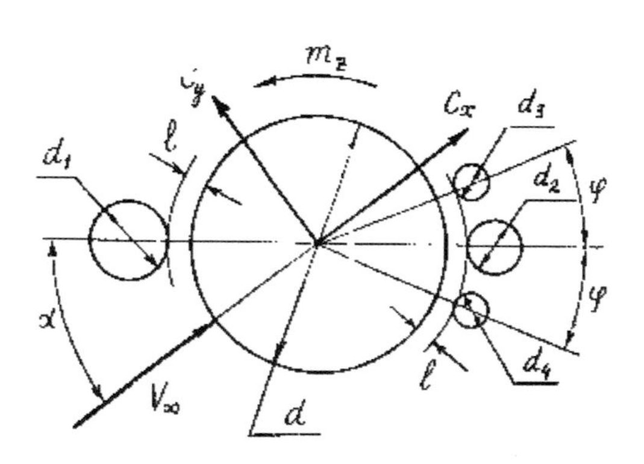

Figure 2. Set #2 Scheme to study l and φ parameters.

There were tested some set lay-out with fixed disposition pipe-lines of selected diameters one of which (set #l) is in Figure 1, and set #2 (Figure 2) to study according to geometrical parameters l and φ. For set #l: $d_1 = 0.33$; $d_2 = 0.22$; $d_3 = 0.1$; $d_4 = 0.06$; $r_1 = 0.79$; $r_2 = 0.74$; $r_3 = 0.66$; $r_4 = 0.62$. For set #2: $d_1 = 0.33$; $d_2 = 0.19$; $d_3 = d_4 = 0.1$; l and φ have being ranged ($\varphi = \varphi_0$ means the pipe-lines d_2, d_3 and d_4 touch on each other). Here and below all linear dimensions are referenced to the diameter of central pipeline d. In the experimental aerodynamic model d diameter was 100 millimetres.

1.1 *The study of hydrodynamic instability of vortex excitation type*

Vortex excitation as it is known (Kazakevich & Grafski 1984) originates owing to periodical eddy separation from smooth surface of stream-lined extent elastic body. Periodical eddy separation form so called Karman vortex trail in the wake of the body. The frequency of f_0 vortex separation is determined by the Strouhal number: $Sh = \frac{f_0 d}{V}$. For each section made fast in the flow the Strouhal number has quite definite value (which will depend upon the angle of incidence too if profile is asymmetrical one). For single cylinder $Sh \approx 0.2$. Owing to periodical vortex separation (either from one side of cylinder or from another one) pressure distribution along its surface gains periodical component therefore cross force Y_D with f_0 frequency arises, and resistance X force gains periodical X_D addition which frequency is $2f_0$:

$$Y_D = Y_{max} sin(2\pi f_0 t),$$

$$X = X_{av} + X_D = X_{av} + X_{max} sin(4\pi f_0 t), \quad (1)$$

where X_{av} – average (static) value of resistance force; Y_{max} and X_{max} – amplitude values of cross and resistance forces periodical components.

If we use aerohydrodynamic indices, according to (1) they will be:

$$C_y = C_{ya} sin(2\pi f_0 t);$$

$$C_x = C_{x0} + C_{xa} sin(4\pi f_0 t), \quad (2)$$

where C_{x0} – average value of resistance force coefficient; C_{ya} and C_{xa} – amplitude values of correspondent coefficients (each coefficient is related to $\rho V^2/2$ and d).

C_x values for set #2 are in Figure 3. As there is torque which m_z coefficient is in Figure 4, the sections of the pipe-line torsionally vibrate comparatively to longitudinal axis.

1.2 *The study of hydrodynamic instability of galloping type*

The loss of hydrodynamic stability of galloping type is the beginning of extent body oscillations in diametrical direction to the flow without twist or with a small one but galloping takes place if section of the body in a definite range of angle of attack has negative gradient of hydrodynamic lift $\partial C_y / \partial \alpha < 0$ (force diametrical to flow) which is rather large as for absolute value.

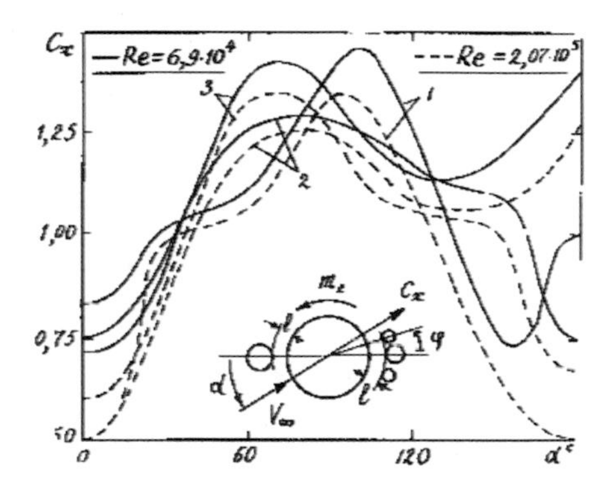

Figure 3. Resistance force coefficient for the set #2: 1 – $\bar{l} = 0.05$; $\varphi = \varphi^\circ$; 2 – $\bar{l} = 0.05$; $\varphi = 30^\circ$; 3 – $\bar{l} = 0.05$; $\varphi = 45^\circ$.

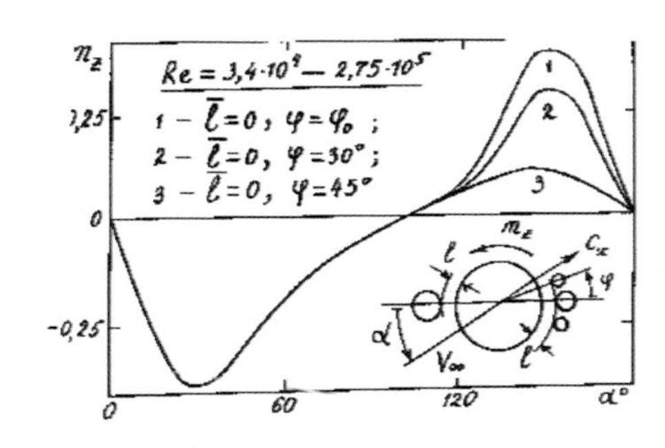

Figure 4. Moment coefficient for the set #2.

The trend of self-excited oscillation regime of type of galloping beginning can be evaluated according to Den-Hartog number (Den-Hartog 1960), that is section losses its stability under given angle of attack, if

$$C_y^\alpha + C_x < 0 , \quad (3)$$

where C_y^α – gradient of cross force coefficient.

Figure 5 shows obtained experimental dependences of cross force coefficient of set upon angle of attack for arrangement #l, and Figure 6 – for three arrangement variants #2. As it is from the chart there are three ranges of angles of attack where derivative $\partial C_y / \partial \alpha$ is negative though condition (3) isn't satisfied in every range. For arrangement #1 (Figure 5, curve 3) derivative $\partial C_y / \partial \alpha$ value in $A\,(\alpha_0 = 80^\circ)$ and $B\,(\alpha_0 = 330^\circ)$ points is $\partial C_y / \partial \alpha \approx -(5-6)$ and as C_x value for this arrangement when α and Re are any (in the range to be explored) is $C_x \approx 0.75 - 1.4$, Den-Hartog number (3) is performed in A and B points and their neighborhoods $\pm 20^\circ$. On the curve #1 at the C point (Figure 3) $\partial C_y / \partial \alpha = -1.2$ and experimental value $C_x = 1.34$, so (3) criterion is not performed.

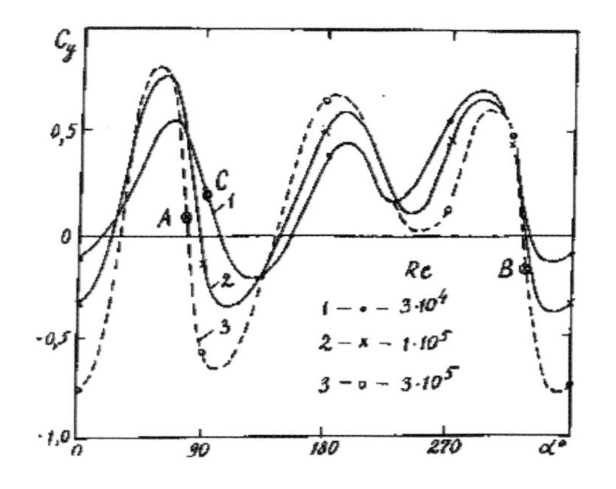

Figure 5. The coefficient of set cross force on the whole for #1 arrangement.

For #2 arrangement (Figure 6) Den-Hartog number is performed, e.g., for curve 3 in the neighborhood of $A\,(\alpha_0 = 60^\circ)$ and $B\,(\alpha_0 = 300^\circ)$ points. In these points $\partial C_y / \partial \alpha \approx -3.5$. Neighborhoods where (3) criterion is performed are $\pm 15^\circ$ approximately. In C point (curve 3) $\partial C_y / \partial \alpha \approx -1.3$, so if $C_x = 1.4$ (3) criterion is not performed.

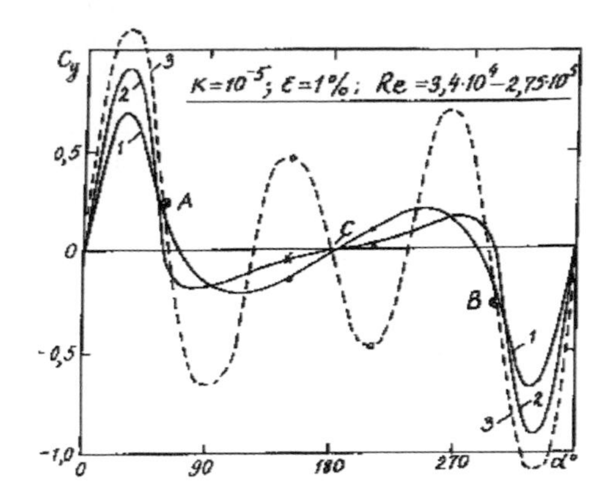

Figure 6. The coefficient of cross force for nr.2 arrangement: 1 – $\bar{l} = 0.05$; $\varphi = \varphi^\circ$; 2 – $\bar{l} = 0.05$; $\varphi = 30^\circ$; 3 – $\bar{l} = 0.05$; $\varphi = 45^\circ$.

1.3 *Dynamic tests of aerohydroelastic self-excited oscillations of pipe-line elements*

To define qualitative and some quantitive characteristics of self-excited oscillations of pipe-line there were performed direct dynamic tests of pipe-line models in air tube. Tests procedure and measuring instruments are described in detail in (Grafski & Kazakevich 1983) and (Goman, Grafski & Kyrychenko 1998). The tests took place with the help of specially developed multicomponent strain measurement scales which main units were mechanical resonant circuits. Their elastic and dissipative properties may be varied during the trial.

The tests have been performed on models of pipelines sets for different variants of arrangements. Each pipe-line itself and the whole set were separate mechanical contours where δ decrement, m mass and ω own frequency may be changed. Abovementioned strain measurement scales permitted theoretically to find the conditions of beginning and disclose the range of various types of aeroelastic instability existence according to their main signs.

Two arrangement sets shown in Figures 1 and 2 had been used for arrangement #1 gave ability to test aeroelastic self-excited oscillations of the pipeline on the whole and every its separate pipe-line; model for arrangement #2 – the whole pipe-line only. The range of Reynolds number was $Re = 3 \cdot 10^4 - 2 \cdot 10^5$.

The results of arrangement nr.1 model tests showed the model on the whole made flexural and torsional vibrations, and some auxiliary pipe-lines –

longitudinal and lateral flexural vibrations. The charts to show amplitudes of those oscillations according to Re and angle of attack are in Figures 7 and 8 (amplitudes a_{xi} longitudinal and a_{yi} lateral vibrations of each auxiliary cylinder are referred to d – diameter of central cylinder). Reynolds number for each pipe-line has been calculated in accordance with its own diameter.

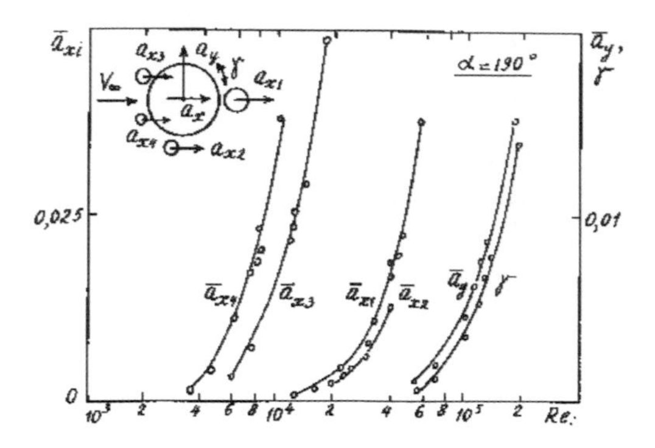

Figure 7. The changes of kinematic parameters of flexural and torsional movement for the whole set and auxiliary pipe-lines #1 arrangement according to Reynolds number.

As it is seen from Figure 7 the loss of aeroelastic stability for the model on the whole took place when $Re = 5 \cdot 10^4$, but for auxiliary pipe-lines in the set it happened much earlier. Figure 8 shows amplitudes of oscillations of pipe-line model as the whole. We can see the amplitude of longitudinal vibrations is rather less than lateral ones. The changes in flexural vibrations (that is their continuous growth from $Re = 4 \cdot 10^4$ to $Re = 2 \cdot 10^5$) means galloping type of self-excited oscillations in examined speed range.

As for arrangement #2, the results discovered two types of aerohydrodynamic instability losses (Figure 9): vortex excitation (A region) and galloping (B region) (the given speed is $V_{pr} = V/\omega_y d$). These three cases are in accordance with following parameters: a) $\bar{l} = 0.5$; $\varphi = 30$, A is a region of vortex excitation ($\alpha - 0° - 180°$), B is a region of galloping ($\alpha - 0° - 180°$); b) $\bar{l} = 0.5$; $\varphi = \varphi_0$, A is a region of vortex excitation ($\alpha = 180° \pm 20°$), B is a region of galloping; c) $\bar{l} = 0$; $\varphi = \varphi_0$, A is a region of vortex excitation ($\alpha = 0° \pm 10°$), B is a region of galloping ($1 - \alpha \cong 0°$, $2 - \alpha = 20° - 180°$). In the region of vortex excitation the amplitude a_{yi} of lateral oscillations firstly increases when speed is picked up and then dies down; in the region of galloping the amplitude of oscillations increases as a rule with speed growth. Given results have been obtained for stiff plates, that's why amplitudes of oscillations turned out to be insignificant.

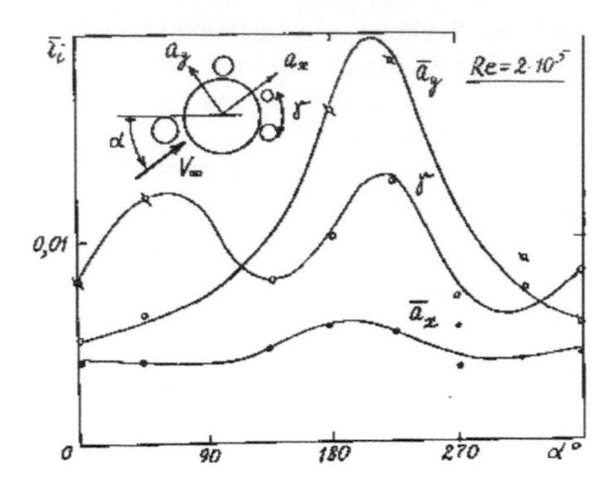

Figure 8. The changes of longitudinal and torsional vibrations of the whole set according to angle of attack (#1 arrangement).

The region of self-excited oscillations of A vortex excitement with a given speed is $V_{pr} = 5 - 10$, or according to Strouhal number $Sh = 0.2 - 0.1$. The loss of stability of galloping type was when $V_{pr} = 14.5$. If galloping takes place both lateral and longitudinal oscillations occur with their own bending frequencies.

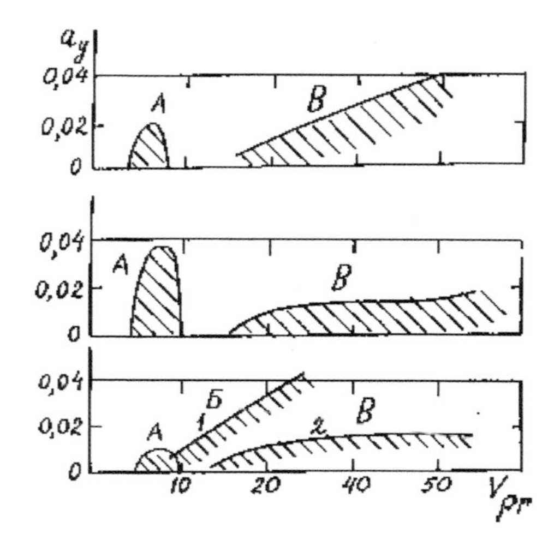

Figure 9. Amplitude dependence of lateral oscillations of the set of speed given.

Instability of type of galloping was revealed in all tested ways of surrounding pipe-lines location. The most stable position as for galloping in the whole tested range of V_{pr} values and all angles of attack turned out to be the arrangement of the set when $l = 0$ (that is auxiliary pipe-lines are snug against central one).

2 CONCLUSIONS

The results of tests concerning the aerohydrodynamic properties of pipe-line elements showed the pipe-line cross-section which consists of pipe-line set with different diameters has inclination to display aerohydroelastic instability of type of vortex excitement and galloping. That's why any specific design needs preliminary careful aerohydroelastic tests.

Maximal possible safety of the pipe-line against hydroeiastic instability beginning in the range of parameter change can be obtained by limiting the working speed of unit transportation lower than minimal stalling speed of vortex excitement beginning.

REFERENCES

Kazakevich, M. & Grafski, I. 1984. *Subharmonic entrainment of aeroelastic vibrations of circular cylinder*. Reports of AS of UkrSSR. A Series, 4.

Den-Hartog, G.P. 1960. *Mechanical vibrations*. Moscow: Phizmatgiz.

Grafski, I. & Kazakevich, M. 1983. *The aerodynamics of badly-flown bodies*. Hand book. Dnepropetrovsk: DSU.

Goman, O., Grafski, I. & Kyrychenko, E. 1998. *Aerodynamic characteristics of immersed construction of the system for submarine extraction*. Dnipropetrovs'k: NMU of Ukraine: Transaction #2.

Geomechanical Processes During Underground Mining – Pivnyak, Bondarenko, Kovalevs'ka & Illiashov (eds)
 ISBN 978-0-415-66174-4

Basic positions of the project of energy-efficient system for solar heat supply

M. Tabachenko & K. Ganushevych
National Mining University, Dnipropetrovs'k, Ukraine

V. Ryabichev
Volodymyr Dahl East Ukrainian National University, Anthracite, Ukraine

ABSTRACT: Schematic solutions during use of solar radiation in heat supply systems are presented. Recommendations for selection and calculation of flat and focusing helium receivers' parameters are given.

In average, annual quantity of solar radiation that falls on the Earth's surface is equal to 2000-2500 kW year / m^2 and 1000-1500 kW year / m^2 in high latitudes areas (Daffi Jr. 1987 & Beckman 1987). Thus, in order to receive heat currents that are sufficient for functioning of the modern energy systems and technological processes, it is necessary to use solar concentrators.

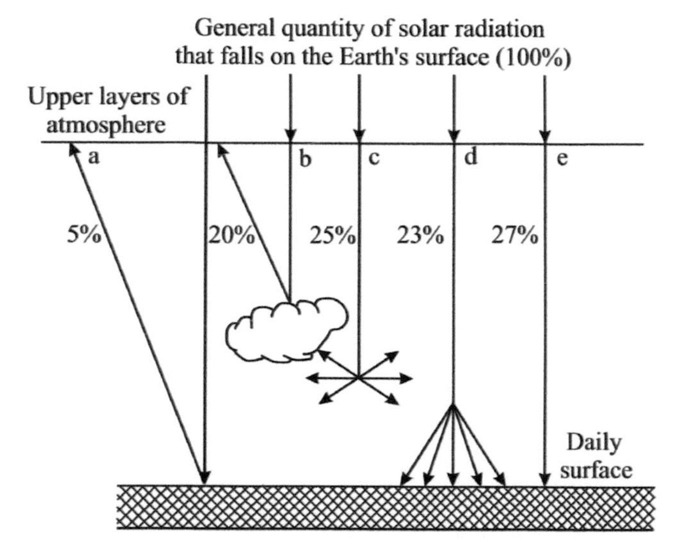

Figure 1. Passing of the solar beams through the Earth's atmosphere: (a) reverberation from the Earth's surface; (b) reverberation by the clouds; (c) absorption by the atmosphere itself; (d) dissipation by the atmosphere with reaching the land surface; (e) emission particle that reaches the Earth's surface.

Specially painted and glassed, they allow to receive low-temperature heat due to solar radiation that can be used for heating and ventilation of the apartments, for water heating etc.

Types of passive solar systems are shown on Figure 2. Types of helium systems are chosen depending on the structures' purposes and existing conditions of heat supply (Table 1).

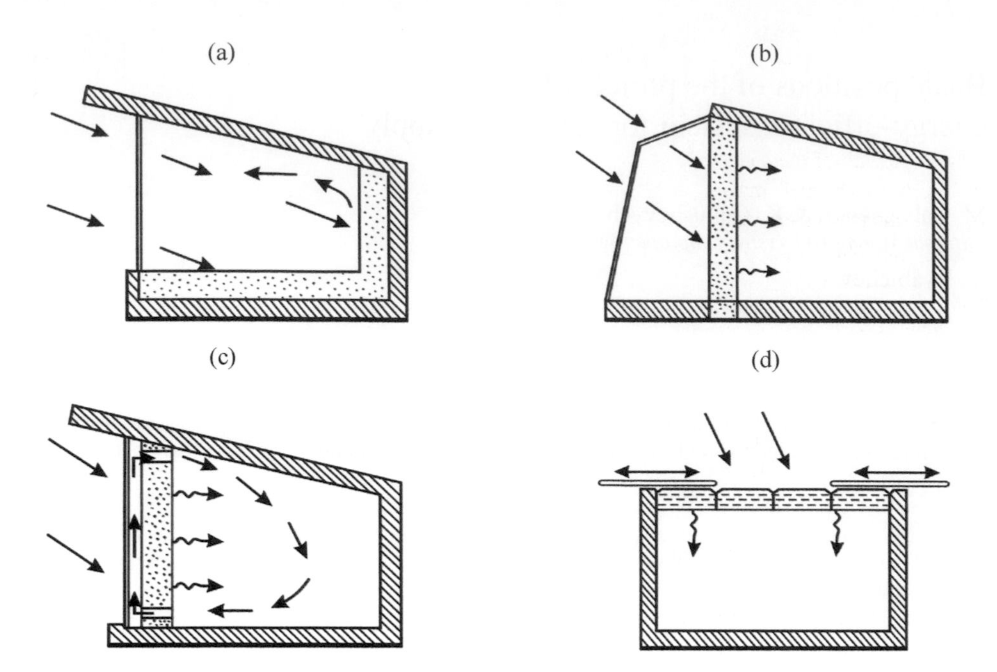

Figure 2. Passive solar systems.

Table 1. Recommendations for selection of helium systems.

Types of the structures	Recommended helium systems
Structures that are connected to constantly functioning energy sources	Full year
Enterprises of social nourishment, hotels, hospitals, sanatoriums, kindergartens, bathhouses, laundries	Seasonal with a backup unit, with 100% supply of hot water
Recreation centers of seasonal operation, pioneer camps	Seasonal with a backup unit that ensure, if necessary, cover of technological load of hot water, thanks to the backup unit
Dwelling houses with the boiler house for heating, summer showers, camps	Seasonal without backup unit

Designing of the systems begin with determination of the needed area of helium receiver that is calculated depending on various variants of the collectors' structures (Daffi Jr. 1987 & Beckman 1987).

Basic accounting equation for the flat collector computation:

$$q_u = q_{abs} - u\left(T_n - T_0\right), \tag{1}$$

where q_u – useful heat quantity that is allocated for the time and square unit (heat current), W / m^2; q_{abs} – absorbing quantity of the heat, W / m^2; u – coefficient of time losses of the collectors, W / (m^2·K); T_n – temperature of the absorbing surface, K; T_0 – temperature of the surrounding wind, K.

Complexity of this equation's application is in the determination of the absorbing plate' average temperature using concepts of the absorber's efficiency f' that is equal to ratio of actual absorbing energy to absorbing energy, in case of the liquor's and the plate's temperatures equality that describes the collector's productivity as the function of the heat carrier's average temperature

$$q_u = f'\left(q_{abs} - u\left(T_T - T_o\right)\right), \tag{2}$$

where f' – function of the construction parameters of absorption that barely depends on the consumtion and temperature of the heat carrier's:

$$f'=\left\{l\left[d+(R-d)f_p\right]^{-1}+\frac{\delta_k}{\lambda_k}+\frac{uR}{d_{in}\cdot\alpha_{in}}\right\}^{-1}, \quad (3)$$

where l –length of the collector, m; d – diameter of the absorber's channel, m; r – radius, distance between the channel's axes, m; δ_k – thickness of the absorber's material, m; λ_k – heat conductivity, W / (m²·K); d_{in} – internal wall's diameter, m; α_{in} – coefficient of heat emission of the internal collector's prop, W / (m²·K); f_p – efficiency of the absorber's rib

$$f_p=\frac{t_h\left[\sqrt{u\lambda_c}(R-d)/2\right]}{iu\lambda c^{-1}(R-d/2)}, \quad (4)$$

where λ_c – heat conductivity of the absorber, W / (m²·K); i – value that is attributed to the 1st section of the accumulator.

Thus, f' for effective absorbers is always greater than 0.9. For example, for the rest of stamped radiators $f'=0.95$, for aluminum – 0.97.

Level of the temperature's distribution along the current is calculated by integrating of accounting equation (1) for the liquid's element

$$\frac{T_{end}-T_{int}}{T_b-T_{int}}=1-exp\left(-uf'F/W_T\right), \quad (5)$$

where T_{end} – end temperature of the collector's heating , K; T_{int} – initial temperature of the heating, K; T_b – average balanced heating temperature, K; F – surface area of the collector, m²; W_T – water equivalent of losses (volume) – multiplication of the liquor volume's loss to density and heat capacity, W / K;

Equation (5) coincides with the notes in symbols of ε – NTU equation of heat conduction between heating current with endless consumptions and temperature T_p and heated current from T_{in} to T_{ext} with number of heat carrying particles $NTU=\frac{kf}{w}$ that is defined by consumption W_T with heat conduction coefficient between the currents u and area fF. So, the efficiency of the helium receiver, as the heat exchanger, is defined from the following equation

$$\varepsilon_g=1-exp\left(-NTu_g\right). \quad (6)$$

According to the equation (1), average temperature of the heat carrier is founded $\overline{T_T}$.

$$\overline{T_T}=T_{int}+\left(T_p-T_o\right)\frac{uF}{W_T}\varepsilon_g. \quad (7)$$

Value $\overline{T_T'}=\left(T_{in}+T_{ext}\right)/2$ is more suitable for the calculations. For this purpose we will use equation of the collector's productivity as its function and introduce factor $\bar{f}$, after this the equation (1) will look like as follows

$$q_u=\bar{f}\left(q_{abs}-u\left(T_T-T_o\right)\right), \quad (8)$$

where $\bar{f}=\frac{W_T}{uF}\cdot\frac{2\varepsilon_g}{2-\varepsilon_g}$.

With $f\approx 1$ and $\frac{W_T}{uF}>2$, value $\bar{f}\approx 1$ and equation (1) turns into (8), and the error does not exceed 2%. Consumption value that ensures set temperature of the collector is calculated from the following equation

$$W_T=f'uF\,ln^{-1}\left[1-\frac{u\left(T_{end}-T_{int}\right)}{q_{abs}-u\left(T_{int}-T_{end}\right)}\right]. \quad (9)$$

Since during computations of helium receivers, most often the heat carrier's temperature at the entry is known, it is convenient to use heat efficiency equation that includes this parameter

$$q_u=f_R\left[q_{abs}-u\left(T_{int}-T_0\right)\right], \quad (10)$$

where f_R – coefficient of heat extraction from the collector that is the ratio of practically received energy to the energy that would be received if temperature of the whole absorbing plate would be equal to temperature of the heat carrier at the entrance

$$f_R=\frac{W_T}{uF}\varepsilon_g. \quad (11)$$

Coefficient of the collector's heat losses is calculated by two components: losses through transparent surface u_c and bottom u_b. If the heat carrier does not move between the absorber and glass

$$u=u_c+u_b. \quad (12)$$

This value for flat helium receivers can be found by the following empirical equation

$$u = \left[\frac{v}{344 / T_d \left[(T_d - T_0) / (v + c) \right]^{0.31}} + \frac{1}{\alpha_c} \right] 2 + \alpha_{st} \frac{F_{sa}}{F} + \to \dots$$

$$\dots \to + \left[\frac{\delta_d}{\lambda_d} + \frac{1}{\alpha_d} \right]^{-1} + \frac{\sigma \left(T_d^2 + T_0 \right) \left(T_d - T_0 \right)}{\left[\xi_d + 0.05 v (1 - \varepsilon_d) \right]^{-1} + (2v + c - 1) / \xi_c - v}, \quad (13)$$

where v –number of glass covers; α_c – coefficient of glass' heat conduction, W / (m^2·K); F_{sa} – surface area of the wall, m^2; δ_d – thickness of the collector's bottom, m; λ_d – heat conduction of the bottom, W / (m^2·K); α_d – coefficient of bottom's heat emission; σ – constant of Stefan-Boltzmann; ξ_d – degree of the surface's darkness.

$$C = \left(1 - 0.04\alpha_c + 5 \cdot 10^{-4} \alpha_c^2 \right) (1 + 0.0588v).$$

The first two figures of the equation (13) determine heat consumption through light-transparent barrier. They depend on slope angle of the collector and its darkness degree

$$\frac{u_c(\beta)}{u_c(45)} = 1 - (\beta - 45)(2.59 - 1.44\xi_d) 10^{-3}. \quad (14)$$

The second and third figures – coefficient of losses through the bottom and walls. In the effective collector, coefficient losses through the bottom is equal to 0.5-1 W / m^2·K. Losses coefficient through walls for a single collector is equal to less than 3% of total losses and in case of their combination – less than 1% and may not be taken into account.

For more complex cases dependence of u on u_c and u_d is defined as follows

$$u = \frac{u_c + u_d}{1 + \alpha_d u_C / \left(\alpha_c \cdot \alpha_{C.d}^g + \alpha_d \alpha_{C.d}^g + \alpha_{d.C} \right)} + \to \dots$$

$$\to \dots + \frac{u_c + u_d}{\alpha_d + \left(\alpha_C \alpha_{d.C}^g + \alpha_d u_C \right) / \alpha_C + \alpha_{C.d}^g}. \quad (15)$$

Important characteristics of the collectors that are needed for calculation of q_{abs} – reduced absorption ability that shows which part of the falling on the collector solar radiation is being absorbed by it. It is determined by multiplication of the coefficients of transparent surface transmittance θ and absorption (a) of the collector (θ_a) separately for straight and diffuse solar radiation. Published and saved meteorological information in most of the cases shows data of intensity of the radiation components on the surface. Thus, coefficient of position q_{abs} should be added to the calculation – ratio of intensity on the horizontal plane. For reverberated and dissipated radiation, their values are determined as follows

$$p^r = sin^2 \beta / 2; \; p^D = cos^2 \beta / 2, \quad (16)$$

where p – ratio of solar radiation's intensity that falls on the collectors' plane to horizontal – coefficient of position; r – density of the current of reverberated solar radiation, W / m^2; D – density of the current of solar radiation's diffuse component, W / m^2; β – slope angle of the collectors to the horizon, degree.

Absorption ability of the collector with diffuse radiation (reverberated and diffused) is taken to be constant that corresponds to the value for the straight line drawn during beam's fall at an angle of 60°. When summing up straight solar radiation which was absorbed by the collector it is necessary to have p^s and $(\theta_a)^s$ values in terms that correspond to solar radiation current's density S. This significantly increases volume of the calculations and creates complexities in calculations directed to determination of long-term collectors' characteristics. In order to facilitate them it is reasonably to use averaged values for the period when slope angle of the solar beam on the collector's surface is $i < 55°$, and p^s and $(\theta_a)^s$ values are for determination of the coefficients (Rabinovich 1982.).

In Table 2, some calculation results are shown for the latitude of 45° according to which the calculations for southern regions of Ukraine can be made (Volevakha & Goysa 1987). Collector's dip angles are shown based on the helium system's work period: summer $\beta \approx \varphi + 15°$, all year round $\beta \approx \varphi$ (φ – angle of geographical latitude, degree).

Table 2. Average monthly values $\bar{p}^s$ and $\bar{\theta}_a$ for collectors of southern orientation $\psi = 45^\circ$ (ψ – azimuth angle).

Month of the year	$\beta = 30^\circ$		$\beta = 45^\circ$		$\beta = 60^\circ$	
	$\bar{p}^s$	θ_a	$\bar{p}^s$	θ_a	$\bar{p}^s$	θ_a
1	2.104	0.733	2.308	0.750	2.580	0.753
2	1.727	0.755	1.924	0.754	1.980	0.753
3	1.419	0.749	1.480	0.754	1.443	0.755
4	1.195	0.752	1.171	0.756	1.074	0.744
5	1.075	0.753	1.000	0.748	0.855	0.734
6	1.020	0.755	0.938	0.747	0.828	0.727
7	1.040	0.757	1.000	0.749	0.850	0.730
8	1.130	0.754	1.095	0.754	1.000	0.740
9	1.303	0.752	1.330	0.755	1.264	0.752
10	1.586	0.750	1.710	0.753	1.747	0.754
11	2.053	0.739	2.392	0.749	2.559	0.756
12	2.255	0.721	2.992	0.742	3.133	0.755

With arbitrary orientation of the building at which the helium receiver is installed, it is necessary to know how the deviation from southern orientation changes intensity of falling on the collector solar radiation. Calculations have shown that for the incline of $\beta \approx \varphi + 15^\circ$ with deviation of up to 10° does not change year's amount of the falling radiation for more than 5%, up to 30° – not more than 10%. Dustiness of the surface and shading of the absorber by walls slightly decrease amount of absorbed radiation. Their precise calculation is complex and not always possible, thus in the most of calculations it is calculated by introduction of reduction coefficient 0.951.

Formulas for calculation of solar radiation's intensity that falls and gets absorbed by the collector of various spatial directions and its rated absorption ability:

$$q_f = p^2 S + p^D D + p^r (S + D)\rho, \tag{17}$$

$$q_{abs} = 0.951(\theta_a)^S p^S S + (\theta_a)^D \left[p^D D + p^r (S + D)\rho\right], \tag{18}$$

$$(\theta_a) = q_{abs} / q_f. \tag{19}$$

Now the efficiency coefficient of helium receiver can be calculated from the formula

$$\eta = q_u / q_f \tag{20}$$

and to write down three equivalent equations for its definition

$$\eta = (\theta_a) - u(\bar{T} - T_0)/q_f, \tag{21}$$

$$\eta = f' \left[(\theta_a) - u(T_T - T_0)/q_f\right], \tag{22}$$

$$\eta = f_R \left[(\theta_a) - u(T_{int} - T_0)/q_f\right]. \tag{23}$$

Concept of technical characteristics of the helium receivers of various types give results of their tests shown on the Figure 3. The results of tests are shown in the shape of points on the straight lines that are described by the equation (21-23). With that, losses coefficient equals to tangent of incline angle in ή coordinate system. $\Delta T / q_f$, the point of intersection with the ordinate corresponds to rated absorption ability.

Table 3. Technical characteristics of helium receivers (averaged data) with average wind speed equal to 5 m / s, $T_n = 305 - 325$ K and $a = 0.95$.

Type of structure	Number of glass surfaces	ε_n	u, | Wt / m²·K;
Flat solar water heater	1	0.95	5.5-7.5
Also	1	0.05	2.8-3.5
-//-	2	0.95	3.2-4.8
-//-	2	0.05	1.8-2.2
Flat solar wind heater	2	0.95	3.6-3.8

As shown on Figure 3, ratio of efficiency of the helium receivers of various types varies from an interval to interval and the right selection of solar heater's type depends on its working conditions that are defined by parameter $\Delta T / q_f$.

For solar water heaters, the following values of losses coefficient u, W / m²·K are established

For single-glass 7.5

For double-glass......4.35

When using equations (20) and (22) for the calculations.

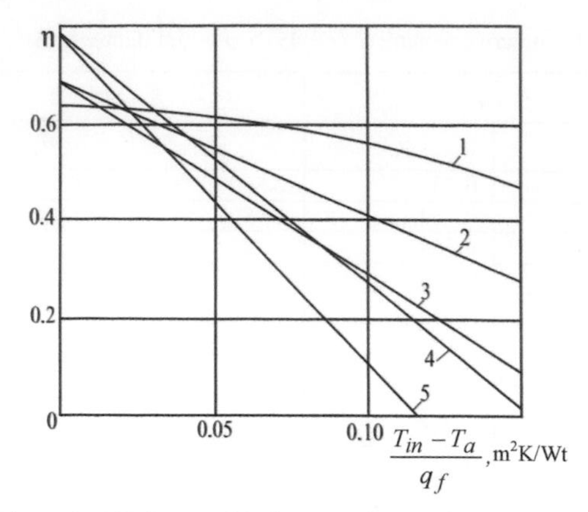

Figure 3. Efficiency of helium receivers of various types: 1 – vacuum connection sleeves; 2 – double-glass selective; 3 – double-glass non-selective; 4 – single-glass selective; 5 – single-glass non-selective.

f' is used for determination of the wind collectors' number, in order to find u, as a rule, formula 13 is used. f' values are found from the following equation

$$f' = \left\{1 + u / \left[\alpha_d + 1/\left(1/\alpha_d + 1/\alpha_{d.d}^r\right)\right]\right\}^{-1}. \quad (24)$$

Shown calculation methods of flat collectors are used for calculations of helium receivers of more complex structures as well. Thus, heat productivity equation for parabola-cylindrical concentrating collector can be presented as follows

$$q_u = f_R\left[q_{abs} - \frac{u}{n}\left(T_{in} - T_0\right)\right], \quad (25)$$

where n – concentration degree that is equal to the ratio between areas of reverberator's beams and the receiver's areas of beams. Coefficient of an external wall's losses.

$$u = \left(\frac{1}{\alpha_b} + \frac{1}{4\sigma\overline{\varepsilon}_d \overline{T}_d^3}\right). \quad (26)$$

Factor of the heat removal, f_r also defined according to (11), and the value of f' – the following formula

$$f' = \left(1 + \frac{ud_{ext}}{\alpha_{int} d_{ext}} + \frac{ud_{ext} \ln \xi_{ext} / d_{int}}{\lambda_T}\right)^{-1}. \quad (27)$$

where d_{ext} – diameter of the external channel, m; d_{int} – the same for the internal wall, m; ξ_{ext} – darkness degree on the external wall.

Quantity of heat that is absorbed by the concentrator is calculated in some other way than for flat receivers. It is connected, in the first place, with the fact that on the receiver only current of the direct solar radiation that falls on the reverberator is concentrated. Current of diffuse radiation that falls directly on the receiver is considerably less and is not taken for calculation for the systems with high level of concentration

$$q_{abs} = Sp^s(\theta_a)\rho\gamma F, \quad (28)$$

where (θ_a) – has the same meaning as for the flat collectors (in case of an unglassed receiver $\theta = 1$); ρ – reverberating ability of the reverberator-reflector, averaged in angles; γ – coefficient of capture that shows the quantity of reverberated current falling on the receiver – characteristics of orientation accuracy of the concentrator and receiver; in general form, its value is calculated for each specific case from the following formula

$$\gamma = \frac{\int_A^B I(\omega)d\omega}{\int_{-\infty}^{\infty} I(\omega)d\omega}. \quad (29)$$

When transferring, 10% of individual consumers to solar heat supply in regions of Ukraine, 0.17 Mt fuel / year can be saved. With effective use of solar radiation with capacity of 4 300 MJ / m² a year, the needs of consumers in heating can be assured for up to 25%, in hot water supply – up to 50%, in conditioning – up to 75%.

Replacement of the traditional boiler houses with the solar units allows to substitute and to save organic fuel, decrease volume of heat emissions and combustion products' emissions into the environment, to solve a line of social problems in regions where there is no centralized energy supply and to involve working potential of unprofitable enterprises.

But during exploitation of various systems and devices which use solar energy, many problems connected with environment protection occur. Use of the low-boiling liquors in solar energy systems and inevitable leaks of these liquors during long-time exploitation of the systems can cause significant pollution of drinking water. Special danger is created by liquors that contain nitrates, chromates that are highly toxic matters.

In order to decrease corrosion of water solar systems that are used for heating and cooling of the apartments and to prevent their freezing, salts with chromites, borites, nitrates, sulphates etc. are added that create serious danger to people's health. Thus, during solar heating and cooling systems' functioning that use rated matters it is necessary to periodically check if there are no leakages of working liquors.

Darkening of the vast territories of land due to the solar concentrators can lead to its degradation. It is also necessary to mark ecological consequences in the air heating station's location region when solar radiation concentrated by glass reverberators passes through it. This is the change of balance, moisture, wind direction, etc. It is necessary to take into account that building of solar energy stations requires vast territories: for example, station of 100 MW-power will occupy area equal to 5 km^2.

In conclusion, it is necessary to mention that unwanted consequences that are caused by use of solar energy systems have local character and they can be avoided if to follow precise safety rules.

REFERENCES

Daffi Jr., A. & Beckman, U.A. 1987. *Heat processes with use of solar energy*. Moscow: Mir: 420.

Rabinovich, M.D. 1982. *Engineering method of solar radiation calculation that falls and get absorbed by the collector*. Helium technique, 4: 7-9.

Volevakha, M.M. & Goysa, N.I. 1987. *Energy resources of Ukraine's climate*. Kyiv: Scientific thought: 132.

Geomechanical Processes During Underground Mining – Pivnyak, Bondarenko, Kovalevs'ka & Illiashov (eds)
© 2012 Taylor & Francis Group, London, ISBN 978-0-415-66174-4

The estimation of effect of the dust-proof respirators' protective efficiency upon the mining workers' dust load

Y. Cheberyachko
National Mining University, Dnipropetrovs'k, Ukraine

ABSTRACT: Here is the methodics of the coefficient's value of dust load which proposed accounting respirator's presence while calculating the volume of dust load. It depends on the air consumption respiration resistance, filters' quality and climate conditions.

1 INTRODUCTION

Lately is greatly increasing the number of professional diseases of workers respirational organs especially at coal industry (Medvedev 2005). Such situation has turned out as a result of the whole rank of some negative factors but the heaviest one is erroneous calculation of value of dust load. The errors appear first of all because of doubtful data of dust concentration in the air of working zone, volume of workers' lungs aeration, wrong estimation of coefficient of delivery unhealthy aerosol into respirator. Thus at the mines the protective devices with different protective efficiency are used whereas this fact isn't accounted while calculation of dust load (Instructions... 2003). Besides the dust-proof DIPRO's type (with the filtering half mask or isolated half mask with shifting filters) is ignored; class of protection (high – FFP3, middle – FFP2, low – FFP1), the volume of respirative resistance to air flow etc. So at working process the filtering cell its resistance to the half mask space because of impenetrability on the obturation stripe (Figure 1).

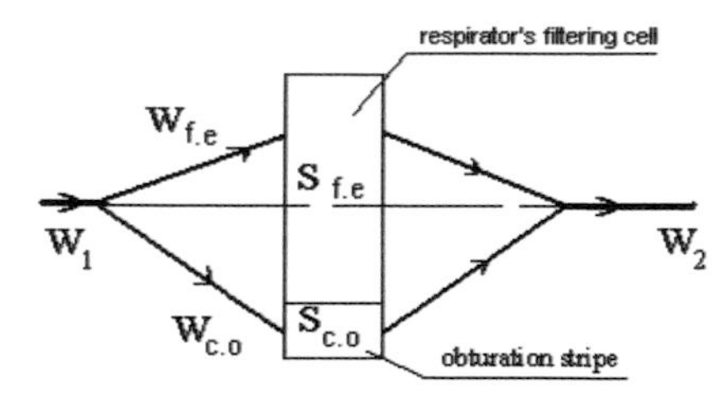

Figure 1. The simplified design (scheme) of aerosol flows in respirators: $W_{f.e.}$ – mass consumption of noxious matters getting on the respirator's filtering cell; $W_{c.o.}$ – mass consumption of unhealthy matters, through respirator's obruration stripe; $S_{f.e.}$ – square of filtering cell; $S_{c.o.}$. – square of obturation stripe.

2 FORMULATING THE PROBLEM

Hence not taking account the causes cited above, while setting dust's amount in the worker's lungs, we get understated, results whereas in reality they are more sizable, that considerably detracts prognosis of arising untime diseases.

For calculation of dust load accounting DIPOR presence class of protection of respirator should be defined and calculated: the coefficient of getting the harmful aerosol bath though filters unhealthy and on obturation stripe; to clear up the value of respiration resistance; to fix the time of exploits, working regime, dispersion content of dust, air temperature and air moisture degrees is quite necessary.

The class of respirator's protection is defined with a coefficient of test-aerosol penetration is defined with depends on lots of factors: the characteristics of filtering material, air supply, dispersional structure of dust, air resistance of DIPOR, climatic conditions at the workings place. This index is determined through experiments while certification tests of readymade device (product), not the filters isolated (as certain producers point out) over special test-aerosols under normal conditions (air temperature 20 °C, relative moisture not over 75%) (Golinko 2008).

$$K_n = \frac{W_2}{W_1} \cdot 100\%, \qquad (1)$$

where W_1 – mass consumption of unhealthy matters before DIPOR; W_2 – mass consumption of unhealthy matters after DIPOR.

According to DSTU 149:2003, DSTU 149:2002 for each protection class following penetration coefficients are fixed: through the filtering half mask: FFP1 – 22%; FFP2 – 8%; FFP3 – 2%; through the shifting filter cells of isolating half mask: P1 – low effectiveness – 20%; P2 – middle effectiveness –

6%; P3 –high effectiveness – 0.05%; through obturation stripe of the isolating half mask coefficient of under ejecting air mustn't exceed 2%.

That is when checking only protection class of DIPOR we realize that while calculation of dust load three coefficients should be introduced for checking DIPOR presence: for FPP1 – 0.22; FFP2 – 0.08; FFP3 – 0.02.

3 MATERIALS UNDER ANALYSIS

However, at the result of experimental researches was found that coefficient of penetrative action of respirator while the process of dusting is changing and depends on many factors. For example, for respirator RPA-TD the growth of coefficient value of penetration after dusting for a certain period of time was fixed (Figure 2).

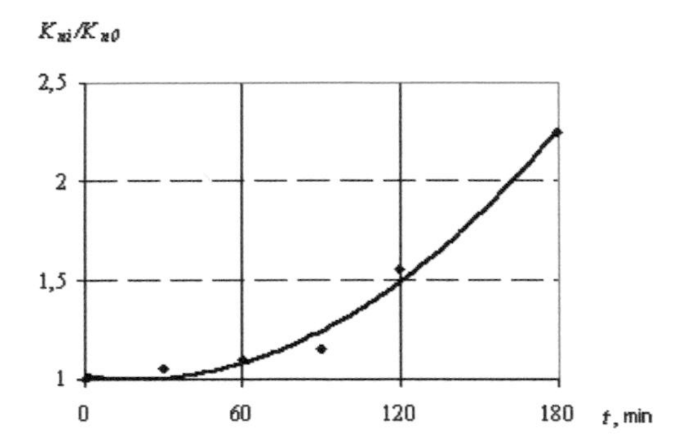

Figure 2. Dependence of relative coefficient of penetration on respirator RPA-TD on the time of dusting.

4 METHOD OF THE PROBLEM SOLVING

Hence, is a worker works in respirator over three hours a day so for the dust load's calculation coefficient accounting respirator's presence will be equal to 0.17.

At the result of research of dependence of protective efficiency respirators from the quality of their filters there was obtained an expression for defining of respirator's penetration coefficient proceeding from values as follows: air supply through DIPOR (Q_1), pressure drop (Δp) filter resistance ($R_{f.e.}$), the meaning of coefficient of filtering.

$$K_n^p = 10^{-\alpha(R_{f.e.}S)} + 1 - 0.8\frac{\rho d^2}{18\eta}\sqrt{\frac{4\pi v^3}{Q - \Delta p / R_{f.e.}}}, \quad (2)$$

where α – coefficient of filtering action m / Pa·s; s – square of filter, m^2; Q – air supply through the respirator; Δp – density of aerosol particles, kg / m^3; η – kinematical air toughness, m^2 / s; v – velocity of aerosol particles, m / s.

The growth of filtering cell's resistance increases the chance of polluted air penetration though respirator. This is conditioned by the redistribution of air flows and increasing ejection of polluted air through pass ability of obturation stripe (Figure 3).

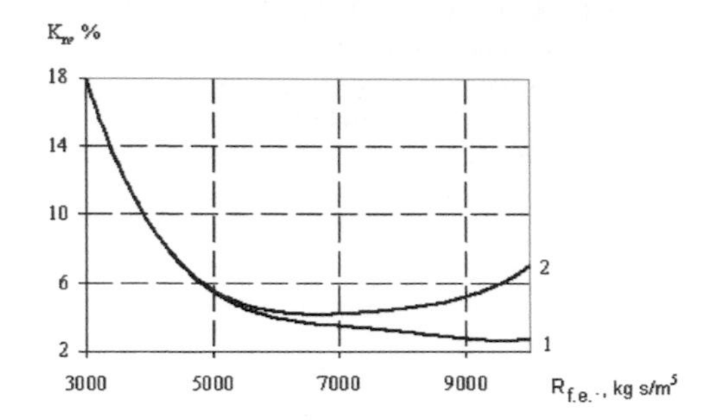

Figure 3. Dependence of penetration coefficient on air flow resistance in filter (1) and respirator with filters (2).

As shown in formula it's evidently clear that value of air supply depending on heaviness of works accomplished influences upon the filtration efficiency immediately raising of filtration speed causes the growth of pressure drop at respirator (Golinko 2008).

As after raising filtration speed the probability of catching aerosol particle by fibre decreases, that increases the coefficient of respirator's penetration (Figure 4).

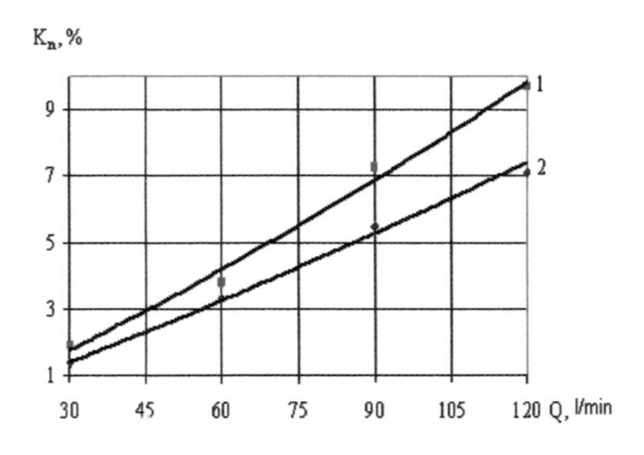

Figure 4. Dependence of penetration coefficient respirator's RPA-TD (1) as in the whole and its filters from eleflen (2) apart from the air supply.

5 RESULTS

Thus, for the respirator RPA-TD of eleflen while performing hard works characterized by the air supply of 120 l / min, the filter's respirative resistance of eleflen will equal about 4000 kg·s / m^3, then the coefficient counting respirator's presence will equal – 0.1.

Particular attention should be paid to the changing if parameters DIPOR under the influence of climatic conditions as in mining workings they differ greatly from normal ones, over which they conduct laboratory researches on stating the principal indices. Experimental researches showed that heightening air temperature and air moisture results in essential pressure drop on respirators provided for the type of filtering material (Figure 5, 6) (Golinko 2008).

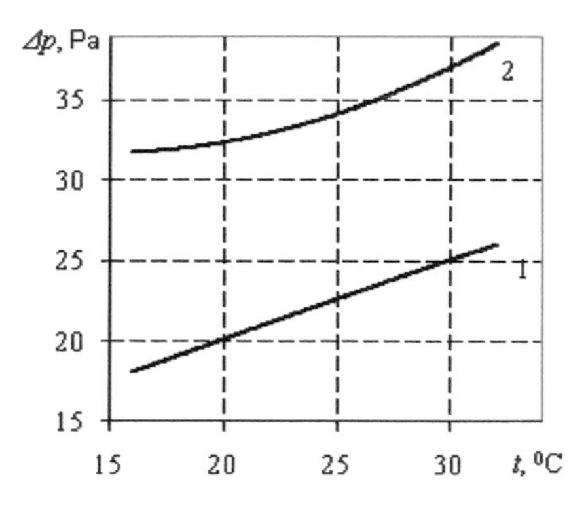

Figure 5 The curves of pressure drop's dependence of respirators on air temperature for the filters: 1 – eleflen; 2 – FPP15-1.5

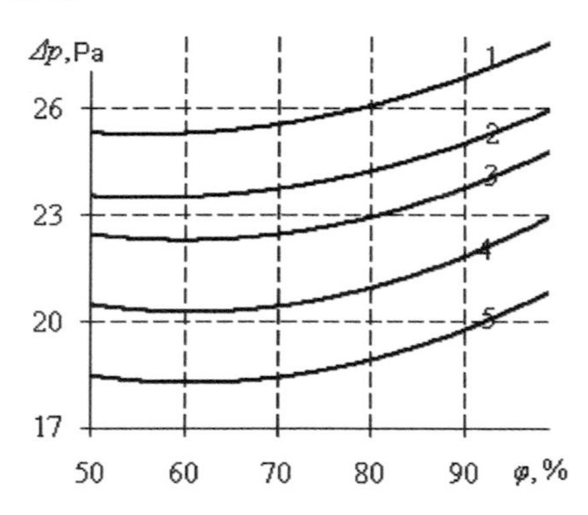

Figure 6. The curves of pressure drop's dependence of respirators on air moisture, filters of eleflen at different air temperature, °C: 1-30; 2-26; 3-24; 4-20; 5-16.

Additional growth of pressure drop on DIPRO can be estimated due to the formula

$$\Delta p = k_1 k_2 R_1 Q_1 , \quad (3)$$

where k_1 – additional pressure drop caused by raising temperature; k_2 – the additional pressure drop caused by increasing air moisture.

Temperature raising results in increasing penetration coefficient of respirator (Figure 7).

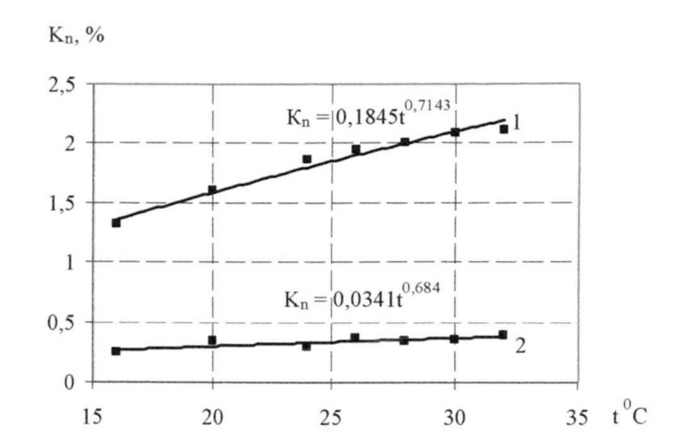

Figure 7. Curves of coefficient's penetration dependence on temperature: air filters of eleflen (1) and filters of FPP15-1.5 (2).

It can be explained by the fact that under the coercion of temperature aerosol particles start quicker moving worsening the action of the main mechanism for their catching that is electrostatic (Petryanov 1984).

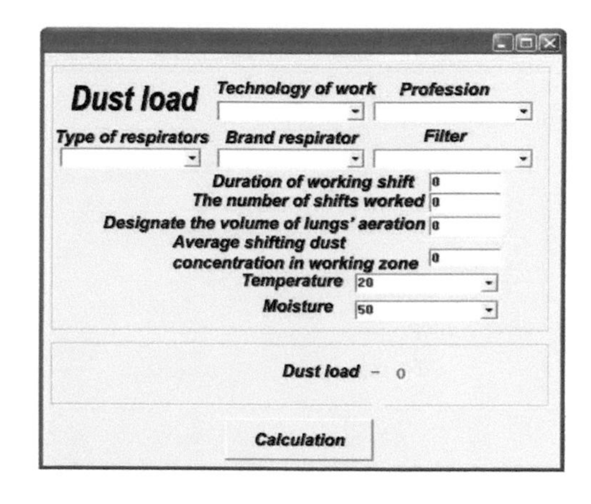

Figure.8. The general view of urnndow program for dust load calculation accounting a respirator's presence and kind of work accomplished.

Hence for the calculating of dust quantity which got into worker's lungs accounting DIPRO presence it's necessary to define a respiration's brand which is used and its characteristics (protection effectiveness of filters used in it, the original pressure drop both on respirator and apart, whether it's possible on filter); duration of working shift; to choose a profession (or designate the volume of lungs' aeration and average shifting dust concentration in working zone); the temperature and moisture of atmosphere within working zone where the respirator's exploitation is provided.

For the dust load estimation accounting the data cited above, the program was designed which lets calculate the dust load meaning past introducing imprint needful data (Figure 8) (Golinko 2009).

6 CONCLUSIONS

Thus the designed program can estimate quickly dust load accounting respirator's work, its characteristics and kinds of works that makes possible to prognoze more exactly the amount of accumulated dust and the probability of disease in respiration organs of workers who use the dust proof respirators.

REFERENCES

Medvedev, E.N., Kashuba O.I., Krivokhizha B.M. & Krutenko, S.A. 2005. *Dusty conditions and the incidence of pneumoconiosis in the coal Ukraine mines*. Makeevka-Donets Basin: MACNII: 205.

Instructions for measuring the concentration of dust in mines and keeping dust loads. 2003. Coll. inst. Rules for safety in coal mines. Zatv. Order of Fuel and Energy Ministry of Ukraine, 662: 151-161.

Golinko, V.I., Cheberyachko, S.I. & Cheberyachko, Y.I. 2008. *The use of respirators and on coal mining enterprises*: A Monograph. Dnipropetrovs'k: National Mining University: 99.

Petryanov, I.V. & Koshcheev, V.S. 1984. *Lepestok (Light respirators).* Moscow: Nauka: 216.

Golinko, V.I. & Cheberyachko, Y.I. 2009. *Develop a load calculation dust miners considering the type of filter respirator*. Bulletin of the Kryvy Rig Technical University. Scientific Papers, 23: 226-229.

Geomechanical Processes During Underground Mining – Pivnyak, Bondarenko, Kovalevs'ka & Illiashov (eds)
© 2012 Taylor & Francis Group, London, ISBN 978-0-415-66174-4

Derivatography as the method of water structure studying on solid mineral surface

V. Biletskyi, T. Shendrik & P. Sergeev
Donetsk National Technical University, Donetsk, Ukraine

ABSTRACT: It is proposed to use derivatographic method for investigation of water pellicle structure on mineral surface. It allows to identify some kinds of moisture, especially to distinguish the pellicle moisture from other types. In case of hydrophilic materials this method allows to select individual species of moisture pellicle (probably, strongly combined and adhesive).

1 INTRODUCTION

The actual problem of the dewatering and fine dispersing materials process intensification is the development of reliable methods for valuation of water structure on solid mineral surface (especially, coal surface).

Thus the question of correlation determination between the amount of superficial moisture which can be mechanically removed, stands especially roughly and that, which is not removable by mechanical methods – filtration, centrifugal etc. The last has presented by the structured (limited) pellicles.

2 UNDERGROUND

Earlier this problem was studied by line of authors (Bochkov 1996; Bejlin 1969; Ivanova 1974; Derivatogramms 1992; Kazanskij 1961 & Deriagin 1989), but qualitative and quantitative estimation of the pellicle moisture remains ambiguous.

Nowadays there is a line of quantitative estimation methods for different types of moisture.

First of all, there are methods for efficiently determination of strongly combined (hygroscopic) moisture, for example according to GOST 8719-70 and its equivalents (Bejlin 1969). As for other moisture forms, the lack of clear bounders between separate moisture types (according to connection of energetic indexes "solid-liquid") leads to difficulty of quantitative appreciation.

Secondly, there is an attempt of moisture types identification and their quantitative determination, which is made in (Bejlin 1969 & Kazanskij 1961) with help of isothermal drying thermo-grams of silica gel. According to M.I. Bejlin (Bejlin 1969) thermo-grams allow to differ various kinds of capillary moisture: "internal capillary moisture" (moisture of capillary escalation); capillary joint moisture and hygroscopic moisture too. By the method what we meant, the identification of strongly combined moisture is difficult in according to non-clear thermo-grams bends of mineral system drying.

Thirdly, nowadays the unique method of moisture determination, which not retire by mechanical methods, is the method of maximal molecular moisture capacity (MMMC) (Bochkov 1996). What is widely used on the theory and in the practice. This method is an integral estimation of row species moisture such as: capillary, internal and partly external, film moisture, which cannot be excised mechanically.

Their differential estimation by MMMC method is impossible. Having a concrete practical meaning, this method is of little use for theoretical analysis of different moisture types.

Thus, our short survey demonstrates the presence the individual methods for estimation of different moisture types and the limits of their usage. At the same time, so named "thin" and "thick" moisture pellicle (anot speaking about their possible elements) cannot be identified by known methods, although exactly they roughly differ by the energy of connection with a hard surface from others (Deriagin 1989).

The goal of this article – The search and investigation of new methodological approaches in identification of different moisture kinds on coal surface. Especially, the valuation of derivatography using possibility as the method of moisture morphology studies on coal surface.

3 EXPOSITION OF BASIC MATERIAL

On the practice, physical-chemical investigations of minerals and especially coal, the method of derivatography is widely used (Ivanova 1974 & Derivatogramms 1992). One of its function is determination of moisture and hydroxides content in minerals (Derivatogramms 1992).

The derivatography usage for estimation of water phase structure on coal surface was offered by us. For this, coal analytical samples of stamp "Ж (fat)" from mine "Samsonivska-Zahidna #1" were investigated with ash content 9.5%. Fresh crashed coal and oxidized on air by 20 °C was used, during one month. Experiments were duplicated for coincidence determination of results. For moisturizing coal had been used water with $pH = 7$.

The model of tests which was probed had such set of characteristics:

Sample 1. Non-oxidized coal, $W^a = 1.2\,\%$;

Sample 2. Non-oxidized moisturizing coal, water:coal = 1:1;

Sample 3. Oxidized coal, water: coal = 1:1.

The investigations were made by derivatigraph of company MOM-Q-1500.

On the Figure 1 initial derivatogramms of coal samples are shown. Endothermic spades of water evacuation of thermogravimethric curves (DTG), reduced to basic line, are shown on Figure 2. Temperature intervals and characteristic points are projected on basic line.

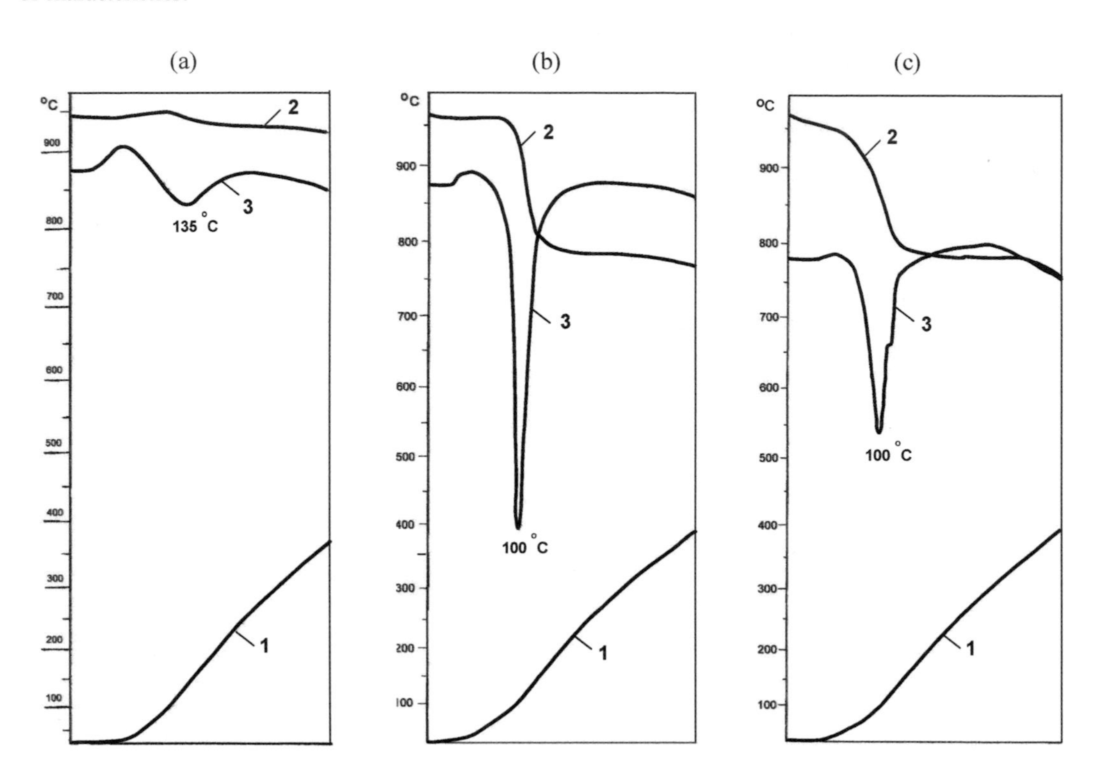

Figure 1. Fragments of coal derivatogramms in the moisture-removing region; a, b, c – corr. samples 1, 2, 3; 1 – temperature curve, 2 – thermo gravimetric (TG) curve, 3 – differential thermo gravimetric (DTG) curve.

How you can see, the maximums of water evacuation for samples sufficiently differ by intensity, configuration, especially extent of symmetry and area under curves. Let's stay on this characteristics more detailed.

Intensity of DTG peaks by its nature determinates the maximal speed of moisture-removing, and area under curve shows the mass of moisture what is removed.

The effect of non-symmetry is connected with the existence of different moisture kinds on coal surface. So as different kinds of moisture have various energy of connection with solid surface, that is why they are removed in different temperature intervals. Especially here, different temperatures take place, which are corresponding to maximal speed of moisture-removing for different kinds of moisture (points on basic line, what are corresponding to tops of peaks). Every kind of moisture IS characterized by its own peak of moisture-removing (in definite temperature intervals) and their overlay leads to non-symmetric peaks on DTG. That is why, to decompose non-symmetric peaks of DTG to Gauss curves allows to identify separate moisture forms. Let's analyze these peaks of moisture-removing, using DTG – curves.

Peaks of moisture-removing for air dry coal (sample 1, Figure 2a) has small intensity, "soft" fronts of DTG, it is practically symmetric. The last (symmetric), obviously, attested to homogeneous water structure, which in this case represents only as pellicle moisture. Small intensity (amplitude) and some plane of peak leads to low rate of moisture-removing. This fact is exactly characterized for strongly combined moisture. Additionally, maximal rate of moisture-removing is observed at higher temperature (135 °C), that also testifies the high-power connection of contacting phases "water-solid".

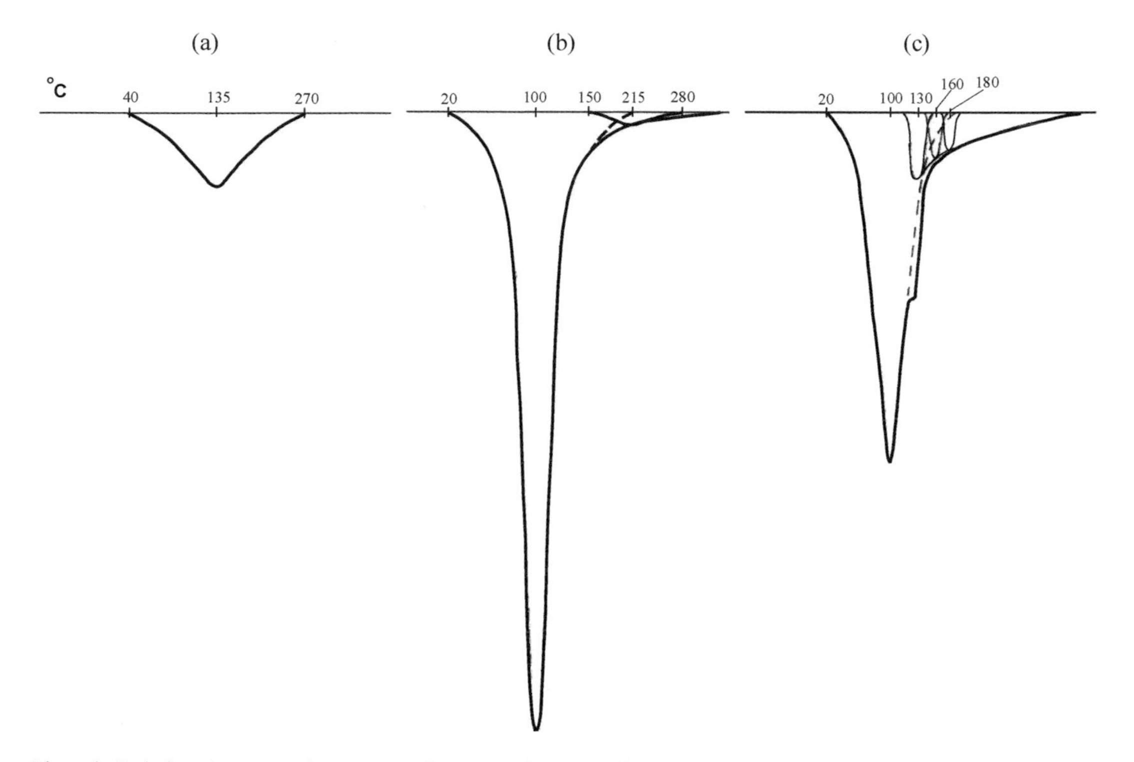

Figure 2. Endothermic curves of water-removing corr. a, b, c – samples 1, 2, 3.

The DTG-peak of non-oxidized moisturizing coal (sample 2, Figure 2b) significantly differ from previous – it is more intensive, non-symmetric in the field of high temperature.

Its decomposition on Gauss-components gives two peaks – main with maximum at 100 °C which obviously corresponds to gravitation and capillary moisture, and small additional peak in the region > 150 °C what is corresponding to moisture pellicle.

The moisture-removing DTG-peak of oxidized moisturizing coal (sample 3, Figure 2b) has more complexes configuration in the region of high temperature.

Decomposition of peak on Gauss-components gives extremes at 130; 160; 180 °C. Obviously, the oxidation process of coal has led to increasing of moisture pellicle by the way of strongly-combined and adhesive moisture. Besides that, in this case moisture pellicle is presented by several layers of strongly-combined moisture (α – i β-films (Deriagin 1989)) and adhesive "thick" pellicle, that is stretched on distance to 1000 nm from solid surface (Deriagin 1989).

4 CONCLUSIONS

1. Derivatographic method allows to identify some kinds of moisture, especially to separate pellicle from other types of moisture.

2. In the case of hydrophilic (hydrophilized) materials, which are characterized by more structural moisture layers on its surface, this method allows to excrete individual types of moisture pellicle (perhaps, adhesive and strongly-combined moisture).

Further, it is expedient to use derivatographic method for investigation influence of various reagents on the system "solid-water". It is actual for physical and chemical methods for dewatering intensification.

At first, it will take the possibility to detect the change relation of different kinds of moisture at physical and chemical influence of reagents.

From other side, fixation of the reagent influence on the system "solid-water" by the help of derivatographic method and interpretation of its results will assist in developing of the DTA-method.

REFERENCES

Bochkov, Y.N. & Zarubin, L.S. 1996. *Estimation of effect of mechanic methods foe dewatering small coal.* Theory and practices of small coal. Moscow: Nauka: 5-20.

Bejlin, M.I. 1969. *Theoretical basis for coal dewatering process.* Moscow. Nedra: 240.

Ivanova, V.P., Kasatov, B.K., Krasavina, T.N. & Rozinova E.L. 1974. *Thermal analysis of minerals and mountain rocks.* Leningrad: Nedra: 400.

Derivatogramms, infrared and Mossbauer spectrum of standard samples of phase state. 1992. St.-Petersburg. Committee of RF on geology and use bowels: 159.

Kazanskij, M.F. 1961. *Temperature influence on station of adsorbed capillary moisture in macro pores of dispersed solids.* Engineering Physical Journal, 3. V.4: 38-43.

Deriagin, B.V., Ovcharenko, F.D. & Churaev, N.V. 1989. *Water in dispersed systems.* Moscow: Chimija: 286.

Geomechanical Processes During Underground Mining – Pivnyak, Bondarenko, Kovalevs'ka & Illiashov (eds)
© 2012 Taylor & Francis Group, London, ISBN 978-0-415-66174-4

Organization of checking system over the state of cross-section of layered working with the use of informative systems

V. Fomychov , V. Pochepov & L. Fomychova
National Mining University, Dnipropetrovs'k, Ukraine

ABSTRACT: Information that is describing the changes in the contour parameters of the working are the basis in the method of making decisions in information system of a mine control at the estimation of the state of layered workings. The decreasing of measuring errors and increasing of the quality analysis of the received data is achieved due to conducting of measurements with the use of automation means and special algorithm of the working cross-section contour design. At this proposed system of state control over the state of cross-section has a low cost, simplicity and high reliability in industrial environments.

1 CURRENCY AND ORGANIZATION OF THE TASK

Ensuring in the possibility of the reuse of the layered workings allows cutting the prime costs of the coal mining. In the conditions of most mines in Ukraine such task requires the difficult approach to the workings protection. Complication of this task decision consists of the necessity of the search in optimum combination of support elements of the layered working. To expose such combination is only possible on the basis in the analysis of the working moving contour, during its exploitation.

The basic problem of providing such high quality analysis is receive of exact measuring information in the chosen points of real working contour. The second most difficult problem is correct determination of the working contour deformation in concerning to its initial state. Both these problems are still unsolved.

In most methods used in natural conditions, the change of the layered working cross-section begins to be fixed from a few centimetres. Mainly such measurements are fulfilled in arch and on soil of working, thus technology of measurements implementation is made in concerning to the certain chosen point which is immobile. For Ukrainian mines conditions, to define such point in the real terms of exploitation in the specific working is not a trivial task, and therefore can lead to significant errors in the results.

Modern development of the management facilities production, allows creating the difficult informative systems, providing a high performance and safety of labour on various industrial objects. Application of such informative systems on mines must be accompanied by the development and introduction in their composition, means of controls and analysis of the state of layered workings, since, such as mine an effective management operation is impossible without the account of affecting geotechnical factors on technological process.

Based on the above-mentioned description, it is necessary to develop the accessible and simultaneously exact method of determination of the underground workings state, at all time of their exploitation. This method must conform to the requirements of the providing possibility of its integration in modern information systems of a production control. In the same time, the using equipment in this method must not be expensive and must answer requirements on accident prevention for coal mining enterprises.

The developed approach in implementation of measurements must provide the different level of automation, which makes it possible to use it in the wide range of the real conditions. And majority of the applied equipment must be made on an enterprise or re-used many times.

2 GENERAL STRUCTURE OF THE SYSTEM

Structurally the system consists of two isolated components: set of the measurements stations and programmatic complex of data analysis. Exchange between these components takes place information as photo received by a digital photo camera with high resolution of matrix. Photographing of the measuring station is made from beforehand certain position with the use of the fixed stand, which provides small angular error, within the limits of 0.1°.

Connection of image to specific measuring station is carried out by the fixation of number of the measuring station by the light-reflecting paint on working arch elements. After checking all measuring stations in manual mode or with use of the automated control stations, the information is transmitted as a set of files in the image recognition contour for the receive of coordinates of measurements points.

Based on the mine working photo that has been received for specific measuring station, massifs with coordinates of point scatterers positions installed on the ends of measuring bolts are determined. After it, through the special algorithm the current type of working contour is built, which is saved in the base of these measurements. On the basis of the received information, the system depending on the put task, executes prognostication of working section changes and analysis of using support efficiency.

The location of the measurements stations, number and place of measurements points on each of the stations, is determined on the basis of preliminary analysis of the tensely-deformed state of the system "roof supports of working-mining massif" (Bondarenko 2006 & Bondarenko 2010). The basic factors of the determining descriptions of the measurements stations are the expected sizes and directions deformation of working contour during conducting of preliminary calculations executable on the stage of the working planning.

The measuring station composition equipment includes anchor-base which is set in accordance with the chosen assembling scheme. The example of such scheme is resulted on a picture 1. For the working with a circular vault there are 22 possible certain points of anchors setting for reflectors (Bondarenko 2007).

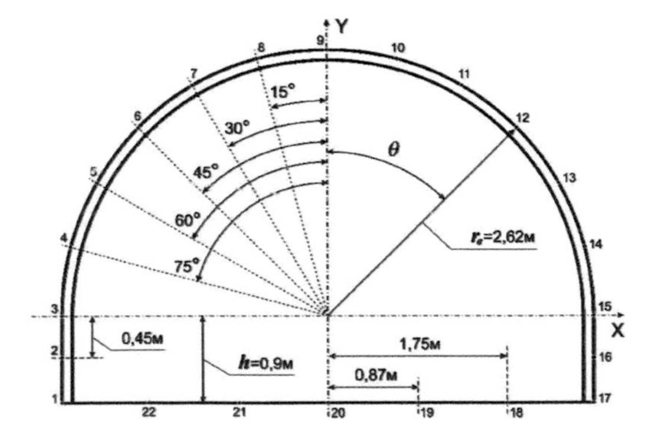

Figure 1. Example of assembling chart of measuring points for making with a circular vault.

Reflectors can be assembled stationary on anchors-bases or for the case of data hand removal, carried and assembled by a mine survey before the implementation of photographing of the measuring station. In addition, the equipment composition portable by a mine survey included a reference line, digital photo camera with high resolution, laser range-finder and stand with the installation system of «horizon». For the variant of automatic control after the measurements stations all of this equipment is replaced by the stationary automatic module assembled in the district of the measuring station.

The eventual point of data collection from all of the measurements stations is a calculable machine, set within the limits of surface production complex of a mine. In the task of this computing machinery is included conducting of final analysis of the got information. Depending on the level of realization of this method this equipment can work in the interactive mode or real-time mode.

The specific element of the measuring station equipment is a complex of anchor-base and reflector. A base is set as an ordinary steel-polymer anchor on the external end of which takes place knot of bayoneting lock. Exactly this knot is used for the reflector installation. A reflector is a steel bar on one end of which a bayoneting lock takes place, and on other – the light-reflecting element of small diameter witch is used for the receipt of light spot at photographing of the measuring station.

For the implementation of the received data measurements analysis the specially developed software is needed. The general structure of such programmatic complex consists of the followings elements: the first module is recognition on the image of technological elements of the measuring station (point reflectors, reference line, mark of the measuring station); the second module is a construction of current working contour from data of the measuring station elements recognition; the third module is an interpolation of change of working contour on the accessible data base; the fourth module is an analysis of applied support efficiency and selection of its parameters for the improvement of the working stability indexes.

3 INSTALLATION AND EXPLOITATION OF THE MEASURING STATION

Assembly of the measuring station elements is reduced to setting of anchors-bases in points of the working contour. These points are chosen in accordance with specifications what were advanced to the analysis of the working contour changes on the basis of the offered method in (Bondarenko 2006 &

Bondarenko 2007). Then on the free ends of these anchors, with help of the bayoneting support, point reflectors are supported.

Right after setting of anchors bases, is carried out photographing of the measuring station. Thus to one of reflectors is supported a ruler, which in the subsequent analysis of pictures will be used for matching the horizon and linear sizes, those were got at measuring.

During the first photographing of the set measuring station a choice and attachment in place of camera position is carried out in relation to the plane anchors location. The photo which was received as a result is accepted as standard and on his basis primary attachment of initial working section and point reflectors is carried out.

The subsequent measurements are executed through the free or clearly specified interval of time. Basic features during conducting of measurements are: necessity in plane placing, the location of standard ruler anchors, and determination of camera exact position in relation to the plane of anchors location (this task decides through a laser range-finder). Then it is possible to execute photographing of object from hands or special stand. The second variant is preferable as increase the quality of the received image and allows controlling the point of survey in space. At photographing the off-site illuminating or flash of camera can be used. In last case measuring quality a few higher, because light spots from reflectors will have more clear border, that at subsequent treatment of image will allow more precisely to define their positions in space.

After completion of the measuring station exploitation, in case of stationary installation of point reflectors, their dismantling is performed in order to continue their usage at other measuring station. Anchor-base is the unique element of measuring station that cannot be repeatedly used, as a type of their installation is not foreseen by possibility of anchors extraction from a mining massif.

4 TREATMENT AND ANALYSIS OF THE RECEIVED DATA

The most difficult implementation of the offered method phase is treatment and analysis of the received images for determination of the working cross-section state.

On the first stage with help of technologies of the image recognition, analysis is performed and transformation of image in the array of data as well, which contains information of the measuring station recognized mark, real correlation of standard ruler length and number of the picture points, on the image of this ruler, and also coordinates of reflectors light spots in concern to the «dead» centre of image. Such point is chosen by an operator arbitrarily at preparation of the system to exploitation and is an object necessarily present on all of pictures of the measurements stations of one mine. By such object, for example, there can be a standard ruler, but in this case it is necessary to dispose it in the same position in concern to the reflectors of the concrete measuring station.

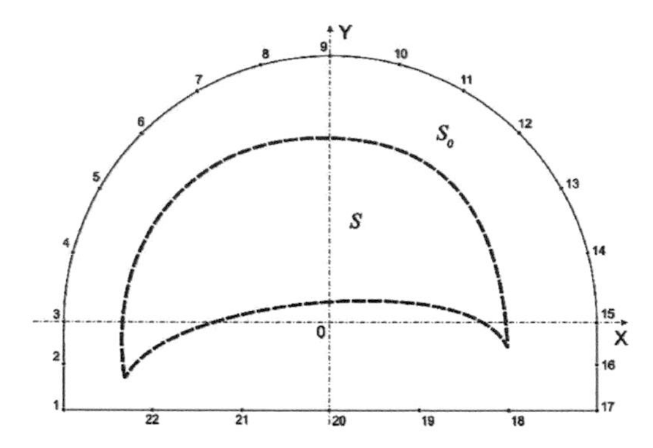

Figure 2. General graphic presentation of conducting technology of measurements data analysis for the working with a circular vault.

On the second stage, binding of the received reflectors coordinates is carried out with the previous results of measurements. Every light spot is identified as a reflector with a specific number and its new coordinates are written in the proper vector of information. At the same time correction of an entire network of received values respectively to the initial coordinate massif is conducted that had been plotted during analysis of template image of measuring station.

Thus, the received data, at first, are written into the initial contour of S_0 (see of Figure 2), and secondly, through the special algorithm, connected in the resulting contour of S .

Now there is possibility of retrograde analysis of the cross-section of working contour development, taking into account influence on this process of various mine-technical factors those were operating in time and space. That is, on the third stage of data processing is becoming possible to extrapolate the conducting of the concrete working within the framework of measures applied on its guard. In the systems control, plotted on the basis of strategies of the «real time», realization of this stage data processing will allow to warn accidents that were caused by the swift rock collapse in the working cavity (Fomychov 2008).

The fourth stage of the measurements data processing is examined and can be used in the informative systems for the search of optimum indexes of different type's combination of support at maintenance of working in the conditions of the certain mine field. That, the received information for the nearby layered working is used as primary indexes during the selection of optimum support descriptions. For this purpose in management information by a mine include the electronic reference book of technologies of the working guard. This reference book contains various variants directed to mine working maintenance, contours of the cross-section of which develop by different scenarios.

5 ADVANTAGES AND DISADVANTAGES OF THE METHOD

It is necessary to take to basic dignities of the offered method: at first, simplicity and, as a result, high exactness of the moving determination in the chosen points of working contour (at the removal of measurements exactness is actually determined correlation of maximal geometrical parameters of working cross-section and linear resolution of the digital camera matrix); secondly, possibility of prognostication of the working contour change and search, with according it, optimum parameters of support; thirdly, low prime cost of the system, possibility of the repeated use practically of all its component.

Problems of realization: at first, development of algorithm construction of the current working section on the results of measurements; secondly, creation on the base of the fixed changes of the working cross-section, the interpolation system of the working contour implementation in the future.

6 CONCLUSIONS

The offered method of cross-section control of the layer working allows with low expenses to provide high exactness of the measuring of the working contour moving and with the use of the specialized mathematical tools conducts the analysis of the got data. On the basis of such analysis the construction of technological contour of the management information by a mine is possible of providing the choice of optimum scheme of working guard.

REFERENCES

Bondarenko, V.I., Kovalevs'ka, I.A., Symanovych, G.A. & Fomychov V.V. 2006. *Computer simulation of the stress-deformed state of small-layered rock mass around the stratal of working*. Book I. Prelimit stage of deformation of the "rock-roof supports". Dnipropetrovs'k: System technologies: 172.

Bondarenko, V.I., Kovalevs'ka I.A., Symanovych G.A. & Fomychov V.V. 2010. *The methodology and results of computer simulation of the stress-strain state of the "layered array-roof supports" of advance working / Mining Magazine*. Moscow: The Publishing house "Ore and Metals". Special edition: 62-66.

Bondarenko, V.I., Kovalevs'ka I.A., Symanovych G.A. & Fomychov V.V. 2007. *System methodology of forecast stability stratal of working in the layered coal-bearing thickness*. Materials of International scientific and practical conference "School of underground development". Dnipropetrovs'k: Art News: 158-165.

Fomychov V.V., Pochepov V.M. & Fomychova L.Ya. 2008. *Information model of mine – the features of engineering and the basic criteria of realization*. Scientific Journal NMU, 11: 37-40.

Geomechanical Processes During Underground Mining – Pivnyak, Bondarenko, Kovalevs'ka & Illiashov (eds)
© 2012 Taylor & Francis Group, London, ISBN 978-0-415-66174-4

Evaluating of metal-resin anchor parameter influence on the support capacity

G. Larionov & N. Larionov
Institute of Geotechnical Mechanics of the National Academy of Sciences of Ukraine, Dnipropetrovs'k, Ukraine

ABSTRACT: The paper is devoted to the important scientific problem solution: anchor spacing obtaining with taking into account the transfer mechanism of "anchor bar-resin-rock" system. The system parameters influence evaluating on strain – stress state is obtained as a result of two task solving. The modified general N.E. Zhukovsky task solving is used for shear borehole surface stress distribution obtaining. Then it is used as boundary condition for space strain – stress state task in anchor vicinity. The way allows gaining anchor space dependence with taking into account the mine opening depth, pretension value, length and anchor diameter and shear stress intensity.

1 INTRODUCTION

The increasing of bolt support problem efficiency is always actually. The problem of interest is the knowledge of parameters which are needed to change for it. An anchor capacity influence parameter evaluating is needed to answer the question. To have possibility to do reinforcement capacity influence estimation of design parameters in resin-metal anchor it is necessary the parameters were included into designed schemes. The achievement of optimum reinforcement capacity is influenced by the following parameters (Canbulat 2008): pretensions rebar values; depth of support, length and diameters of an anchor and boreholes, bolt and borehole profiles; the performance of the resin, bar and rock, stress rock intensity and anchors space. The design schemes variety are used in mechanical engineering. Thus an idea to use one of them for resin – metal anchor parameters estimation is appeared.

2 FORMULATING THE PROBLEM

As known, to the considerable efforts transfer in weak durability environment and to reduce the contact surface "rebar – the weak durability material" efforts the intermediate shell with periodical ribs on an external and internal surfaces are used. Such designed schemes have received wide scope of application in various branches of manufacture and in a life. Resin-metal anchors which are wide used for opening supporting by the nature are very similar to such construction. So, considerable anchor loadings are distributed on low level durability surrounding rocks (in comparison with rebar material durability) by means of fixing mix shell (FMS). It is evident that loading will be distributed on the bigger area and fitting conditions for loading transfer to surrounding will be created. But to use the well-known design scheme and to do the design parameter estimation FMS must have the threading on external surface. As was found that as usual two-prong bits are used to bore hole for anchor in the rock. During the boring are present simultaneously two almost constant efforts: torque and thrust ones. Thus almost-periodic ribs are appeared on the borehole surface. After bar arrangement and FMS hardening corresponding surface was formed. Thus, the surface almost-periodic ribs may be taken as a thread and the system "bar-FMS-rock" can be quite reproduced the offered M.E. Zhukovsky design scheme.

3 SOLVING THE PROBLEM

The contact surface efforts for this scheme may be obtained with the solving of generalized M.E. Zhukovsky task (Figure 1).

The right segment means a rebar body, middle – a FMS one and the left – massive of rocks (Figure 1). Contact rib segments mean bar and FMS thread accordingly. The bar and FMS thread efforts denote by $p_0, p_1, ..., p_n$, and FMS and rock thread by – $t_0, t_1, ..., t_n$. Further as an anchor field we will understand a part of a body which is founded between two ribs.

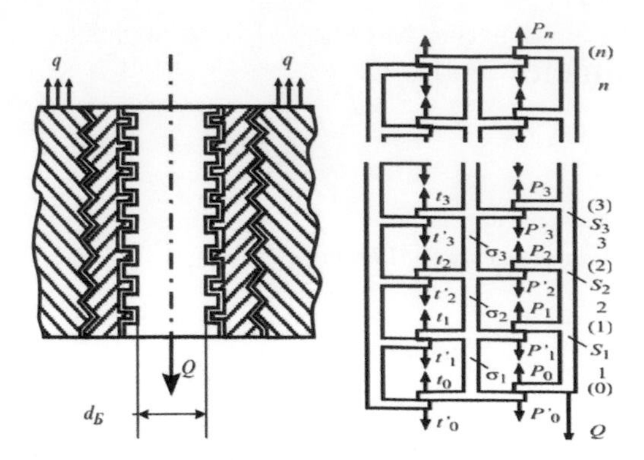

Figure 1. The loading and design "bar-FMS-rock" system scheme.

After updating of the generalized N.E. Zhukovsky task (Larionov 2011) and using the simple, but complicated transformations, we will obtain expression for efforts t_k and p_k :

$$t_k = \frac{\lambda_1\lambda_3 Q}{8sh\frac{\beta_1+\beta_2}{2}sh\frac{\beta_1-\beta_2}{2}}\left[\frac{e^{-(k+1)\beta_2}}{sh\frac{\beta_2}{2}} - \frac{e^{-\left(k+\frac{1}{2}\right)\beta_1}}{sh\frac{\beta_1}{2}}\right];$$

$$p_k = \frac{\lambda_1 Q}{2sh\frac{\beta_1+\beta_2}{2}sh\frac{\beta_1-\beta_2}{2}} \times \left[sh\frac{\beta_1}{2}e^{-(k+1)\beta_1} - sh\frac{\beta_2}{2}e^{-\left(k+\frac{1}{2}\right)\beta_2}\right] + t_k. \quad (1)$$

where

$$\lambda_1+\lambda_2+\lambda_3+\lambda_1\lambda_3 = 4ch\beta_1 ch\beta_2 - 2(ch\beta_1 + ch\beta_2);$$

$$ch\beta_1 = 1+\frac{\lambda_1+\lambda_2+\lambda_3}{4} + \sqrt{\left(1+\frac{\lambda_1+\lambda_2+\lambda_3}{4}\right)^2 - \frac{\lambda_1\lambda_2}{4}};$$

$$ch\beta_2 = 1+\frac{\lambda_1+\lambda_2+\lambda_3}{4} - \sqrt{\left(1+\frac{\lambda_1+\lambda_2+\lambda_3}{4}\right)^2 - \frac{\lambda_1\lambda_2}{4}}.$$

To define the stress-strain state in a anchor vicinity the known approach (Larionov 2011) to the solving of similar problems is used (Figure 2).

The facts of anchor arrangement closely to stop face give a sure in elasticity methods using and the stress-strain state defining in anchor vicinity. It gives the grounds to believe that rocks at a face had not time to receive plastic deformations and are in an elastic condition. Suppose that environment is homogeneous and isotropic, and anchor fastening is carried out in borehole depth on distance which considerably exceeds its diameter.

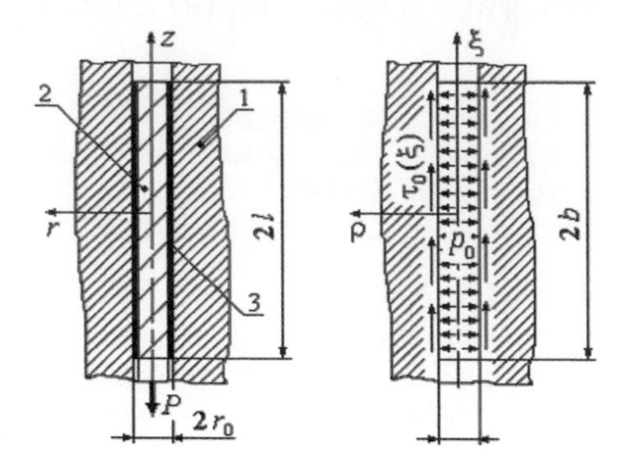

Figure 2. To statement of elasticity problem: 1 – rock, 2 – an anchor, 3 – a fixing mix.

Normal surface stresses p_0 is supposed to consider as constant and equal to certain depth mountain pressure. Shear stress as show last researches (Dey, Aziz & Indraratna 2001) essentially changes along an anchor length. Pressure distribution in anchor vicinity is described with the basic equations of the elasticity theory, which in axe symmetric case look like (Larionov 2011)

$$\mu\left(\nabla^2 U - \frac{U}{r^2}\right) + (\lambda+\mu)\frac{\partial}{\partial r}\left[\frac{1}{r}\frac{\partial}{\partial r}(rU) + \frac{\partial W}{\partial z}\right] = 0;$$
$$\mu\nabla^2 W + (\lambda+\mu)\frac{\partial}{\partial z}\left[\frac{1}{r}\frac{\partial}{\partial r}(rU) + \frac{\partial W}{\partial z}\right] = 0, \quad (2)$$

where U, W – radial and axial components of deflection vector; μ, λ – Lama's coefficient; r, z – cylindrical coordinates.

$\nabla^2 = \frac{\partial^2}{\partial r^2} + \frac{1}{r}\frac{\partial}{\partial r} + \frac{1}{r^2}\frac{\partial^2}{\partial \varphi^2} + \frac{\partial^2}{\partial z^2}$ Laplas operator.

Having made replacement of variables $p = r/r_0$, $\varsigma = z/r_0$, we obtain the equations (2) in Papkovicha-Nejber form:

$$U = 4(1-\nu)B_r - \frac{\partial}{\partial \rho}\left(\rho B_r + \varsigma B_z + \frac{1}{r_0}B_0\right),$$

$$W = 4(1-\nu)B_z - \frac{\partial}{\partial \varsigma}\left(\rho B_r + \varsigma B_z + \frac{1}{r_0}B_0\right),$$

where B_0 – harmonious scalar; B_r, B_z – harmonious vector components which satisfy to the Laplas equations:

Besides, harmonious functions $B_0(\rho,\varsigma)$, $B_r(\rho,\varsigma)$, $B_z(\rho,\varsigma)$ should be chosen so that to satisfy to boundary conditions:

$$\sigma_r\big|_{\rho=1} = \begin{cases} -p_0, & |\varsigma| \le b; \\ 0, & |\varsigma| > b; \end{cases}$$
$$\tau_{rz}\big|_{\rho=1} = \begin{cases} -\tau_0(\varsigma), & |\varsigma| \le b; \\ 0, & |\varsigma| > b, \end{cases} \qquad (3)$$

where σ_r, τ_{rz} – radial and shear stresses tensor components; $b = l / r_0$ – relative length of FMS section; $p_0 = \gamma H$ – hydrostatic rock pressure in H depth. Anchor borehole surface shear stresses will be defined as square-law function of axial coordinate (Larionov 2011):

$$\tau\big|_{\rho=1} = \tau(\varsigma) = a_0 + a_1\varsigma + a_2\varsigma^2;$$

$$\tau\big|_{l=0} = \tau_1; \quad \tau\big|_{l=L} = \tau_2,$$

where τ_1, τ_2 – shear stresses at the beginning and the end of a FMS anchor section ($\tau_1 > \tau_2$). Then general solution decision will look like:

$$\sigma_r = -\frac{2}{\pi}\int_0^\infty \Bigg[\left((3-2v)K_0(\beta\rho) + \left(4(1-v)\frac{1}{\rho} + \beta\rho\right)K_1(\beta\rho)\right) \times$$
$$\times\left((C_1 p_0 \sin(\beta b) - C_2 a_2(\beta))\cos(\beta\varsigma) + C_2 a_1(\beta)\sin(\beta\varsigma) + \left(\beta K_0(\beta\rho) + \frac{1}{\rho}K_1(\beta\rho)\right)\right) \times$$
$$\times\left((D_1 p_0 \sin(\beta b) - D_2 a_2(\beta))\cos(\beta\varsigma) + D_2 a_1(\beta)\sin(\beta\varsigma)\right)\Bigg]d\beta;$$

$$\sigma_\varphi = \frac{2}{\pi}\int_0^\infty \Bigg[\left((1-2v)\frac{\rho}{\beta}K_0(\beta\rho) + 4(1-v)\frac{1}{\beta\rho}K_1(\beta\rho)\right) \times$$
$$\times\left((C_1 p_0 \sin(\beta b) - C_2 a_2(\beta))\cos(\beta\varsigma) + C_2 a_1(\beta)\sin(\beta\varsigma)\right) +$$
$$+ \frac{1}{\rho}K_1(\beta\rho)\left((D_1 p_0 \sin(\beta b) - D_2 a_2(\beta))\cos(\beta\varsigma) + D_2 a_1(\beta)\sin(\beta\varsigma)\right)\Bigg]d\beta;$$

$$\sigma_z = \frac{2}{\pi}\int_0^\infty \Big[(\beta\rho K_1(\beta\rho) - 2vK_0(\beta\rho))\left((C_1 p_0 \sin(\beta b) - C_2 a_2(\beta))\cos(\beta\varsigma) + C_2 a_1(\beta)\sin(\beta\varsigma)\right) +$$
$$+ \beta K_0(\beta\rho)\left((D_1 p_0 \sin(\beta b) - D_2 a_2(\beta))\cos(\beta\varsigma) + D_2 a_1(\beta)\sin(\beta\varsigma)\right)\Big]d\beta;$$

$$\tau_{rz} = \frac{2}{\pi}\int_0^\infty \Big[(\beta\rho K_0(\beta\rho) + 2(1-v)K_1(\beta\rho))\left((C_1 p_0 \sin(\beta b) - C_2 a_2(\beta))\cos(\beta\varsigma) + C_2 a_1(\beta)\sin(\beta\varsigma)\right) +$$
$$+ \beta K_1(\beta\rho)\left(-(D_1 p_0 \sin(\beta b) - D_2 a_2(\beta))\sin(\beta\varsigma) + D_2 a_1(\beta)\cos(\beta\varsigma)\right)\Big]d\beta.$$

Thus the expressions for deflections in radial direction and along a symmetry axis take a form:

$$U = \frac{r_0}{\pi\mu_0}\int_0^\infty \Bigg[\left(4(1-v)\frac{\rho}{\beta}K_1(\beta\rho)/\beta + \rho K_0(\beta\rho)\right) \times$$
$$\left((C_1 p_0 \sin(\beta b) - C_2 a_2(\beta))\cos(\beta\varsigma) + C_2 a_1(\beta)\sin(\beta\varsigma)\right) +$$
$$+ K_1(\beta\rho)\left((D_1 p_0 \sin(\beta b) - D_2 a_2(\beta))\cos(\beta\varsigma) + D_2 a_1(\beta)\sin(\beta\varsigma)\right)\Bigg]d\beta;$$

$$W = \frac{r_0}{\pi\mu_0}\int_0^{\infty}[(\rho K_1(\beta\rho))((C_1 p_0 \sin(\beta b) - C_2 a_2(\beta))\sin(\beta\varsigma) - C_2 a_1(\beta))\cos(\beta\varsigma) +$$

$$+ K_0(\beta\rho)((D_1 p_0 \sin(\beta b) - D_2 a_2(\beta))\sin(\beta\varsigma) - D_2 a_1(\beta)\cos(\beta\varsigma))]d\beta.$$

For stress-strain state obtaining in anchor vicinity with taking into account anchor design parameters the consecutive decision of two tasks was used. The first task solve has allowed to define efforts and shear stresses on FMS contact surfaces (the generalized N.E. Zhukovsky modified task solve is used). Using of shear stress surface distribution as a boundary condition for the second task, mentioned above, has allowed defining stress state tensor components in anchor vicinity (Larionov 2011). Comparison of the received results with finite element method (FEM) statement ones (Figure 3) demonstrates satisfactory accuracy for engineering calculations.

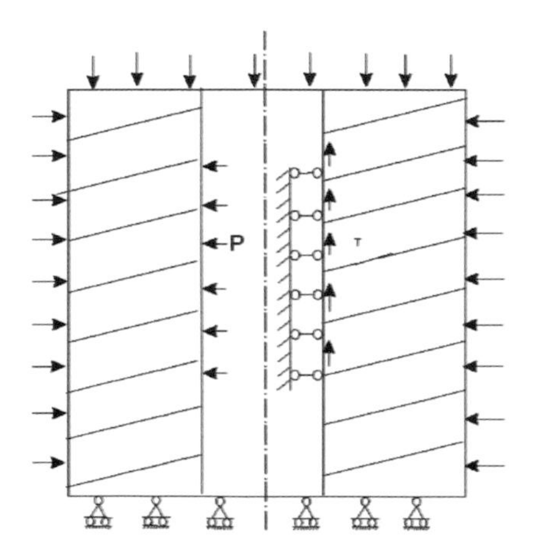

Figure 3. combined the loading and design FEM scheme.

The reconstruction of the analytical form function when the tabular form function exists for the design anchor parameters evaluating is needed. The reconstruction of the analytical form function is made with mesh value function methods as usual. When the function value obtaining time expense and number of design parameters are small the approximation methods are used as usual. However in case of complicated and multivariable task when using the methods is time expensive, it is impossible to do reconstruction even. Using the analytical reconstructed form function to design parameter influence evaluating has two demands. These demands are opposite. From the one hand, the reconstruction function is to be closed to the mesh value function data, from another to supply possibility for the design anchor parameters evaluating.

The new sequence approximation method (Larionov 2011) is proposed for solving the problem. The method allow to approximate reconstruction in analytical form function when it exist in table form in vicinity point $M(X_0), X_0 = X(x_1^0, x_2^0, ..., x_n^0)$, $X_0 \in \overline{D}$ and $\overline{D}$ – domain of its definition. There are theorem have formulated and top limit error estimation for analytical form function representation as univariable function product in the paper. Function representation is proposed as:

$$F(x_1, x_2, x, ..., x_n) \approx \alpha_n g_1(x_1) g_2(x_2) .. g_n(x_n),$$

where $g_1(x_1), g_2(x_2), ..., g_n(x_n)$ – approximation functions of the lines which caused by crossing the function surface by planes parallel to co-ordinate ones, and which pass through $M(X_0) \in \overline{D}$, α_n – approximation factor. The demands to the functions $g_1(x_1), g_2(x_2), ..., g_n(x_n)$ are continuity with the continuously first derivative differentiable. So, as evident, it is enough wide class of functions. But all work, as you see, have done for function representation in point vicinity. What to do for expand the results in domain of its definition? As a using experience demonstrated, solutions of real tasks allow representing them in whole domain of its definition. In spite of the fact that the errors of the function representation raised up to domain boundary errors do not exceed 5-7%, what is sufficient for mine engineer. Thus influence parameter evaluating method consists of approximate solution reconstruction (when it exists in number table form) in form of exponent functions and its index exponent comparisons. The greater index value, the stronger function parameter influence.

Using of the modified generalized M.E. Zhukovsky task for carrying out of practical parameter calculations is connected with some problems of computing character (Larionov 2011). The matter is that dependences between contact efforts are described with hyperbolic functions (1) that are not convenient. For convenience of using such dependences in practice it would be presented in more simple form (exponent form functions would allow not only to do calculations in more simple way, but to do design parameters degree influence of contact efforts estimating too). Considering special importance of FMS layer in the loadings transfer mechanism from anchor to rock, we will consider the effort value in FMS as goal function:

$$\sigma_{vy}^{p} = \frac{1}{L}\int_{0}^{L}\sigma_{vt}(\xi)d\xi ,$$

where

$$\sigma_{vt} = \sigma(L,q,d_a,d_{vt},E_a,E_{vt},h_a,h_{vt}) = \sum_{i=1}^{k-1}(p_i - t_i);$$

L – FMS length; q – bar pretension value; d_a, d_{vt} – rebar and borehole diameters; E_a, E_{vt} – elasticity bar and FMS materials modulus; h_a, h_{vt} – periodical bar and FMS surface rib steps. We will try to estimate of design parameters degree influence of FMS efforts. Developed method is used to solve this task. The contact system "bar-FMS" efforts – p_k and "FMS-rock" efforts – t_k are obtained with formulas (1). As a result the dependence between σ_{vt} and parameters q, d_a, d_{vt}, E_a, E_v, h_a, h_{vt} and L are obtained (4):

$$\sigma_{vt} = a_L q \frac{d_{vt}^{3.94531}}{d_a^{4.08821}} \frac{E_{vt}^{0.861906}}{E_a^{0.455547}} \frac{h_a^{2.54292}}{h_{vt}^{1.56405}} \frac{1}{L^{0.998028}}, \quad (4)$$

where a_L – approximation factor. Relative errors function distribution (4) for 19 mm diameter rebar demonstrated in Figure 4.

(a)

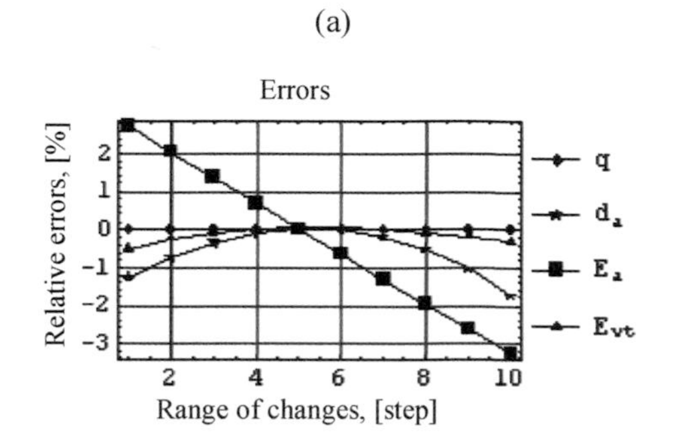

(b)

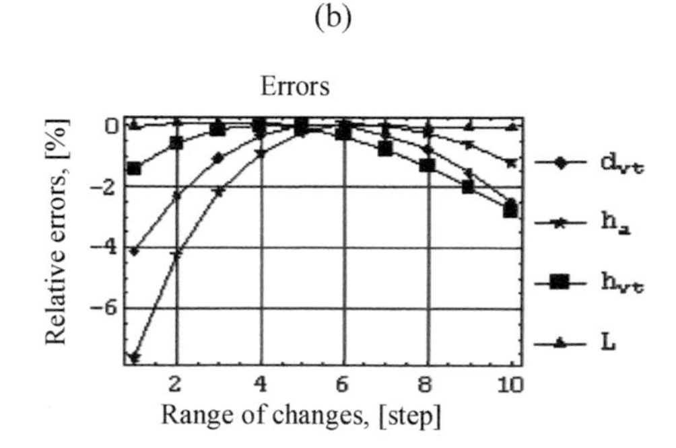

Figure 4. Relative function (4) errors distribution in base point vicinity.

More convenient formula form (4) is present as:

$$\sigma_{vt} = \varphi(q,d_a,d_{vt},E_a,E_{vt},h_a,h_{vt},L) = a_L q \frac{d_{vt}^4}{d_a^4} \frac{E_{vt}^1}{E_a^1} \frac{h_a^3}{h_{vt}^2} \frac{1}{L}. \quad (5)$$

It is necessary to notice that the problem of a classes approximation functions choice is one of the major problems not only in applied mathematics, but in technology applications also. How to limit a class of approximating product functions? As works (Ivahnenko & Urachkovskiy 1987) testify, the variation factor of approximating functions cannot be represented as criterion of its choice. As established the criterion for the class choice limit is to be external to the problem. Thus, as criterion which limits a class choice of approximating functions would be selected a dimension of initial function. The function representation form in the formula (5) though has the greater errors in comparison with (4), however completely corresponds to the theory of dimensions. Use of this theory for restriction of approximating function classes is the extremely important for an estimation of parameters influence on original function.

The dependence between parameters: initial pretension values P, opening supporting depth H, working FMS length L, rebar d_a and borehole d_{vt} diameters, shear stress intensity I_p^r and influence anchor radius ρ have obtained with using the theory (Larionov 2011). Influence anchor radius $\rho = f(P,H,L,d_{vt},I_p^r,d_a)$ is a continue function of the parameters. The dependence between parameters and anchor space is obtained as:

$$\rho(P,H,L,I_p^r,d_a) = a_{d_{vt}} \sqrt[3]{\frac{PHd_a}{LI_p^r}},$$

where a_{da} – approximation factor.

Maximum boundary relative error value between the formula and applied number algorithm using do not exceed 5%. Minimum boundaries relative error value is founded in choose point vicinity. It will be noted that influence anchor radius is proportional to anchor rigidity system that corresponds with conclusion (Mark 2000). In accordance with it influence anchor radius is opposite proportional anchor length that to confirm the theory regularity.

4 CONCLUSIONS

1. There new sequence approximation method is proposed to found the dependences between system parameters confirm the efficiency. Relative boundary errors do not exceed 5-7% what allow to use approximate reconstruction in analytical form function for engineer computation.

2. FMS efforts in depending on parameters q, d_a, d_{vt}, E_a, E_{vt}, h_a, h_{vt}, l are obtained.

3. Anchor space dependence between parameters: initial pretension values, opening supporting depth, working FMS length, rebar and borehole diameters, shear stress intensity and influence anchor radius have obtained.

REFRENCES

Ismet Canbulat. 2008. *Evaluation and design of optimum support systems in South African collieries using the probabilistic design approach.* Dissertation submitted to the Fuculty of Engineering Built Environment and Technology for the degree Philosophy Doctor. Pretoria: University of Pretiria: 340.

Larionov, G.I. 2011. *The anchor design parameter evaluating.* Dnipropetrovs'k: National metallurgy academy of Ukraine: 286.

A. Dey, N.I. Aziz & B. Indraratna. 2001. *Shear behaviour of infilled bolted join*t. AIMS (Aachen International Mining Symposia): 323-341.

Ivahnenko, A.G. & Urachkovskiy, U.P. 1987. *Compound system modeling with experiment data.* Moscow: Radio and communication: 118.

Mark, C. 2000. *Design of roof bolt systems. New technology for coal mine roof support.* Pittsburgh, PA. U.S., DHHS, NIOSH: 156.

Geomechanical Processes During Underground Mining – Pivnyak, Bondarenko, Kovalevs'ka & Illiashov (eds)
 ISBN 978-0-415-66174-4

Prospects of plant waste using for energy

A. Pavlychenko & O. Borysovs'ka
National Mining University, Dnipropetrovs'k, Ukraine

ABSTRACT: The necessity to identification and scientific justification of environmentally safe methods of treatment of fallen leaves and other plant waste to more energy are grounded. Arguments proving the necessity of removing fallen leaves from the city are given. Ecological effects of plant waste burning in cities are analyzed. The ecologically optimum methods of utilization of fallen leaves are validated. Environmental, ecological, economic and social aspects of this direction of research are marked out.

1 INTRODUCTION

In autumn in our country it is accepted to burn fallen leaves, in spring – last year's grass. Air in cities in the peak of the spring and autumn burning becomes heavy and bitter, that results in the increase of respiratory diseases of its inhabitants. The situation is complicated by the features of seasonal weather conditions: windless weather, increasing number of foggy days, high humidity. This ensures that the smoke does not rise but spreads like a low above the ground, covering not only the territory of burning, but also adjoining areas. A significant number of cases of fallen leaves burning is observed in the cities, the atmosphere of which and are already polluted by industrial emissions and motor transport.

Note that the collected leaves in most cases get in containers with domestic garbage and transported to landfills or burned in the places of accumulation. Transporting of leaves to landfills increases pressure on public services as well as area for disposal of garbage grows. And because of leaves burning within the limits of city a range of pollutants gets in the air which accumulate in the leaves during the vegetation period.

Thus, the combustion of one ton of plant waste is released about 9 kg of micro particles of smoke in the air. A dust, nitric oxides, carbon monoxide, heavy metals and a number of carcinogenic compounds are presented in their composition. As leaves in nature are wet, in most cases they do not burn, but simply smolder and that promotes a large number of hydrocarbons. Some of them, such as aldehydes and ketones, cause irritation of eyes, nose, throat and lungs. Considerable part in the smoke of burnt leaves is made by polycyclic aromatic hydrocarbons, most of which are carcinogens that can cause cancer. In addition, dioxins are coming in the air with smoke, which are one of the most poisonous substances for human.

2 FORMULATING THE PROBLEM

That is why there is a need in determination and scientific justification of environmentally safe methods of treatment of fallen leaves and other plant waste to get additional power resources.

The use of fallen leaves will allow a city to get additional electric energy and bioorganic fertilizer, which can be brought in the soils of parks, gardens and other green areas of the city and these measures can improve the ecological condition of green spaces and their resistance to unfavorable factors.

3 GENERAL PART

The leaves are renewable organic raw material, which accumulates annually in the city and can be used for energy and fertilizer needs for the city. Ecologically safe utilization of fallen leaves will improve the ecological, economic and social situation in the city.

Cleaning plant leavings is the time-consuming work, which for decades from year to year is executed by gardens service wipers and citizens on Saturdays cleaning day. The need for removal of leaves arises out of rules of exploitation of the green planting, and that is why specialists of landscaping services insist on conducting this work type, bringing over next arguments.

The first group of arguments – health of the green planting. Urban green spaces get the whole complex of pollutants that penetrate the leaves from the contaminated soil, water, and also from atmospheric air.

A motor transport contaminates city soils with oil

and heavy metals. Due to frequent misoperation of systems for surface water collecting pollutants from highways enter the area of open ground – that on soils with the green planting.

Greater part of surface of soil in town is asphalted, and it means that such a toxic substance as benz-a-piren that is excreted by asphalt also accumulates in soils of green areas.

Salt which is generously covers a city in winter time also gets in soil. So keep these leaves and allow them to be included later in the soil is more harmful for plants. It is known that heavy metals and many other pollutants can reduce the life expectancy of trees and cause weakening resistance to diseases and pests.

The second group of arguments is sanitary-hygienic. Most of the fallen leaves very quickly become a mixture of leaves and different garbage that many citizens set up under trees and bushes. The remains of food, plastic bags, dog excrement continue to accumulate actively all winter long and to mix with the leaves that creates a favorable habitat for rats and dangerous infections that they spread.

The third group of arguments is maintenance of lawns. Good lawns in city are needed both from the aesthetic and an environmental point of view. The roots of lawn grass densely interlaced with each other clamp soil and prevent polluted soil to turn into a toxic dust that the wind carries the face of pedestrians. Most lawn grasses are able to live on contaminated soil and clean it, sucking the pollutants from the soil and accumulating them in their green parts. At haying seasons these toxic matters are accumulated in leaves, they are "packaged" and ready for utilization, and a lawn continues to grow and to execute the work of living filter again. However, it touches exactly to the well-groomed lawn which consists of the special lawn grass and which is regularly handled in compliance with the technology, but not place which we name a „lawn", while it is a populated by weed herbs. Creation and maintenance of the real city lawn is the time-consuming and costly work and what is the most important it requires a professional approach. Lawn has strict requirements for the one who takes care after him and one of the mandatory conditions – lawn should be cleared of fallen autumn leaves. Lawn with fallen leaves strongly loses in quality and rots during the winter. Professional gardeners say – a tree and lawn are always in a conflict. For example, trees need nutrients from their own leaves, and a lawn is stable when it is clean and regularly trimmed. They are habitants of different ecosystems. Most used in parks trees are the forest inhabitants thereby forest soil is always ideal and usual for them, and this type of soil is created over the years of fallen leaves by many microorganisms and animals, bacterial and other processes. But a real lawn is quite rare phenomenon in a natural environment.

Urban environment needs trees and lawns. Very often they coexist on the same plot of land and people forced to resolve their conflict, of course, while reducing the environmental comfort for both.

Thus, removal of fallen leaves from territory of city is necessary, but instead of collecting and exportation there are numerous cases instances of their combustion. Damage from such actions is heterogeneous and extremely dangerous.

Besides the direct danger to human health, incineration of leaves and dry grass results in such threats:

1. Wintering insects burn in a dry leaves, such as ladybugs. Aphids which are the catch of ladybugs remain to winter on the stage of egg on branches. Combustion of leaves in autumn creates conditions for the development of aphids in spring.

2. Incineration of leaves results in destruction of the ground cover, in fact vegetable remains burn down directly, soil-forming microorganisms perish. In addition, they die from the heavy metals formed at burning.

3. At normal terms, when leaves rot through, the necessary for the development of plant substances go back into soil. Parts of the leaves which slowly decompose, structure the soil, improving its quality. The gradual destruction of fallen leaves creates conditions for development of soil microflora and fauna, which, on the one hand, executes the work of processing the leaves, on the other hand, prevents the development of pathogenic organisms to trees (fungal and bacterial diseases of trees). City trees are prone to various diseases.

Only part of the seedlings takes root in urban environments and grows up to "adult" age. One of main reasons is quality of city soils. No one prepares the soil in the forest – nature does it with the annually got material annually – fallen leaves. The forest litter is a factory for the production of soil that nourishes trees and protects their health. And ash appears at incineration of leaves. Despite the generally accepted opinion ash is very poor fertilizer and thereby burning the leaves each year result in the increasing soil depletion.

4. On natural areas and lawns fire destroys seeds and roots of herbaceous plants, damages the lower parts of trees and bushes and the tops of their roots.

5. Destruction of natural litterfall leads to an increase of soil freezing in 2-4 times.

6. Smoke from fires in foggy days can form smog and hang up in the air for long time. In this case visibility on the roads is deteriorating, that results in the increase frequency of accidents. In addition

smog causes significant decline of health.

7. Smoke-filled cities use much more energy for illumination

An additional problem is that usually a lot of different garbage burns and with leaves that substantially increases air pollution. During the combustion of a plastic bag to 70 various compounds free oneself in air, most of which are poisonous to humans. They usually cause throat irritation, cough. Turbid black smoke from smoldering plastic waste contains carcinogenic polycyclic hydrocarbons. At burning of rubber carcinogenic soot and oxides of sulphur also appear and they cause respirator diseases. Epithelium of the mucous membrane of the respiratory tracts, that constantly gets irritated by smoke, unable to resist microbes. People with bronchitis, bronchial asthma, rhinitis or tonsillitis suffer the most.

Wooden fibreboard, woodchip board and plywood quite often get to the bonfires. These materials contain resins, which include formaldehyde and they can be painted with oil paint that contains lead.

Thus, the problem of fallen leaves in the city is very difficult. From one side, they must be removed from territory of city, from the other side their removal and burning causes a range of environmental impacts. There is no single solution to this problem. Cleaning the leaves must be selective that is decision should be made for each green planting object separately considering all its features.

In the system of landscape management all the objects are divided into classes and categories according to their purpose and location in urban building. There are five classes of objects, which lay different claims to maintenance and intensity of care. The highest allowances are established for objects of the first service class (green planting of the city setting, the most responsible for the location and value, the most visited city parks, gardens, squares, areas next to public and historic buildings, major street highways). The second class – is objects of district designation: parks, gardens, squares, boulevards, streets, roads and driveways. The third class includes the green planting of local value: gardens, boulevards, public gardens, streets and passages, landscaping and gardens intra neighborhoods. Landscapes and historic parks, greening objects of various departments, schools, hospitals, preschool establishments are in the forth class. The fifth class includes parks and forests within the city limits.

Rules of fallen leaves cleaning, as well as other norms of retaining, are set depending on which class the green object is. Fallen leaves can be left only in parks, urban forests and partly in the landscape and historical parks. The classification of green planting correct provides the correct measures, but balks the cases where an individual approach is needed. Making decision depends on the territories owners and exploiting organizations, which can automatically take advantage of rules and norms that are established for the class of object, but may submit proposals on the application of other approaches of green planting maintenance, taking into account features of each greenery class.

I class. City green planting, most responsible for a location and value, are the parade places of city, and they must have lawns of high aesthetic value. Autumn leaf cleaning is required for them.

The most visited city parks – more difficult question. Large-sized parks often have more visited parts (parade and entrance parts, areas with high recreational load), where it is necessary to have well-groomed lawns and to conduct the autumn cleaning up of leaves. Also they have the less visited areas, where there no grass under the trees, often with the areas of forest flora. Leaving the leaves under the trees in these areas gives only favorable results – improving the health of trees, soil quality, habitats of the birds and squirrels. In the small intensively visited parks that are close to residential neighborhoods, subway stations and other heavily visited sites needed cleaning the leaves from sanitary reasons, even if they don't have ordered lawns.

Cleaning leaves in the gardens and squares of urban value is inevitably – often they have lawns cover that needs cleaning. In addition, the attendance of such objects is high, that creates the need for regular cleaning of the territory only because of the garbage that people remains. Other objects of class I – the most important part of city streets and avenues – also require cleaning fallen leaves, because the quality of grass coverage is important both from aesthetically and an environmental point of view.

II class. Parks, gardens and squares of district purpose strongly differ one from other by the load and its distribution inside the object. Individual approach must be applied here. In areas, not adjoining to transport highways and intensively visited places, in " quiet" areas which are used mainly for walks and recreation of people (with a small number of dogs), leaving leaves under trees will be rather useful, than harmful. Often in such areas lawns are out of high value and it is possible to donate their appearance in favor of trees. Boulevards, streets and green spaces passages of district importance, with lawns covering must be cleaned from leaves from the aesthetic and environmental reasons.

Class III. In gardens, parks and street greening should be applied the same approach of class II. At maintenance of territories of the interblock planting of greenery, gardens and public gardens of local value electoral approach to separate areas on removed and not removed from the leaves is important. Based on the aesthetic, environmental, health and performance features of objects and especially the different parts of them, one can identify areas where there is no need to remove leaves. Careful consideration of this matter will help improve the health of green space and get the savings funds, which traditionally is not enough for the maintenance of green spaces.

Class IV. Objects of this class are very diverse in their purpose. Landscape and historic parks can have areas of natural soil formation that is without autumn cleaning the leaves, the choice of these areas entirely determined by the characteristics of parks and their parts. That applies to the green space in the departments. Often these areas allow to conduct zoning and split "front" (cleaned) and "natural" area (without cleaning leaves). Quite different approach is needed to such objects as hospitals area and child's establishments. Considerations of cleanliness and health security are the most important there. Leaves should be removed necessarily.

V class. Existing norms do not include cleaning fallen leaves in forest parks and urban forests and traditionally it doesn't perform. They do not have a lawn actually, they have a herbage. A small area around the commercial facilities that have a large number of garbage may be an exception – cleaning the leaves there is associated with sanitation, as well as environmental considerations.

Thus, depending on specific conditions, fallen leaves can be either removed, or abandoned by green areas, but in any case it can not burn in the city – it is dangerous for environment and prohibited by law. Leaves deleted from the territory of the city, in our opinion, it is impractical to export to landfill, because it can be utilized, that is applied with a benefit.

Fallen leaves, which accumulate annually in the cities, refer to the renewable plant resources on which it is possible to get various valuable products. Thus, the methods of fallen leaves utilization are:

1. Composting in order to obtain bio-organic fertilizer.

2. Processing into fuel briquettes and pellets.

3. Reception of activated carbon, biologically active substances and other products.

Composting of fallen leaves is the safest way of their utilization in relation to the environment. Annual cleaning leaves in parks in 20 years results in the decrease of wood growth to 50%. The same happens on the small holdings, in the gardens. Composting is the decomposition of plant remains without access of oxygen to form compost - more or less homogeneous organic matter that can be used as fertilizer. The most valuable feature of compost is a big content of chemical compounds that are in great necessity for plants. The fact that the trees as they have large and deep root system deliver necessary chemicals in sufficient quantity and optimal proportions to their bodies, including the leaves, and they collect them from large areas or deep layers of the soil. Ash that remains after the burning of leaves, despite to popular belief, does not contain several elements and can not be a good mineral fertilizer.

For composting leaves are laid in heaps 2 m wide and up to 1.7 m high. The layer of soil 25 cm thickness piles in basis of heap. Each of leaves layer must not exceed 30 cm. It is possible to throw down bird dung, food tailings on the same heap. Each of layers is covers with earth. During a summer compost is shoveled 2 or 3 times. Compost is ready when it turned into a homogeneous dark crumbly mass. In the summer allocation compost ripens for 2-3 months, in the autumn - after 6-8. Similarly, composting can be done in trench with depth to 1 m and 1.5 m wide. Trench composting is more comfortable because compost is evenly moistened and does not dry up.

Compost can be used as a fertilizer in a year after the allocation. Its useful properties will be kept for 4 years. It is useful to apply compost as basal fertilization for trees and bushes.

Recycling of fallen leaves into fuel briquettes and pellets. Humanity requires energy, and the need for it increases every year. At the same time, the supplies of traditional natural fuels (oil, coal, gas, etc.) are limited. The supplies of nuclear fuel are also limited – uranium and thorium, from which it is possible to get plutonium in reactors. Reserves of thermonuclear fuel – hydrogen – are practically inexhaustible, but controlled thermonuclear reactions are not mastered yet and it is not known when they will be used for industrial energy generation in pure form. There are two ways remain: using of untraditional renewal energy sources and severe economy at the cost of energy.

The new forms of primary energy are solar, geothermal, wind and bioenergy. Unlike fossil fuels these forms of energy are not limited by the geologically accumulated supplies, that is their use and consumption does not lead to the inevitable exhausting. In addition, indicated schemes of energy transformation can be connected by one term "ekoenergy", because these methods of clean energy obtaining do not result in contamination of environment.

Bioenergy is based on biomass obtaining, which is used directly or after appropriate treatment as fuel. Using the energy properties of biological resources has a large value for the economy of

Ukraine.

Biomass includes: the remains of agricultural plants, wastes of stock-raising (manure), wood processing and forestry residues, organic part of municipal solid waste and fallen leaves of big cities.

For more than ten years fuel briquettes and pellets are popular and economical fuel that used in many countries. Briquettes are produced from dry sawdust of coniferous and deciduous species, sunflower peelings, straw, leaves, etc. The technology of wood fuel pellets is based on the process of compression of shredded organic waste under high pressure when heated, and lignin is the connecting element. Plant cells contain it, i.e. wood pellets contain no harmful substances, including glues.

In European countries for a long time branches of trees, burned grass and dry leaves are converted into small pellets – granules with a peanut nut size. Pellets are poured in the bunker, and automatically supplied to the boilers that work like gas-producing boilers while receiving gas with properties similar to natural, but three times cheaper.

Equipment that can work on such pellets already is in Kyiv – he is produced by the Czech company "Werner". During the last two years the special caldrons from "Werner" already successfully work on the hothouses of communal enterprise of green planting maintenance of the Desnyanskij district of Kyiv.

Reception activated carbon. In modern scientific literature there is only some information about the possibilities of rational use of fallen leaves. Basically it is used in non-processed form as fuel, feed additives or bio-organic fertilizer. At the same time directions of utilization of fallen leaves can be diversified by adding this raw material the physical and chemical processing with a receipt on its basis of absorbent carbon, biologically active matters and other products.

4 CONCLUSIONS

This direction of investigation has ecological, environmental protection, also social character, and it's very difficult to examine the question of economic effect, because it says, especially about health population and environment.

The expected economic effect from introduction of the system of vegetable wastes utilization on city territory can be next:

– provide clean energy through the use of renewable vegetable raw materials;

– receipt of bio-oganic fertilizers from vegetable wastes;

– energy savings for illumination of the city, because for smoky cities use much more energy for lighting;

– reducing the area of land that are allocated to the landfills of solid waste, by exclusion of fallen ingress leaves and other plant waste to municipal solid waste;

– reduce the number of respiratory disease in autumn, and thus reduce the number of medical certificates.

The complex use of fallen leaves will allow: to reduce the level of contamination of atmospheric air due to the refusal of burning of leaves, to prevent the destruction of seeds and roots of herbaceous plants, underbodies of trees and bushes in the burning of leaves, to prevent destruction and degradation of soil by burning plant residues; to get additional source of clean energy for the needs of the city, to get a clean fertilizer that can be used to improve green spaces of the city, and create favorable environmental conditions for the population.

Thus, different approaches to the utilization of renewable plant waste into useful products and energy are promising and promote the effective solution of environmental and economic problems, namely the improvement of environment in urban areas and finding additional sources of raw materials and energy.

REFERENCES

Goryshina, T.K. 1991. *Plant in the city.* Leningrad.: Publishing house of Lenynhrad University: 152.

Fendyur, L.M. 2001. *Greening urban areas.* Zaporozhye: ZSU: 32.

Kucheryavyi, V.P. 2005. *Planting settlements: a textbook.* Lviv: Svit: 456.

Kucheryavyi, V.P. 2003. *Plant melioration: school-book.* Lviv: Svit: 540.

Nikolaevskaya, I.A. 1990. *Beautification of the cities: textbook for construction technical schools.* Moscow: The Higher School: 159.

Gabrel, M.M. 2004. *Spatial organization of urban systems.* Kiev: Publishing House of ACC: 400.

Urban Development. Planning and development of urban and rural settlements. The state building norms of Ukraine (DBN. A. 2. 2.-1-95). 2002. Kiev: State consumer standard of Ukraine.

Lesyshyna, Y.A. 2007. *Inhibitors and activated carbon based on fallen leaves.* Author's abstract of thesis ... cand. chem. sciences: 02.00.13. Donetsk: Institute of Physical Organic Chemistry: 22.

Geomechanical Processes During Underground Mining – Pivnyak, Bondarenko, Kovalevs'ka & Illiashov (eds)
 ISBN 978-0-415-66174-4

New method for justification the technological parameters of coal gasification in the test setting

V. Falshtynskyy, R. Dychkovskyy, V. Lozynskyy & P. Saik
National Mining University, Dnipropetrovs'k, Ukraine

ABSTRACT: Implementations of test researches enable considerably to decrease expenses at the estimation of the efficiency of underground coal gasification for concrete mining and geological conditions. The new method of test researches is offered with the simulation of underground conditions of the real mine. Dependences of the material and thermal balance are proposed. The basic technological parameters of gasification are determined during the transmission results to natural conditions.

1 INTRODUCTION

The development of borehole underground coal gasification (BUCG) technology with transformation into the enterprises of power-chemical complex are interested: mining, chemical industry and power engineering. Setting the parameters of gasification of the coal seam and their working off in experimental and natural conditions requires the involvement the diversified scientific schools and experts in various scientific fields.

The complexity of the experiment will allow to confirm a number of analytical solutions and bring them to the real recommendations un oder to design the power-chemical enterprises for production and complex processing the coal seams in the place of its occurrence.

The main objectives of stand researches are (Kolokolov 2000; Falshtynskyi 2009 & Savostianov 2007)

– establishment the functional dependence of changes in composition of BUCG gases with growth of an outgassed space by constant consumption of blowing reagents and the combined supply scheme taking into account the increased humidity (44-67%) of layer and ash-content of coal (36-42%);

– establishment the borders' area of heating up breeds of the immediate roof over an outgassed coal layer;

– establishment the parameters generating the reactionary channel;

– determination the inertness of a coal layer at its firing in the combined feed system;

– determination the movement speed of feed point of blowing;

– definition the expenses, losses and pressure of blasting and gas with growth of an outgassed space;

– definition the qualitative composition of condensate and fly-ash at underground gasification.

2 DEVELOPMENT THE CRITERIA FOR SIMILARITY AND CALCULATION OF MATERIAL AND HEAT BALANCE OF GASIFICATION

The process of BUCG is carried out by the criteria of similarity in scales 1:10, 1:20...40. The basis of natural conditions at the test experimental installation shows the geological structure of the rockmass, technological parameters of an underground gas generator, physical speeds and process kinetics. It will allow to establish the mechanism of roof rocks deformation withe growth of an outgassed space at various speeds of the fire wall advance. According to the developed mining-geological conditions, elemental composition of the modeled coal seam and parameters of an underground gas generator csn be determined the material and heat balance of the process of underground gasification in conditions close to nature (Table 1, 2, 3, 4).

Mathematical modeling of material-heat balance of BUCG process at the test installation was provided with the software product of "mathbalanse BUCG" (Falchtyskyy 2012). It was developed by staff of the National Mining University. It provides an algorithm for calculating the parameters of the borehole underground coal gasification process associated with thermochemical conversions of solid fuel into gaseous state and condensate in the conditions of elemental composition of the coal seam, external inflow of water and heat balance of the underground gas generator.

Table 1. Parameters of material balance of BUCG process.

Type of blowing	Yield of BCG gases from gas generator, %							The lowest warmth of BCG gase combustion	Humidity of BUCG gas
	H_2	CH_4	CO	N_2	H_2S	CO_2	O_2	MJ/m^3	g/m^3
Aerial	4.15	3.88	10.26	77.27	0.22	3.68	0.54	2.86	206
Steam-air O_2, N_2, steam	9.29	7.43	9.05	63.86	0.29	9.30	0.78	3.91	311
oxygen +steam O_2, N_2 steam	15.21	12.79	5.02	45.51	0.33	19.98	1.16	7.26	309
oxygen + carbon dioxide + steam O_2, CO_2, N_2, steam	18.34	14.18	8.7	0.27	0.31	57.06	1.14	6.32	287
Oxygenous O_2, N_2	8.16	6.53	22.45	21.46	0.32	39.89	1.19	7.51	195
Oxygen + carbon dioxide O_2, CO_2, N_2	8.27	6.83	26.37	11.73	0.34	45.31	1.15	6.68	236

Table 2. The main chemical products yield in the modeling of BUCG process.

Types of blowing mixtures	The BUCG chemical products yield (t)			
	Coal pitch	Raw benzol	Ammonia	Sulfur
O_2N_2	63.7	20.48	32.8	1.02
H_2O (steam) $+O_2N_2$	74.9	43.7	57.3	1.02
O_2 (30-62%) N_2	86.2	44.1	66.5	1.2
O_2 + steam	61.2	36.2	67.4	1.04
CO_2+O_2	91.3	45.3	58.1	1.1
$CO_2+O_2+H_2O$ (steam)	83.6	38.2	62.5	1.12

Table 3. Geometric and quantitative model parameters of the underground gas generator at the BUCG area.

Type of blowing	Parameters of blowing, m^3/h	Speed of outgasing, m / day	Intensity of the BUCG process, kg / h	Maximum temperature of the process, °C	Pressure in the reaction pipe, MPa	Yield of the generator gas, m^3/kg	Temperature of the на generator gas at the yield, °C
BUCG area #1 (layer C_6^1, $m = 0.9$ m, mark Г) Solenovsk deposit							
Aerial	134.5	1.1	22.9	977	0.15-0.17	2.14	178
Steam-air O_2 H_2O (steam) N_2	172.4 36.2 56.9 79.3	1.18	23.6	961	0.16-0.19	2.23	185
Oxygenous O_2 N_2	126.7 78.6 48.1	1.05	21.2	1060	0.2-0.35	1.82	218
oxygen + steam O_2 H_2O (steam) N_2	123.6 55.6 39.8 28.2	1	19.8	1012	0.26-0.3	1.88	207

Continuation of Table 3.

Type of blowing	Parame-ters of blowing, m^3/h	Speed of outgasing, m / day	Intensity of the BUCG process, kg / h	Maximum temperature of the proc-ess, °C	Pressure in the reaction pipe, MPa	Yield of the generator gas, m^3/kg	Temperature of the на ge-nerator gas at the yield, °C
BUCG area #1 (layer C_6^1, $m = 0.9$ m, mark Г) Solenovsk deposit							
oxygen + CO_2 O_2 CO_2 N_2	121.2 75.2 18.2 27.8	1	19.8	1029	0.25-0.3	1.74	215
oxygen + CO_2 + steam O_2 CO_2 H_2O (stean)	118.5 42.9 18.2 57.4	1	19.8	1018	0.25-0.3	1.77	210
Geometrical parameters of the model							
Power of the coal seam, m	Thickness of the immediate roof, m	Thickness of the main roof, m	length of the reac-tion pipe, m	Diameter of the pipe, m	Diameter of the working boreholes, m	Power of percarbonic layer, m	Length of the working boreholes, m
0.23	0.8	1.5	1.9	0.2	0.1	0.1	2.0

Table 4. Heat balance of underground gasification.

Indicators	Composition of the blowing					
	Aerial		Oxygenous		$O_2 + CO_2$ +steam	
	$\frac{MJ}{kg}$	%	$\frac{MJ}{kg}$	%	$\frac{MJ}{kg}$	%
1	**2**	**3**	**4**	**5**	**6**	**7**
Heating of combustion on working fuel	23.454	95.836	23.454	88.854	23.454	87.365
Heat content of the massif in the oxidation zone (combustion)	0.701	2.864	1.154	4.372	1.275	4.749
Heat content of blowing	0.318	1.299	1.788	6.774	2.117	7,886
Sum total:	**24.473**	**100**	**26.396**	**100**	**26.846**	**100**
Heating of combustion of the BUCG gas	3.610	14.762	5.310	21.036	3.610	13.534
Heat losses: 1. Heating of ash and slag, *MJ*	0.312	1.276	0.312	1.236	0.312	1.170
Heating and evaporation of moisture (water inflow and humidity of coal and rocks), *MJ*	3.708	15.163	2.634	10.435	2.634	9.875
Heating of roof, footwall), *MJ*	9.206	37.645	7.983	31.625	7.602	28.500
Heat content of dry generator gas	7.619	31.155	9.004	35.669	12.516	46.922
Sum total:	**24.455**	**100**	**25.243**	**100**	**26.674**	**100**
The temperature of generator gas at the gas-outlet holes, °C	143		187		175	

Continuation of Table 4.

Indicators	Oxygen + CO_2		Air+steam		O_2 + steam	
	$\frac{MJ}{kg}$	%	$\frac{MJ}{kg}$	%	$\frac{MJ}{kg}$	%
1	**8**	**9**	**10**	**11**	**12**	**13**
Heating of combustion on working fuel	23.454	87.528	23.454	93.020	23.454	87.486
Heat content of the massif in the oxidation zone (combustion)	1.268	4.732	0.653	2.590	1.258	4.692
Heat content of blowing	2.074	7.740	1.107	4.390	2.097	7.822
Sum total:	**26.796**	**100**	**25.214**	**100**	**26.809**	**100**
Heating of combustion of the BUCG gas	3.610	13.555	3.610	14.343	3.610	13.500
Heat losses: 1. Heating of ash and slag, *MJ*	0.312	1.172	0.312	1.240	0.312	1.167
Heating and evaporation of moisture (water inflow and humidity of coal and rocks), *MJ*	2.946	11.062	4.847	19.258	2.634	9.850
Heating of roof, footwall), *MJ*	8.946	33.591	7.410	29.441	8.857	33.123
Heat content of dry generator gas	10.818	40.620	8.990	35.719	11.327	42.360
Sum total:	**26.632**	**100**	**25.169**	**100**	**26.740**	**100**
The temperature of generator gas at the gas-outlet holes, °C	179		165		169	

3 DESIGN FEATURES OF THE TEST INSTALLATION

Experimental test installation represents the metal capacity in the size of $2.0 \times 2.2 \times 1.8$ m in which on the scale of 1:10, 1:40 the coal seam and containing rocks are modeled (Figure 1). For the simulation on the test installation of natural conditions are established: geometrical parameters, scales of time, speeds, density and pressure, large-scale factors of temperatures, heat emission and heat conductivity, indicators of hydromechanical, geometrical and mechanical similarity and criteria of a homochronous. The received geometrical and quantitative parameters of an underground gas generator model and material-heat balance of the coal gasification process on the BUCG areas taking into account results of calculation the criteria of similarity of experiment (Tables 1, 2, 3 and 4).

The air supply network consists of two mobile compressor installations, one of which working, another – reserve, an air collector, an air reducer, a gas meter, a nipple, cork cranes, mobile pipelines, heat resisting nozzles, the steam generator and receivers.

The mobile compressor installation SO-76 (TU 22-5871-94) made at the Vilnius plant, is intended for receiving the compressed air necessary for giving blowing mixture in a model of an underground gas generator.

Mobile compressor installation with productivity of 30 m^3 / hour consists of the compressor, receivers, a moisture separator, the air filter, a regulator of pressure, a safety valve, the electric motor, the pipeline, the starter and a protection. All units of the compressor are assembled on receivers connected consistently and supplied by wheels and a hand-rail for movement. Connection of the compressor and the engine is carried out by V-belt transfer. The compressor is piston, two-cylindrical, one-stage, simple action with compulsory air cooling from a pulley flywheel with the blades, given to rotation by cranked shaft through a driving belt.

Air arriving in the compressor is cleared in the air filter, representing the grid reeled up in some layers and on a branch pipe goes to soaking-up cavity of the block of cylinders. Alignment of a pulsation of air which arises at back and forth motion of the piston of the compressor, partial clarification of air from water and oil occurs in a receiver. Receivers are connected among themselves. Volume of receivers is 22 litres.

Purification of the air before giving into the network is carried out in an oil separator made in the form of a welded cylinder with a glass, filled with copper-plated tubes. Separated condensate flows down in a receiver and through a draining opening periodically merges.

Necessary working pressure (within 0.15-0.35 MPa) is established by a pressure regulator.

For prevention of increase the pressure above 0.7 MPa the safety valve on receivers is monted. The compressor and receivers are connected by the blasting pipeline. Pressure of air is supervised by a manometer.

(a)

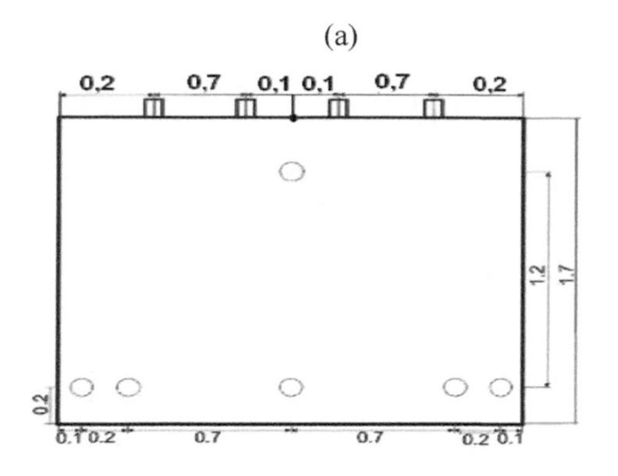

(b)

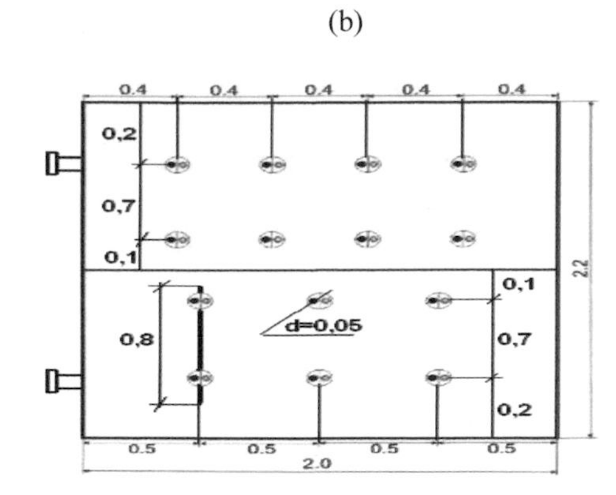

Figure 1. The BUCG test bench (general view): (a) front view, (b) the top view.

The drive of the compressor is carried out from the 4 AM 100 three-phase asynchronous electric motor, capacity of 4 kW, tension 380 In with frequency of rotation of 2880 rpm.

The PNV-30U2 actuator (rated current 10 A, tension 380 V). The direction rotation of a cranked shaft (from a flywheel) is counter-clockwise.

The one-stage bottled gas reducers (BKO-25-2) which are let out on TU 26-05090-87 are applied to pressure decline of the air arriving from an air collector and automatic maintenance of the set working pressure.

Pressure decline in a reducer occurs by its one-stage expansion about passing through a gap between a saddle and the valve in the chamber of working pressure. From a receiver gas, having passed the filter, arrives in the chamber of a high pressure.

At rotation of the regulating screw clockwise the effort of a press spring is transferred through a membrane and a pusher to reducing valve. The flexible hose in length of 6 m is attached to a reducer for a possibility of moving of an airsubmitting nozzle. Movement of a nozzle is made by the metal pipe in length of 4 m for carrying out experiments in a long box of installation.

The hose incorporates to a pipe by the special adapter. In the end of the pipe the heat resisting nozzle is fastened. Blowing from the nozzle moves through the punched tube by which the well is equipped.

The accounting of the arrived blowing and the received gases is made by converters of expense IRVIS-K300. Counters are intended for the volume accounting of nonaggressive gases. Pressure of gas shouldnt exceed 1.7 MPa.

Removal of gases from the model is carried out on the pipe connected to the condensate sump. On the pipe there is a latch and the union for gas sampling. Purification of gases is carried out in a tank filter by the admission of gases through three layers of mineral cotton wool. The scheme of the outgasing pipeline is presented on Figure 2.

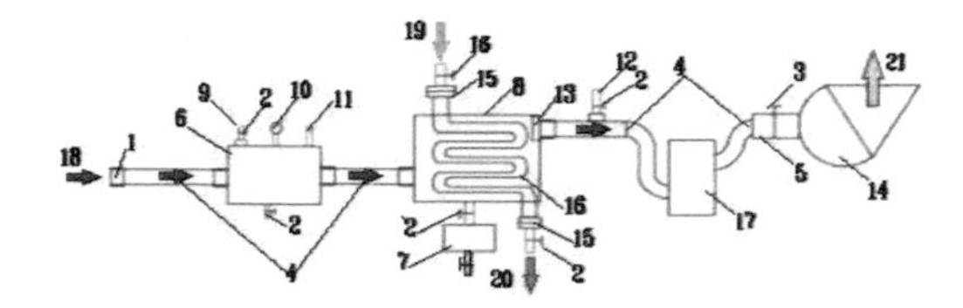

Figure 2. The Scheme of the outgasing pipeline: 1 – an adapter; 2 – crane; 3 – shiber; 4 – pipeline ($d = 50$ mm), $L_1 = 800$ mm, $L_2 = 500$ mm, $L_1 = 400$ mm; 5 – pipeline $d = 80-100$ mm, $L_1 = 500$ mm; 6 – receiver (the broad tank), $h = 250$ mm, $b = 300$ mm, $L = 300$ mm; 7 – tank of drain $h = 150$ mm, $b = 150$ mm, $L = 150$ mm; 8 – the tank for collecting condensate and cooling of gas, $h = 750$ mm, $b = 600$ mm, $L = 600$ mm; 9 – the crane for gas sampling; 10 – manometer; 11 – thermometer (thermocouple); 12 – the crane for gas sampling by a gas analyzer in a line mode; 13 – filter; 14 – ventilator; 15 – connections; 16 – recuperator the tubular spiral; 17 – limy filter; 18 – generating gas; 19 – cold water; 20 – hot water; 21 – the gas filter.

For condensate selection from the gas in addition to the scheme of an outgasing network it was developed the earlier to establish a tank-extender (Figure 3).

Electric consumers of installation work at tension of 220/380 V. The three-phase alternating current is brought to installation by a mine flexible screened cable of the GRShE 3×16+1×10. The seet has having three veins by section of 16 mm. The cable maintains 105 A. Feed of voltage is carried out to the actuator of PVI 63 type calculated on connection of the electric motor of the maximum capacity of 55 kW and rated current of 63 A (150 A).

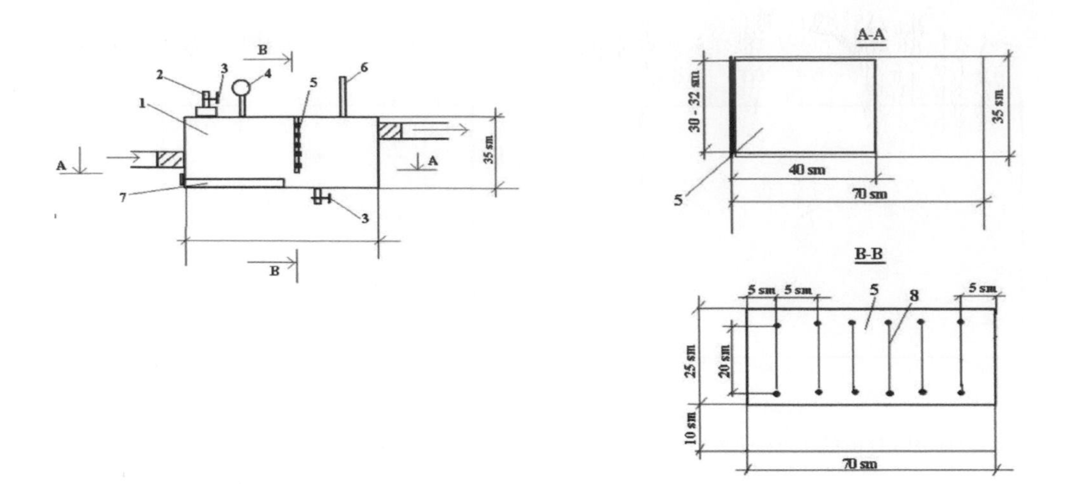

Figure 3. General viev of the tank-extender: 1 – the tank, 2 – the carbine for sampling of gases, 3 – the faucet, 4 – the manometer, 5 – the grate, 6 – the thermometer, 7 – the tray, 8 – the grate crack (width 1 – 1.2 sm).

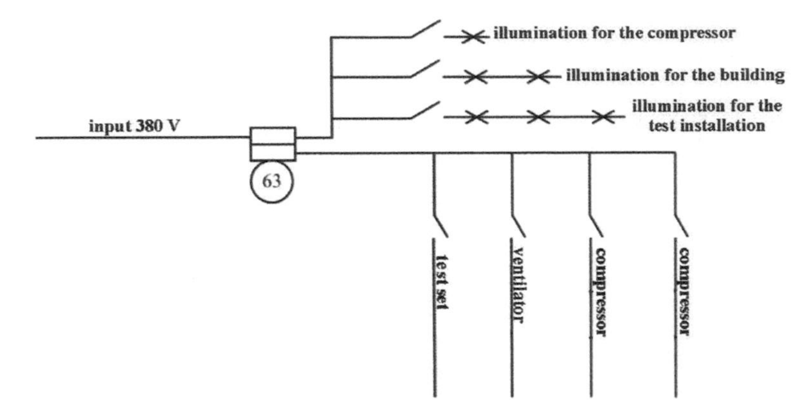

Figure 4. Scheme of power supply of the test installation.

From the actuator to consumers there is a flexible cable. External lighting is carried out by RN 100 lamps. The scheme of power supply of bench installation is provided on Figure 4.

Automatic turning on of the compressor is carried out by an electrocontact showing manometer which turns on the compressor at pressure decrease in a receiver to 0,6 MPa and disconnects at pressure increase to 0,8 MPa. Compressors work alternately, operating time of one compressor shouldn't exceed 12 hours.

For determination of concentration being formed the gases received in model of an underground gas generator, Casboard-3200L equipped with the interface, a chromatograph "Poisk 2" and gas analyzers VH-170. Technical characteristic of a gas analyzer of Casboard-3200L are provided in Table 5.

Table 5. Technical characteristics of a gas analyzer Casboard-3200L.

Matter	Method of measurement	Range of measurement	Accuracy	Inaccuracy
CO_2	Infra-red sensor	50%	$\leq 2\%$ width of scale	$\leq 2\%$
CH_4	Infra-red sensor	100%	$\leq 2\%$ width of scale	$\leq 2\%$
H_2S	Electro-mechanical sensor	traces	$\leq 3\%$ width of scale	$\leq 2\%$
O_2	Electro-mechanical sensor	25%	$\leq 3\%$ width of scale	$\leq 2\%$

The system of temperature-sensitive elements is applied to determination of parameters of a field with converters of a signal equipped with the interface.

Flowmeters IRVIS on a blowing and outgasing pipe of the stand of an underground gas generator are also equipped with interfaces and laptops that provides continuous control and information accumulation of an experiment.

Detecting of components of an analyzed the mix of gases is carried out by the combined detector on heat conductivity and warmth of combustion.

Management of process of gasification is carried out from a platform with blowing giving on the flexible active pipeline with a heat resisting nozzle, a variation pressure in an oxidation zone (exothermic processes) reactionary channel and combined (pulsing) removal of generating gases from a restoration zone (endothermic processes) gas generator with ensuring balance of physical speeds and kinetics of reactions.

4 INVESTIGATION THE WORKING REGIME OF THE STAND SETTING ON FREE MODE

Before the process started on the stand setting it is necessary to conduct the tests of blowing serve in perforated pipeline with the purpose of their use in the experimental stand setting. In the gasgenerator model is necessary to pawn a coal seam and containing rocks (Figure 5).

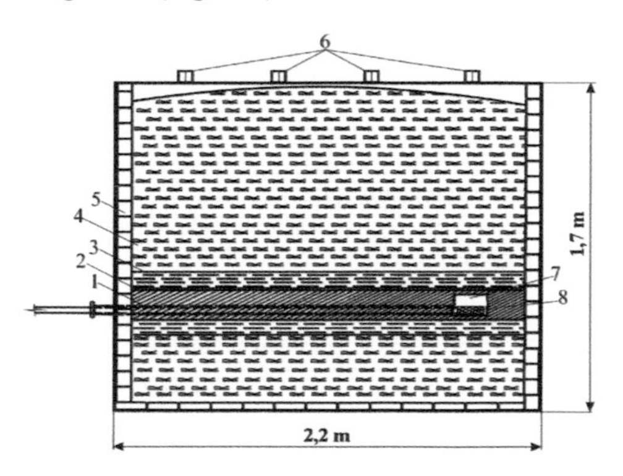

Figure 5. Model of gasgenerator with coal seam and containing rocks: 1 – coal seam; 2 – steel sheet; 3 – clay; 4 – siltstone; 5 – isolation material; 6 – heat couple; 7 – reactionary channel; 8 – the perforated heat-resistant attachment.

It is expedient to conduct an experiment with the moving point of blowing serve and pick up another analogue of gas. Verification the setting on free mode are made as follows. A compressor is included, in receiver air swing to pressure 0.6 MPa. Air is given through a flexible pipeline and attachment in a model. Smoke exhauster was included for air sucking from a model. Flowmeter fix the amount of air, given and exhaust from a model. After model drying is planned conducting the first series of experiments in the hot mode at a forcing method.

5 ICONDUCTING AN EXPERIMENTS ON THE STAND SETTING IN THE HOT MODE

For testing on the experimental stand setting in NMU a testing methods are developed on the design exploitation features of blowing boreholes with the controlling point on the surface stand setting. Designed methods foreseeing preparation the model to the experiment by forming soil layer, coal seam, forming reactionary channel in a coal seam: for the blowing serve (pipeline moving with heat-resistant attachment), channel of gasification and gasoutlet channel, setting the perforated pipeline in the channel for blowing serve, input a pipeline with attachment in the perforated pipe, forming the rocks of roof and coal seam ignition.

During the experiment by the change of pressure and blowing charges, setting hatches on the steady mode of gasification.

During work of setting the followings parameters are determined: speed of attachment moving (sm / h), pressure of blowing serve (MPa), blowing charges (m^3 / min), composition of outgoing gases, temperature in the channel of gasification and gasoutlet channel (°C).

It is expedient to begin an experiment with determination the length of combustion face. Five holes are foreseen in the model corps of underground gasgenerator. The first series of experiments are conducted at length of combustion face – 1.7 m in the forcing mode.

Ignition carried out by the burning hot pieces of coal and blowing enriched by oxygen (O_2 – to 42%, blowing charge 0.2-0.3 m^3 / min, 3 oxygen cylinder by volume 40 l). Necessary coke volumes are 20-30 kg on one experiment. Temperature of self-ignition for Seleznevsk coal are made by a 315-328 °C.

The blowing serve during ignition is carried out by the combined method. Festering compressor and with work of smoke exhaust for rarefaction creation in a reactionary channel. After ignition of reactionary channel passing of the mode of blowing serve is carried out to the forcing method with the increase of pressure from 0.18 to 0.35 MPa.

Planned expense of blowing on one experiment consist 3228 m^3, with the receipt of gas in the volume of 6908 m^3 generator gas. During an experiment there will be gasified 1100 kg of coal.

During the conduction of experiment on the model setting BUCG are determined:

– charges of blowing on gasification (0.65-0.95 m^3 / min) on flowmeter IRVIS with the conclusion of testimonies on a transformer is equipped an interface;

– parameters of rock mass moving as far as coal seam gasified, by reference sensors point with the conclusion of testimonies on a transformer is equipped an interface;

– pressure of festering (0.15-0.35 MPa) on manometers;

– temperature in the combustion chanel, in a coal seam and rocks of roof (400-1100 °C) – temperature detectors with the conclusion of testimonies on a transformer is equipped an interface;

– amount of appearing gas (1.4-2.03 m^3 / min) on flowmeter;

– pressure on the output of gasoutlet pipeline 00.3 MPa) – by a manometer;

– gas composition, 20-32% – portable and stationary gas analyzer, equipped an interface and chromatograph;

– calorific value of gas, 2.18-3.56 MJ / m^3 – by a calculation a way.

During work of stand instituted control after character of motion of combustion face and redistribution of temperatures in a coal seam make during all of experiment through temperature detectors with the input of information on an interface.

Current control of gas composition is made a stationary and portable gas analyzer. Control generator gas extraction on a gas-freeing pipe and his analysis on chromatograph "Poisk 2" carry out with an interval at 1-3 hours, in accordance with GOST 14920-79.

Character of rock mass moving as far as pushing of combustion face of model of gasgenerator is fixed reference point sensors with the input of information on an interface.

After completion of gasification process heat-resistant attachments and pipes take out, unseal a model and determine mass of remaining coal, ash and runback composition from a tank for ash collection and tank-cooler.

Motion of measurings fixed on the personal computer in the program "Magazine of supervisions and registration of experimental information" Results of researches are designed protocol, containing information about a sequence, character and results of researches

6 CONCLUSIONS

On results an experiment the parameters of mathematical design of underground gasgenerator and process of coal gasification are corrected, the adjusted parameters are recommendations to planning of the borehole underground coal gasification station.

REFERENCES

Kolokolov, O.V. 2000. *Theory and practice of thermochemical technology of extraction and processing of coal.* Monograph. Dnipropetrovs'k: NMU: 281.

Falshtynskyi, V.S. 2009. *Improvement the technologii of borehole underground coal gasification.* Monograph. Dnipropetrovs'k: NMU: 131.

Savostianov, O.V., Falshtynskyi, V.S. & Dychkovskyi, R.O. 2007. *Mechanizm of rockmass behaviour at borehole underground coal gasification.* Dnipropetrovs'k: Naukowyi Visnyk NMU, 10: 12-16.

Falchtyskyy V.S., Dychkovskyi R.O., Lozynskyi V.G. & Saik P.B. 2012. *Research an Adaptation Processes of the System "Rock and Coal Massif – Underground Gasgenerator" on Stand Setting. Instytut Gospodarki Surowcami miniralnymi i energiją Polskiej akademii nauk.* Krakow: Szkoła Eksploatacji podziemnej: 241-254.

Geomechanical Processes During Underground Mining – Pivnyak, Bondarenko, Kovalevs'ka & Illiashov (eds)
 ISBN 978-0-415-66174-4

The mechanism of stress formation in the rock massif around the mine working with intersection of it by stoping

O. Kuzmenko
National Mining University, Dnipropetrovs'k, Ukraine

O. Kozlov & O. Haylo
LLC "Krasnolimanskoye", Rodinskoye, Ukraine

ABSTRACT: Showed the mechanism of formation the technological zones and the stress state of massif at the location of mine working on the surface of coal seam on the way of stoping. Identified specific areas, that impact on operation of working face, that are creating difficulties and needs for additional costs for the transition mine working by highly mechanized working face.

1 INTRODUCTION

Currently, the intensification of mining on the Donbas mines and other coal basins of the world are achieved through the introduction of high-performance mechanized complexes with a great resource. Their application yielded excellent results in the working face in coal seams, which occur in good geological conditions where there is a possibility to increase the size of the stoping column.

In geological conditions of Donbas Basin the effective use of intensification of stoping technology is constrained by natural and anthropogenic formation of structures of country rock. However, most of the longwall faces are working in difficult conditions that lead to greater downtime and loss of working time. Costs for the purchase of complexes are not justified because of the inconsistency of geological conditions with the technical requirements of this class of mechanisms. To achieve good results is difficult because coal reserves do not correspond with the technology, which provides for mechanized systems. Reserves of coal seam, suitable for develop by high-mechanized complexes that do not correspond with criterion of readiness for stoping.

Increasing the length of the longwall face and stoping columns to reduce losses due to reinstalling systems, lead to increase the probability of occurrence within the stoping field of geological faults and mine workings that belong to the intersecting by complex-mechanized longwall. The presence of mine workings on the way of longwall face movement is one of anthropogenic structures. The passage the mine working by stoping is requiring additional costs and makes it difficult, but it does not lose work on the stratum and the complex is not subject to the dismantling and assembly work. It is important to maintain high efficiency equipment.

In these areas dramatically increased rock pressure on the support, deteriorating condition of rocks and there is observe an increased release of methane gas. There is folding, crumpled coal and country rocks, the false roof and the ground.

2 FORMULATING THE PROBLEM

The passage by stoping of the mine working, located in the plane of the stratum, requires an assessment of the reliability of the state of technological areas, which provide normal working conditions and execution of the preliminary activities for its formation with desired strength properties as well.

These measures include assessment of mine working deformation, creation of anthropogenic massif over the depth with elasto-yielding characteristics by installing additional timbering, for making to anthropogenic zone the shapes of acceptable parameters for the inscribing of geometrical dimensions of the sections of mechanized roof support.

It is known that the mine workings are located on the way of longwall face are passed in advance and completed their life enough to be subjected to deformation under the effect of rock pressure, as in conjugation, and the width of the stoping column.

In the coal mines of the Donbas the losses of the areas of preparatory mine working reach 60-70%, depending on the structure and depth. More than 40% of mine workings are repaired before handing in exploitation, 52% mine workings are deformed and 20% are in poor condition. A specially should be noted the unsatisfactory state of coupling, where

80% do not correspond with the technological requirements.

In these circumstances, it is important to know the mechanism of formation of strain zones according to the overlay the dynamic zone of bearing pressure at the working face in the area of the bearing pressure of mine working.

To evaluate the effect of overlay the support area of working face, first is necessary to determine the degree of deformation of the mine working, when it is in a coal pillar, outside the influence of mining. This information becomes the starting point for the choice of ways to create an anthropogenic massif over the depth to elasto-yielding characteristics, timbering and run-time of technologic processes, prior to enter the working face in the anthropogenic zone.

Under the anthropogenic zone means a combination of unloaded areas of the country rocks in the massif around a mine working and its own space (Figure 1). In this environment, established rock pressure, formed a system of cracks. A unique gas regime is formed. In addition, the support of the mine working has been spent the resource of the time and there is a mine working available without regard to its use during the transition cleaning works. The support is constructed without taking into account of it use during passage the cleaning works and has spent part of the resource of time. Massif, surrounding the mine working, can be reinforced or weakened. There may be elements of anchoring, plugging stones or backfill wells used for the decontamination of the country rocks. In any case, the massif has changed its natural condition, and subjected to erosion processes from the mine atmosphere. Mine working is filled of materials of the support.

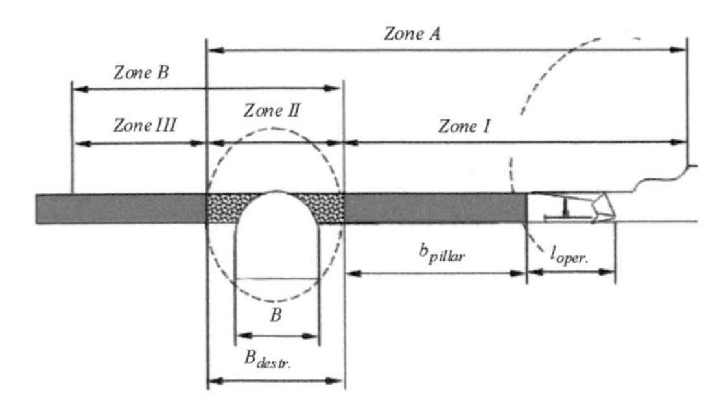

Figure 1. Scheme of anthropogenic zones disposition in the plane of the layer on the way of working face movement.

3 MATERIALS UNDER ANALYSIS

Based on the geomechanical state of the country rocks, anthropogenic zones can be divided over the degree of influence on the stability of the massif in the zone of influence of the working face. Zone A is characterized independent state of the country rocks around the stoping mine working and working face.

It is known that in the massif around a single mine working creates several zones of varying degrees of disturbance and stress-strain state (Kuzmenko 2001; Vynogradov 1989; Calculating...1990; Sazhin 1968; Samodelkina 2003; Skipochka 2003; Skipochka 2006 & Skipochka 2002). Discharge zone, exorbitant strain and fracture zone are located at different distances from the contour of mine working and a different influence on the stability of the massif. Under the conditions of occurrence of flat coal seams in Donets Basin tangible value of the physical and mechanical changes in the state of the massif occur at the distance of 3-5 of width of the mine working ($B_{destr.}$).

In controlling the roof by complete laying of worked out space the part of load up to 25-30% of the overlying rocks perceives backfill massif. On thin flat formation of Donbas the control of roof realize by complete destroy of the country rocks. With this method, the load on the hanging out roof rocks is passed in greater extent on the protective construction and stoping support and stoping mine working. If there are between the working face and the coal pillar mine working the part of hanging out rocks is passing the load on it.

The peculiarity of the passage by working face of mine working is that in a period of time is changing the conditions of formation of the load on the fasteners elements of mine working and mechanized support by incremental of supported area of stoped-out space. Pillar can be destroyed, that is leading to the dynamic increment of load and possible mine dump formation in the technological area.

Along the length of longwall face the increment of the load on the mechanized support section of the convergence of the country rocks in the workspace will be the differential nature due to changes in of the coal pillar width (b). On the conjugation with mine working the hanging out rocks under the worked out space creates the pressure on the pillar. Pillar crushed under loads exceeding its carrying capacity, which depends on the physic-mechanical properties of the coal seam and its geometrical dimensions.

The total area of the load formation on the massif and the anomalous rock pressure the length of the lava is within the total length of the zone I and zone B. This rock pressure is opposed to the active resistance of the mechanical support counter along the width of the operating space of longwall face (loperating) and compressive strength of the coal pillar, located in different stress-strain state.

Mine working is carried out ahead of time by the moment of its transition by stoping. Consequently, it is deformed during exploitation, reducing the cross-section. At the sides of the mine working the part of the coal massif is destroying, losing their original carrying capacity.

It is known, that the zone of rock pressure consists of two branches – increasing the voltage (a) to peak and decline (b) in the depth of the massif to the geostatic stress value (γH).

When you remove the working face from the mine working, zones of the rock pressure is not contacting with one another (Figure 2a) and the formation of the loads on the support occurs, regardless of their relative position in the massif.

The magnitude of face influence and mine working depends on the physic-mechanical properties of coal massif and country rocks, as well as from it structure. Its length ($l_{def.}$) depending on the stoping bed thickness can be 12-20 m, and the workings deformation of the zone of influence of working face reaches several tens of meters.

Thus, the deformation of the mine working can be enhanced even at a distance from the longwall face through the overlay zone of influence of the working face in a stationary bearing zone. The contact of the bearing pressure zone that is forming ahead of the working face, and the bearing zone around passed mine working occurs downstream of their branches. From this point begins the formation of a coal pillar width (b_{pillar}), its loads and mine working condition in front of face (Figure 2b).

The descending branch of the distribution diagram of rock pressure zones overlapped, and coal pillar undergoes of dynamic alternating loads on the part of the moving longwall face and hanging out rock, over the stoped-out area. The geomechanical nature of the interaction of mine working support with the country rocks is changing. Under the influence of the total load coal pillar is deformed, the natural fracture cracks reveal, gas regime change, as in the massif, and mine working.

With longwall face moving towards the mine working the pillar width (b_{pillar}) becomes equal to the length of the zone of bearing pressure ($b_1 + a_1$) and the load on the pillar is forming by hanging out of the roof rocks, as over the mine working and working space on the longwall face and the worked out area (Figure 2b.) Under these conditions, the sections of mechanized roof supports reaction cannot be neglected. The support works in compliant mode and increases the carrying capacity of the coal pillar near the mine working.

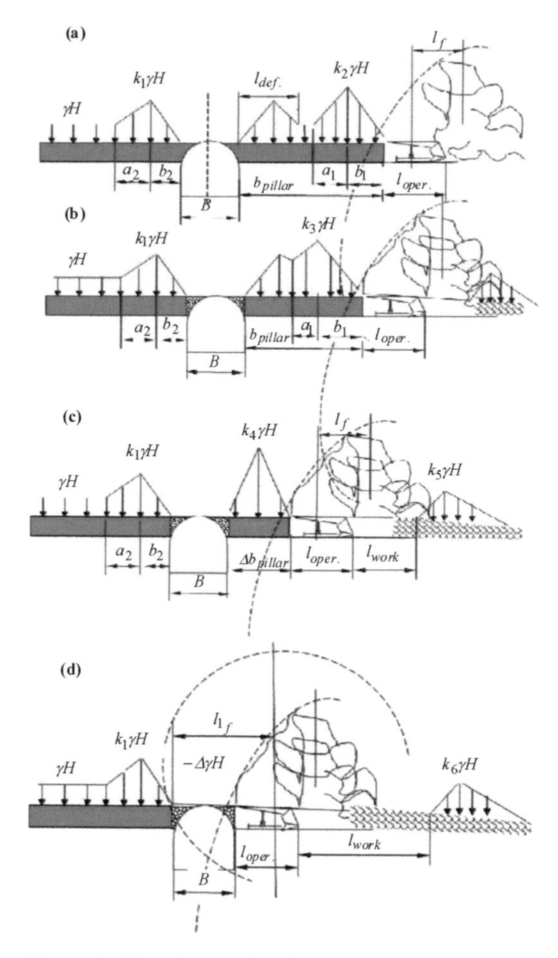

Figure 2. The scheme of rock pressure formation in the massif with the overlay of a bearing zone of clearing works on passing mine working.

The connection of mine working with the workspace of longwall face dramatically increases the area of exposed rock of the roof. The bearing zone disappears without encountering the pillar support. Working space of longwall face is among the cavity of passaging mine working and stoped-out space. The half-span of collapse zone rocks is increasing (l_{1f}), moved ahead of the working space of longwall face. Practically, stoping are not carried out, excavation of coal is absent, and thus ends the formation of anthropogenic zone, part of which is necessary to prepare in advance to successfully overcome the mining operations (Figure 2d).

Taking the width of intersected mine working (B) equal to the width of working space ($l_{oper.}$) without reducing the carrying capacity of the coal from the opposite side of the mine working value of span of hanging out of the roof rocks to increase by 2 times. Rigidly clamped support for the overhanging roof rocks above the coal powered supports is coal massif.

At a certain power and strength of lithological differences given the magnitude of exposure can be considered as the hanging console, supported by flexible support in the form of mechanized roof supports sections. The value of rocks hanging may be different depending on the position of mine working and the step of failure ($L_{det\,er.}$) of the immediate and main roof. With failure of rocks before the approach the longwall face to the mine working the value of hanging should not exceed half a step of failure closer to ratio:

$$(l_f + l_{oper.} + B) < 0.5 L_{det\,er.},$$

where l_f – step of roof rock failure, m; $l_{oper.}$ – width of the operating space of longwall face, m; B – width of mine working, m; $L_{det\,er.}$ – magnitude of face and mine working influence zone; m

Step of failure of rocks can be equal to $L_{det\,er.} = l_f + l_{oper.} + B$, which is highly undesirable from the standpoint of the formation of load on the support. If this happens, the roof rocks of coal massif will be cutting and possible fixing of support on rigid base.

Depending on the position of the line bottom to the longitudinal axis of the mine working of this quantity can be gradually reduced to reach values of the length of the lava as it move towards the mine working of or increase simultaneously along the entire length. The latter is the most adverse effect, since the sections lining both perceive strain throughout the area of rock above the workspace. We must take into account the fact that the country rocks around the mine working of deformed, have developed a system of cracks within the layers and as a result have lost their original capacity. It may be a situation when a section of lining loaded with rock blocks formed by layers of broken lithological differences. Stratification of rocks will be developed along the length of the longwall face as far as reducing the width and length of a coal pillar, increasing thus the load on the section of roof supports.

4 CONCLUSIONS

On the path of the longwall face are anthropogenic areas that have an impact on the stress-strain state of the country rocks and require further study the nature of the formation of the load on the support in the zone of influence of stoping.

The increment of the load on the section of roof support will depend on the angle between the line of longwall face with the longitudinal axis of mine working and solidity of the roof rocks of the area of technological area, and on the velocity of subsidence of the roof rocks in the workspace.

REFERENCES

Kuzmenko, O.M. 2001. *Investigation of the stress-strain state of rocks around the working face of mine working at intense fracturing area*. Vol. 2. Geotechnology at the turn of the XXI century. Donetsk: DUNPGO: 102-107.

Vynogradov, V.V. 1989. *Geomechanics of state management of the massif at the mine workings*. Kyiv: Naukova Dumka: 192.

Calculating methods in fracture mechanics. 1990. Ed. S. Atluri. Moscow: Mir: 392.

Sazhin, V.S. 1968. *Elastic-plastic distribution of strain around mine workings of various shapes*. Moscow: Nauka: 194.

Samodelkina, N. 2003. *A method for taking into account the rheological properties of rocks in the finite element analysis of geomechanical processes*. Technical Physics. probl. razrab. helpful. fossils, 3: 14-20.

Skipochka, S.I., Mukhin, A.V. & Kuklin, V.Y. 2003. *Peculiarities of coal-rock massif geomechanics at high production rates of the stope*. Dnipropetrovs'k: Geotechnical mechanics: Col. sc. works, IGTM NASU, 41: 16-22.

Skipochka, S.I., Usachenko, B.M. & Kuklin, V.Y. 2006. *Elements of coal-rock massif geomechanics during high rates of longwall advance*. Dnipropetrovs'k: PE "Lira LTD": 248.

Skipochka, S.I., Mukhin, A.V. & Chervatyuk, V.G. 2002. *Geomechanics of extraction drifts protection in unstable rocks*. Dnipropetrovs'k: 126.

Geomechanical Processes During Underground Mining – Pivnyak, Bondarenko, Kovalevs'ka & Illiashov (eds)
© 2012 Taylor & Francis Group, London, ISBN 978-0-415-66174-4

The need for modelling the deposit in conditions of Polish coal mines basing on the experiences of LW Bogdanka S.A.

T. Janik & J. Nycz
LW Bogdanka S.A., Poland

D. Galica
Mineral and Energy Economic Research Institute Polish Academy of Sciences, Poland

ABSTRACT: The principle situation in the base of technical, technological and geological changing of Polish mining industry is presented. It is proposed to made the simulation model of Lubelski Wegiel "Bogdanka" S.A. coal mining field using IT system. This allows making the planning the mining works in changing stress geological conditions. The creating geological database and later digital model of the deposit was made by Minex software. Input dates ware assumed every three months by available information. Necessary conclusions and economical evaluations were done.

1 INTRODUCTION

The role of mining geologist in the mining process is to inter alia gather, process and share information about the deposit to fulfill the needs of the mine. Independently from the complexity of deposit's structure, the scope of actions undertaken by mining geologists is the same. With the advance of exploitation the increment of information increases with time. To meet the requirements put before the geological department it is necessary to introduce additional IT tool, namely programme for gathering, processing and interpreting the geological information.

The development of IT tools gives the modern mining geology better capacities to interpret geological data obtained during recognizing, developing and mining of the deposit. The amount of geological data is a derivative of the deposit structure, its size, time of exploitation and size of production. Together with advance in mining a need for more effective management of geological information occurs. These kind of possibilities are offered by many broadly marketed IT tools used for modeling and visualization of spatial geological data. The amount of geological data piles up with time so it becomes problematic to effectively manage the information about the deposit, especially with shortages in the workforce which are observed recently among geological departments of the mines. Digital technique comes with help in these situations. It offers specialist software for better management of information. The only constraints in this field are the capabilities of the software and the skills of the person who uses it. In modern mining one may observe a trend to put more responsibilities on the geological departments of the mine. Introduction of new exploitation technologies (ploughs), reaching for still thinner seams and the increase of depth of hard coal exploitation requires the mining geologist to have higher qualifications which guarantee that he is able to make decisions quickly. In this situation the digital technique gives incommensurable possibilities in comparison to traditional methods of interpretation of geological data.

2 MODELLING OF THE DEPOSIT AS A PART OF IT SYSTEM FOR ASSISTING DECISION MAKING PROCESS DURING PREPARATION PHASE OF MINING

Taking into the account the aforementioned conditionings and coming towards the new challenges related to optimisation of planning and controlling the exploitation as well as the mining process in Lubelski Wegiel "Bogdanka" S.A. an IT system for assisting decision making process has been implemented. Project consortium consisted of Mineral and Energy Economy Research Institute of Polish Academy of Sciences (MEERI PAS) from Krakow and Przedsiębiorstwo Robót Geologiczno-Wiertniczych Ltd. (PRGW). The goal of digitising the mine design process and deposit management was particularly focused on integration of works conducted in the mining division through development of adequate way of information flow and allowing access to the information to the mine's engineers (Dyczko & Kłos 2008).

Key elements of the system were implemented in the surveying and geological department of LW "Bogdanka" S.A. These include: numerical mining map, geological database and digital model of the deposit. The functioning of the system in this area was initially based on the following IT tools (Dokumentacja projektowa ... 2007):

– MicroStation (produced by Bentley, basic graphical environment);

– Minex Horizon (by Gemcom, geological database, deposit model);

– I/Mine Modeller (by GMSI, modelling and visualization of some of the geological data);

– SoftMine (by PRGW, the tools for creating numerical mining map and assisting data implementation to the geological database);

– MS SQL Server 2005 (by Microsoft, database).

During the further development of the system a decision was made to resign from using I/Mine Modeller and to replace it with Surfer.

The solutions implemented in the surveying and geological department are only a part of the system. However, other elements of the system (scheduling of preparatory and exploitation works in particular) are dependent on the quality and availability of the data from the department.

3 FROM GEOLOGICAL DATABASE TO A DIGITAL MODEL OF BOGDANKA'S DEPOSIT

The basic task at this stage was to create geological database and later digital model of the deposit for Minex (Figure 1). The first stage of the project began in the middle of 2007 and included digitisation of ca. 2300 geological profiles (4000 at present), 97 profiles of surface drillholes, 83 profiles of underground long exploratory boreholes and 503 coal chemical analyses from the headings. Basing on this data digital models of two seams (382 and 385/2) were created.

The geological database of the system includes particularly geographic information (drillholes and profiles coordinates), litho- and stratigraphic data as well as quality information. Hydrogeological and geotechnical data has been collected. Because of the specific requirements of Minex it was decided that dedicated SoftMine tools will be used for gathering geological data and facilitating implementing data into the geological database of the programme. Verification and proper interpretation of data has been an important step during the analysis of the huge amounts of geological information. It was also assumed that every three months (together with the advance of the mining works) the database will be updated with newly available information.

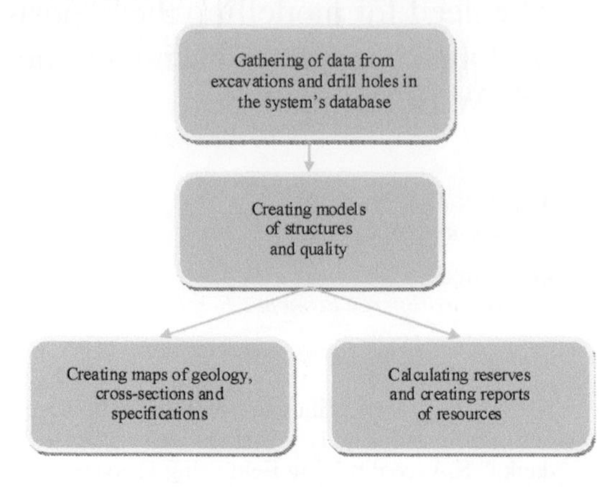

Figure 1. Working scheme of the system with geological database and model of the deposit.

The choice of Minex as a basic tool for deposit modelling was related to the specific structure of the deposit The software for deposit modelling is sometimes dedicated for different types of deposits, but two main types should be highlighted (Siata 2008):

– programmes that use grid models;

– programmes that use block models.

In the case of stratigraphic deposit (as in the case of LW "Bogdanka" S.A.), especially when the mining does not comprise the entire thickness of the seam, it is justified to use a tool based on a grid model.

The grid model is created through interpolation of values of chosen parameters in assumed triangle or rectangular grid. The grid model may contain different kind of geological parameters: structural, qualitative, hydrogeological and other. These models are created independently and usually depict the parameters' values averaged for the whole thickness of the seam (Figure 2).

The composition of a single coal seam model in LW "Bogdanka" S.A. contains following kinds of grids:

– structural: floor and roof, thickness of the seam, thickness of the partings, model of seam's floor inclination;

– qualitative: models of sulphur, ash and calorific value as well as the density model.

Apart from the aforementioned, the geological model in Minex can be edited when it comes to cross-sections, geological profiles of the excavations, profiles of the surface drillholes or penetrometer measures. The first stage of the project was completed in September 2008. The geological model is updated every quarter through feeding Minex database and subsequently preparing new set of grid models for particular seams.

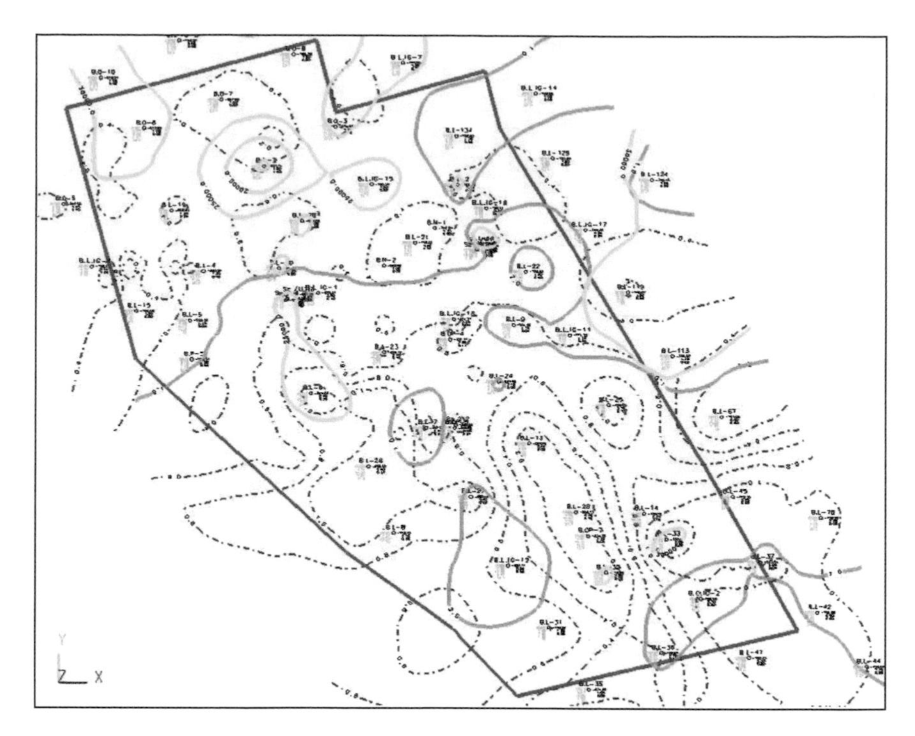

Figure 2. An example of isoline map with geological drill holes generated on the basis of few different Minex grid models.

The important problem associated with the use of block model is that Minex's database has a closed character. The application is therefore secured from unauthorized or accidental interference into the geological database structure. The data gathered in the application's database is easily accessible, it can be analysed, sorted and filtered. However, exchange of the geological data with other programmes used in the surveying and geological department is difficult. The problem of low interoperability with various tools is not however affecting the quality of prepared model. Despite the fact, it is important during day-to-day geologist's work. Therefore in the next stage of the project it is planned to prepare an independent and superior geological database for the system. Due to this it will be possible to use Minex as well as other IT tools (like Microstation, Surfer etc.) as client applications for the main database of the system.

4 ESTIMATING AND REPORTING RESERVES

In the complete geological model complex data that characterise the deposit are contained. Therefore the model contains full data about the quality and reserves of the mineral in the analysed deposit. Depending on the legislation in force and user's requirements these reserves may be calculated and reported in miscellaneous ways. The programme used for deposit modelling should not only allow to calculate volumes and masses of the mineral, but should also show average values of structural and quality parameters within the chosen areas. It should be possible to estimate the changes in reserves resulting from the conducted mining works, improvement in geological information or changes in the panelling design of the deposit (Siata 2008).

In the case of a stratigraphic deposit of LW "Bogdanka" calculating of reserves in Minex is performed on the basis of seam models which are created in advance. Certain computational areas are prepared for which reserves are calculated. The borderlines of the computational areas (Figure 3) can be imported from Microstation where they are drawn as closed polygons in advance and saved as .dxf files. Alternatively they may be created in Minex.

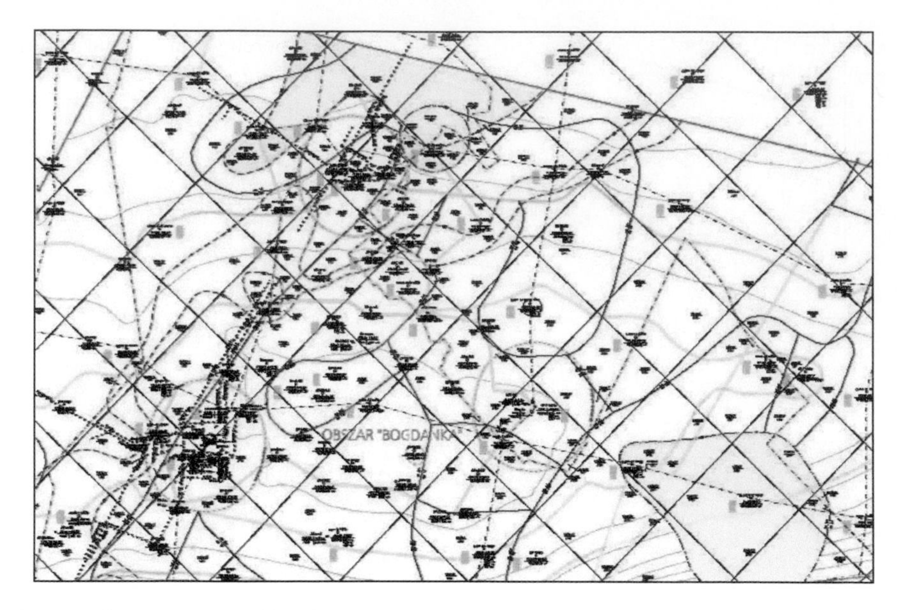

Figure 3. The map of reserves prepared in Minex and Microstation.

The calculating of reserves can be done for particular polygons or for the entire computational area. After the calculations are finished reports are created (Figure 4) in which information about the reserves and quality of coal in the computational plots may be found.

Estimation of reserves with the use of the geological model of the deposit is less time-consuming and more precise. Calculating the reserves and setting mean parameters in the plots requires only changing of the computational plot shape (i.e. to reduce the plot size by the area that has already been mined). Therefore during estimation of reserves it is important to constantly update the geological model with new geological information obtained with the advance of mining.

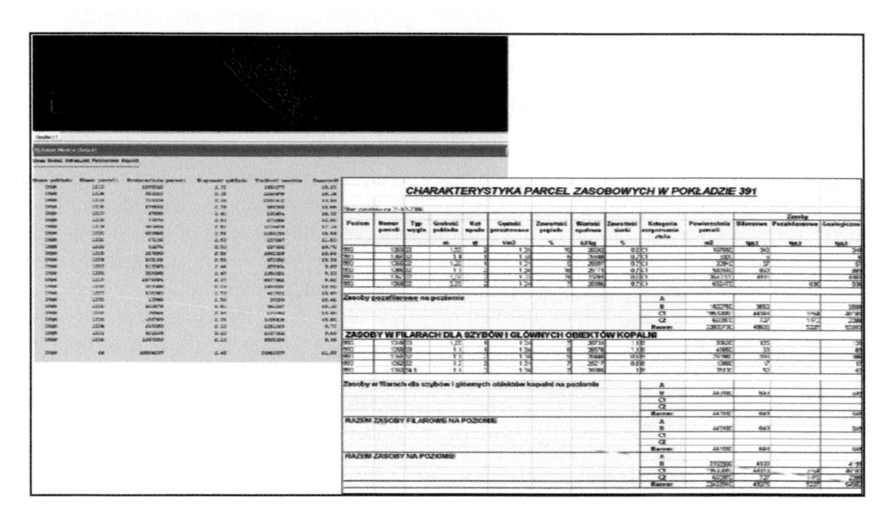

Figure 4. Exemplary reserves reports.

5 THE USE OF MODELS FOR DRAWING GEOLOGICAL PROFILES AND CROSS-SECTIONS

Geological profiles in Minex may edited basing on any chosen type of information. Surface drillholes (Figure 5), heading profiles, profiles from long exploratory boreholes may be used together or separately. There is also a possibility to generate quick correlational specifications in any chosen direction which is a huge support to the geologist when analysing the multi-seam stratigraphy (Figure 5).

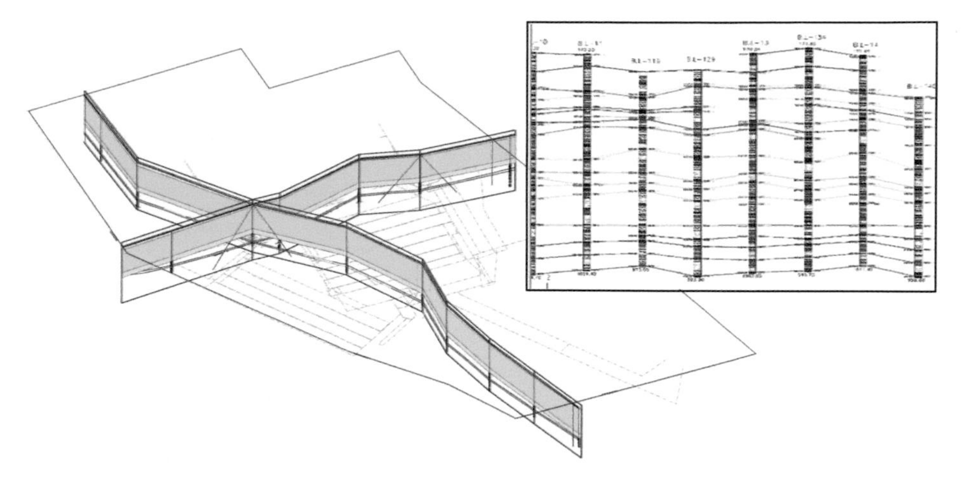

Figure 5. The main cross-sections through the deposit based on the surface boreholes and an example of presenting correlational configuration of seams.

The diversity of options which the tool provides does not however come with the quality of generated visualizations and it is therefore necessary to often use additional software like MicroStation for getting better quality of the final drawings.

6 DIGITAL MODEL OF THE DEPOSIT AS AN ELEMENT OF PRODUCTION LANNING

The person that prepares the geological model should be aware that despite its significant cognitive value, the model itself is only a part of the system for assisting decision making process for production and mine planning.

Deposit modelling in LW "Bogdanka" S.A. is a necessary and fundamental element of a bigger system which purpose is to optimally and rationally manage the deposit. The model is a source of much relevant information for other users of the system.

Structural, quality, hydrogeological and geotechnical data is used mainly for the purpose of preparing optimal exploitation designs and subsequently the conducting of the mining works themselves. The geological model of the deposit is also a source of date for production scheduling in MineSched (Figure 6).

The quality information gathered in the geological model together with the tool for production scheduling that is able to use this kind of data allow to create prognoses of quality and parameters of the winning in the perspective of designed schedules based on a full range of information that comes from the geological exploration. Together with the advance of works over the development of the system and broadening its users' knowledge it will be possible to prepare schedules which take the maximal convergence of assumed quality parameters into account.

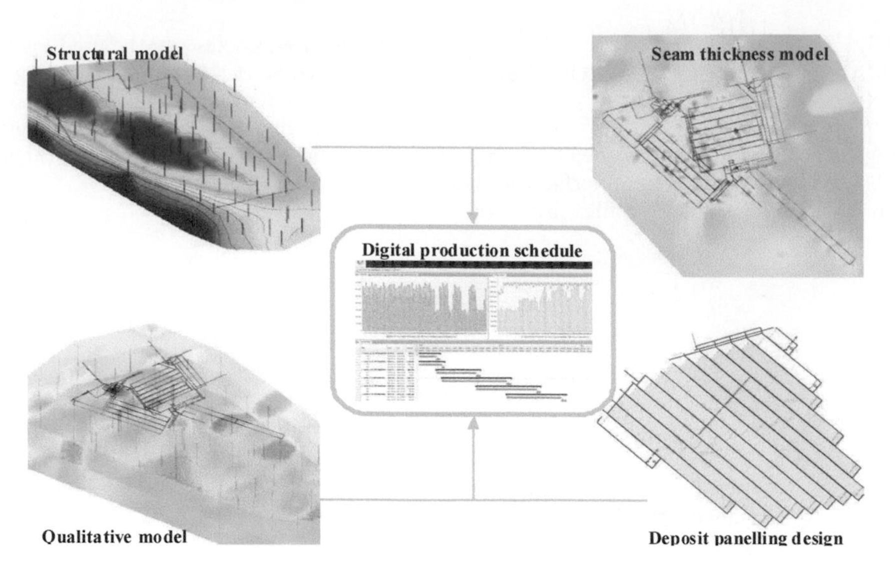

Figure 6. Production schedule based on the geological model.

7 CONCLUSIONS

The implemented numerical model of the deposit has changed the previous method of work of the mining geologist by adding the obligation to update the geological database, continually study to broaden his knowledge and participate in trainings. Following advantages of the created model may be listed:

– gathering all information about the deposit in one place;

– speed of access and edition of geological information in the form of maps, cross-sections and profiles;

– the possibility of correlating various geological information through editing a few different kinds of geological information on one map;

– freedom of editing different kinds of information;

– the possibility of sharing maps, cross-sections and profiles with employees of other departments after saving the map in .pdf, .dwg or .dxf file formats;

– the possibility of using the prepared models in other system tools (Surpac, MineSched).

Following disadvantages of the system can be listed:

– low interoperability of Minex and related time-consuming creation of source files during the updating of the geological model of the deposit;

– the necessity to use Microstation to improve some of the visual results of operations performed in Minex.

The aforementioned problems will be a subject of subsequent works over the system. However even its present state of development allows to see the great potential which lies in the professional and conscious use of such tools.

To sum up, it is necessary to mention that:

1. Still more geological data is created and stored in digital form in relation to general computerization of all aspects of life and work. This data may be used during the creation of geological database without processing if they are archived in standard database applications.

2. The traditional methods of managing geological information are difficult to maintain in the conditions of the need for quick decision making related to the exploitation problems associated with geological structure of the deposit or variability of its parameters.

3. The more complicated the deposit's structure and the larger the mining area the more difficult it becomes to manage the geological information.

4. Standing before still higher requirements regarding the fast interpretation of information and new exploitation techniques, the mining geologist is forced to use modern IT tools for improving his performance and quality of his work.

5. The advancement of civilisation and computerisation of all aspects of social and economic life gives a reason to assume that use of complex digital geological data processing in underground mines will become necessary and is only a matter of time.

REFERENCES

Dyczko, A. & Kłos, M. 2008. *System wspomagania decyzji w procesie przygotowania złoża do eksploatacji w kopalni węgla kamiennego „Bogdanka" S.A. – założenia, funkcjonalność i przepływ danych*. Materiały Szkoły Eksploatacji Podziemnej. Kraków: Wydawnictwo IGSMiE PAN.

Siata, E. 2008. *Model geologiczny złoża i jego rola w zarządzaniu produkcją*. Materiały Szkoły Eksploatacji Podziemnej. Kraków: Wydawnictwo IGSMiE PAN.

Dokumentacja projektowa wdrożenia. 2007. *System zarządzania złożem LW "Bogdanka" S.A.* Puchaczów.

Geomechanical Processes During Underground Mining – Pivnyak, Bondarenko, Kovalevs'ka & Illiashov (eds)
 ISBN 978-0-415-66174-4

Deposit model as a first step in mining production scheduling

A. Dyczko, D. Galica & S. Sypniowski
Mineral and Energy Economic Research Institute Polish Academy of Sciences, Poland

ABSTRACT: Development of IT tools gives the modern mining industry greater possibilities of interpretation of geological data acquired during exploration, access and mining of a deposit. The amount of geological data is a differential of the deposit's structure, its size, time of exploitation and production rate. With the advance of the exploitation, a need for more efficient management of geological information emerges. The article presents key differences between two elementary methods of modelling: grid models and three-dimensional block models, as well as their influence on the method of scheduling mining production. The entire discussion has been based on practical experiences gathered by the authors during deposit modelling conducted in LW Bogdanka SA and KGHM PolskaMiedź SA.

1 INTRODUCTION

A typical development process of every mine may be presented in a few consecutive steps:

1. Mineral exploration
2. Deposit recognition
3. Development of a model of the deposit
4. Development and implementation of a mining plan (deposit development plan and mining operation plan)
5. Liquidation of the mining plant, reclamation and future development of the post-mining terrain.

One needs to take into account that steps 2-4 influence one another. When creating a model of the deposit, one may reach a conclusion that some parts are not explored well enough, so it will be necessary to conduct some additional drillings (back to step 2). When mining the deposit, we get current geological data which need to be implemented into the model developed earlier, so the users are sure that they are working on the latest available information (back to step 3). The updated model of the deposit may give a hint about how to drive the future mining excavations in order to reach optimal degree of deposit's use, avoid natural hazards etc.

2 MODEL OF THE DEPOSIT

The digital model of the deposit includes modelling of structural surfaces, tectonic disturbances, thicknesses, hydrogeological conditions or quality parameters of the mineral. The information comes from observations made in boreholes and in currently driven mining excavations and is gathered in a geological database in relevant tables (Nieć et al. 1999).

Modelling in general is an empirical method of studying complex phenomena or processes basing on construction of models which are their simplified mental (concept) picture, vivid or expressed through mathematical formulas. During the construction of a model, the features of the studied object or process which are found to be irrelevant, from the point of view of the research goal, are eliminated.

The terms "model", "modelling" became popular in natural sciences, in deposit geology they were first used by D. Cox and D. Singer and spread quickly (D. Cox et al. 1986).

Using models in deposit geology has a long tradition. Every classification of deposits is in fact a classification of their models.

The basic features which characterise a model of a deposit (Nieć 1990) are:

– type of surrounding rocks and (if relevant) their specific petrographic features, facies formation, age etc.;

– geological conditions, especially geotectonic, of occurrence;

– the factor controlling the distribution of orebodies;

– mineral composition of the deposit (content, texture and structure of the ore);

– metamorphosis of the surrounding rocks and typical features of the geochemical halos.

The constructed models, presented as contour lines maps, allow for verification of data gathered in the database and their correction in a case of errors or low reliability. It is also possible to modify the model accordingly to other available geological information (Nieć et al. 1999). The knowledge of deposit's geometry, meaning the knowledge of its geometrical model, variability structure and general structural features defined by the descriptive model is important for the design of deposit's exploration (especially for defining the density of borehole grid), designing the mine access and choosing the mining method (Nieć 1990).

Basing on the models, one may calculate the deposit's reserves or its quality parameters in any part of the deposit. Research of this type has been also conducted in Poland, especially in Upper Silesian Coal Basin. An example of deposit's variability and thickness of coal seams (no. 118 and 207) between Jaworzno and Janina Mining Areas has been shown in Figure 1 (Mucha et al. 2010).

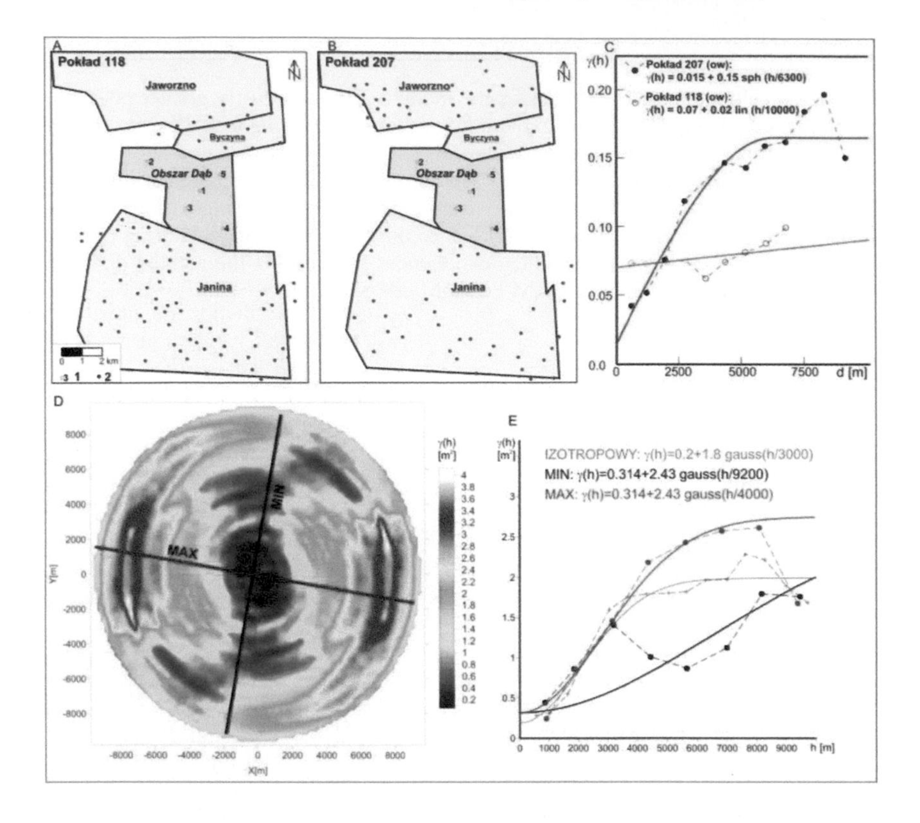

Figure 1. Deposit's variability and thickness of coal seams (no. 118 and 207) between Jaworzno and Janina Mining Areas. Explanation to the figure: A, B – drillhole distribution, C – isotropic relative semivariograms of thickness and their models, D – indicatrix of the seam 207 variability, E – directional semivariograms of thickness of seam 207 and their models (MIN – direction of minimal variability, MAX – direction of maximal variability) (Mucha et al. 2010).

Modelling of hydrogeological conditions allows for tracking their changeability in time resulting from the exploitation. Use of kriging allows for determination of reserves estimation accuracy and their classification accordingly to their accuracy.

The further part of the article presents key differences between two elementary methods of modelling: grid models and three-dimensional block models, as well as their influence on the method of scheduling mining production. The final part of the paper discusses the practical experiences gathered by the authors during deposit modelling conducted in LW Bogdanka SA and KGHM PolskaMiedź SA.

2.1 *Grid models*

According to Reed (2007) the term "modelling" refers to the process of creating a spatial distribution of numbers, created through calculations or estimations. The estimated parameters may include: thickness of the deposit, metal content in the orebody or any other feature of the deposit useful for determination of its parameters. The distribution of number may be two- or three-dimensional. In the case of two-dimensional models (so-called grid models), the dependent variable Z is a function of location determined by X and Y coordinates.

These types of models are used for modelling topography, deposits and other geological layers with seam structure, water-bearing layers etc.

The data for the model is most often taken during field research (i.e. collecting samples, measuring points etc.), then imported to the modelling programme (one of the most common applications used for simple calculations and data visualisation in 2D is Surfer created by Golden Software) and after the calculations are performed in a regular grid of coordinates one may display or print it in numerous ways. The most common ways of displaying the data include contour-line maps or gradient maps where the intensity of a given feature is shown using colours. An exemplary grid model has been shown in Figure 2.

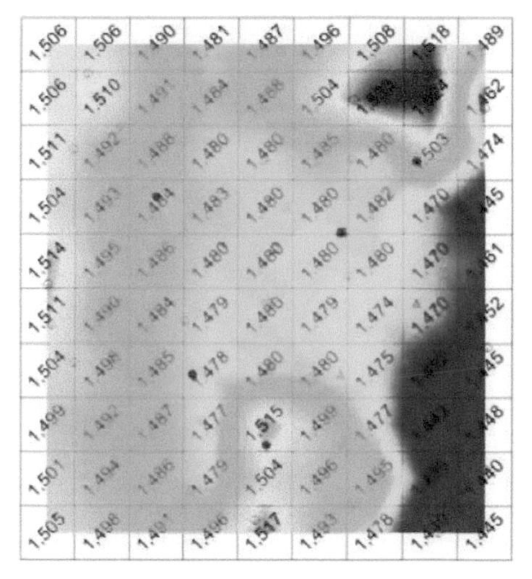

2D contour map with labeled cells

Figure 2. Exemplary two-dimensional grid model (Reed 2007).

The key difference between grid models and three-dimensional block models (described further in the article) may be taken to the fact that the modelled surface in 2D cannot be folded or bent around its axis which means that any given point with determined X and Y coordinates may have only one value of Z (Reed 2007). In a block model such limitation does not exist – subsequent blocks which have the same X , Y coordinates may have different values of Z , therefore creating a column of elements laying one atop the other.

From the practical point of view it is worth mentioning that the difference between the speed of creating a two- and three-dimensional models is considerable. While grid models are created very fast, the calculations necessary for creation of a block model may take many hours, if not days, even on a fast computer. In the industry, one may often see combined models – for example, the floor and roof of the deposit are grid models and the information about content of a useful component in the deposit between these two surfaces is stored in a block model.

2.2 *Block models*

The block model is basically a three-dimensional database describing the structure of the deposit. The model is constructed of thousands (sometimes hundreds of thousands or millions) of cuboids, out of which every one represents a fixed volume of the deposit and contains information about its structure. Their amount is practically limitless: the content of particular elements and chemical compounds, text information, geotechnical parameters etc. – the content of information assigned to a particular block is dependent only on the user and the data available to him.

However, one must remember that every additional piece of information multiplied by the number of blocks significantly increases the size of the model and simultaneously extends the time necessary for performing any calculations with its use. Nothing stands in the way to put any parameters in the model defined by user calculations. The best examples here are the costs of mining or the value of the mineral in a given block. Having this information, we de facto own a three-dimensional economic model of the deposit.

The block model is always created basing on the information gathered in the geological database which in turn are obtained from exploratory drillholes during geological works. It is created with use of geostatistical algorithms, such as kriging, inverse distance or nearest neighbour. The basic advantage of a model over a geological database is the fact that the model contains information about the entire deposit and not only about the spot where drilling took place.

However, the data in the model is only as good as the input information. One cannot count on getting reliable geological information in block of a few or a dozen or so meters when the distances between the exploratory boreholes are in the order of magnitude reaching a few hundred meters. The software producers usually recommend that the size of a block should be not less than half of average distance between the grids of the geological boreholes which are the source data for the creation of the model. The geological model can be rarely created only once and used for the many following years of mine's activity.

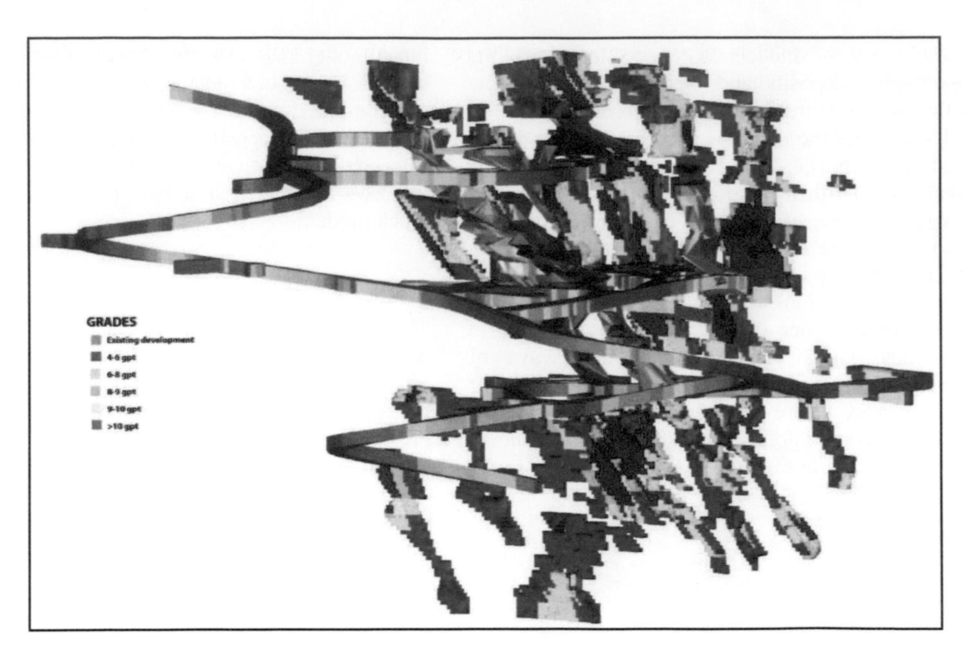

Figure 3. An example of a block model in Komis gold mine owned by Golden Band Resources Inc. (http://www.goldenbandresources.com).

When choosing the software used in the mining plant, it is important to pay attention to the speed and ease of updating the existing model.

It is also possible to perform calculations necessary for determination of volume of the overburden's or in a given part of the deposit. For this purpose so-called constraints are used. This means that for example it is possible to calculate a volume between two surfaces (i.e. a floor and a roof), but only within certain borders (i.e. in a particular plot) or to calculate the volume of the deposit, but only fulfilling the specified quality criteria (i.e. concerning content of a particular useful component above a determined percentage threshold). The constrains may be arbitrarily modified by the user and the results of the calculations can be exported to text files of .pdfs.

It is worth emphasising that the 2D and 3D models have different applications. 2D models are well-suited for seam-structured deposits, created as a result of sedimentation (i.e. hard coal and lignite), often with significant extents. They also allow for better modelling of geological disturbances, such as faults or handling deposits which contain seams with large and small thicknesses. Three-dimensional block models can in turn be applied to ore deposits, vein deposits created as a result of intrusions, such as copper, gold or nickel deposits.

3 PRACTICAL APPLICATION OF DEPOSIT MODELLING IN THE CONDITIONS OF POLISH UNDERGROUND MINES

3.1 *Grid model of LW Bogdanka's deposit*

In the middle of 2007, a project concerning establishing of an IT system for decision supporting in LW Bogdanka was launched. The employees of Mineral and Energy Economy Research Institute of Polish Academy of Sciences were the originators of the idea. The first stage of the work included implementation of data from around 2300 geological profiles (currently around 4000), 97 surface drillholes, 83 profiles from exploratory headings and 503 chemical analyses of coal from the headings, upon which digital models of two seams (382 and 385/2) mined by the company were generated (Janik, Nycz & Galica 2011).

The geological database of the system included mainly geographic data (coordinates and profiles), litho-stratigraphic and quality data. Hydrogeological and geotechnical data has also been gathered. Verification and proper interpretation of such huge amounts of data (mainly seam correlation) were an important step during this stage. Furthermore it was assumed that every quarter, together with the advance of mining, the geological database will be updated with new information.

During the concept phase, grid models were chosen for modelling of the deposit which is nothing extraordinary when one considers the type of the deposit mined by LW Bogdanka, especially the fact that the mining takes place at the entire thickness of the seam.

An example of a contour-line map with geological drillholes generated on the base of a few different grid models of LW Bogdanka mine has been presented in Figure 4.

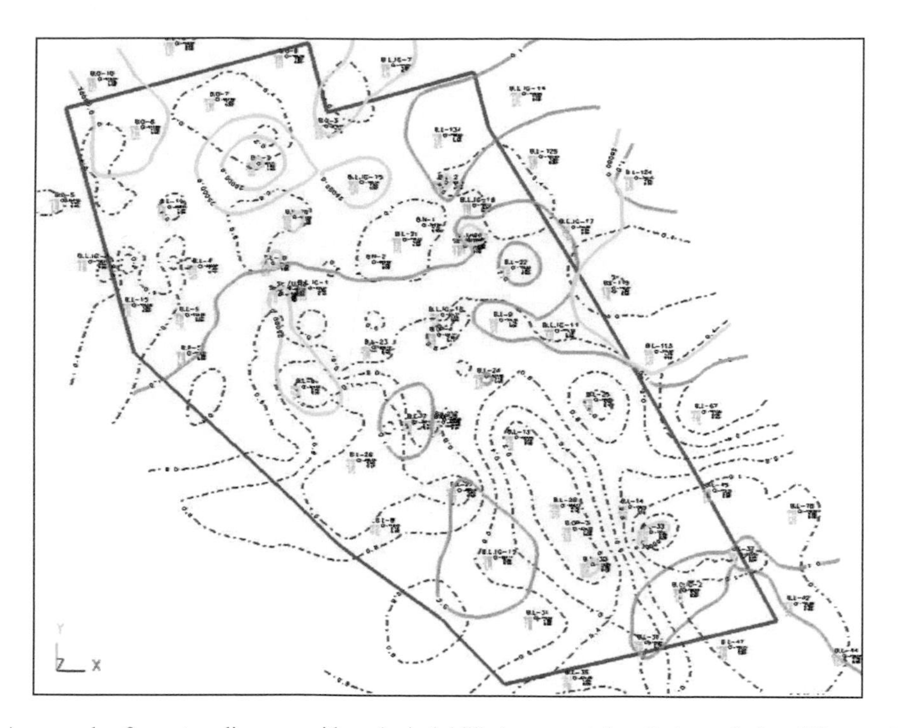

Figure 4. An example of a contour-line map with geological drillholes generated on the base of a few different grid models in Minex (Janik, Nycz & Galica 2011).

The grid model is created through interpolation of values of chosen parameters in a predetermined triangular or rectangular network. The grid model may contain grids (networks) all kinds of parameters: structural, quality, hydrogeological and others.

These models are created independently and usually show values of parameters averaged for the entire thickness of a seam.

The grid model of a single seam in Bogdanka includes the following grids:

– structural: models of floor and roof, thickness, thickness of the partings, inclination of the seam floor, quality: content of sulphur, ash and heating value as well as density model.

Apart from those parameters, the model allows for editing sections, geological profiles of excavations and profiles of surface or geotechnical boreholes. The first stage of the implementation has been finalised in September 2008. At the moment the geological model of LW Bogdanka's deposit is updated every quarter through introduction of new information to the database and then preparation of a new set of geological grids of particular seams (Janik, Nycz & Galica 2011).

3.2 *The block model of Głogów Głęboki – perspective areas of mining in KGHM PolskaMiedź SA*

In 2005 the employees of MEERI PAS were participating in a project called "Multi-variant assessment of accessing and mining of Głogów Głęboki-Przemysłowy (GG-P), with risk analysis of the venture". The project included preparation of analysis and verification of resource base of GG-P deposit with construction of a digital model of the deposit and a preliminary production schedule. For this purpose software produced by Surpac Minex Group was used (later the company becomes a part of Gemcom Software). The choice of this particular software for modelling was predetermined by the fact that it is owned by KGHM Polska Miedź SA. The database and model of the deposit were prepared with the use of Surpac Vision, while scheduling was performed in MineSched.

The digital model was based on data gathered from 53 boreholes located in GG-P mining area. The boreholes which were the source of information on GG-P were drilled in a grid of relatively large distances between one another (1.5 × 1.5 km) which makes predictions of any geological dislocations impossible (Kicki, Dyczko et al. 2005).

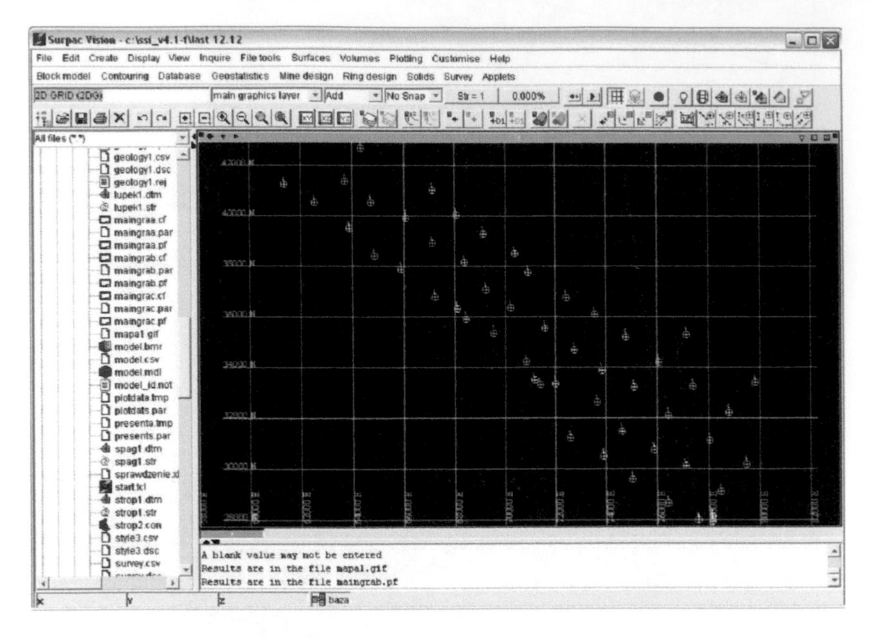

Figure 5. Distribution and location of the drillholes in space (Kicki, Dyczko et al. 2005).

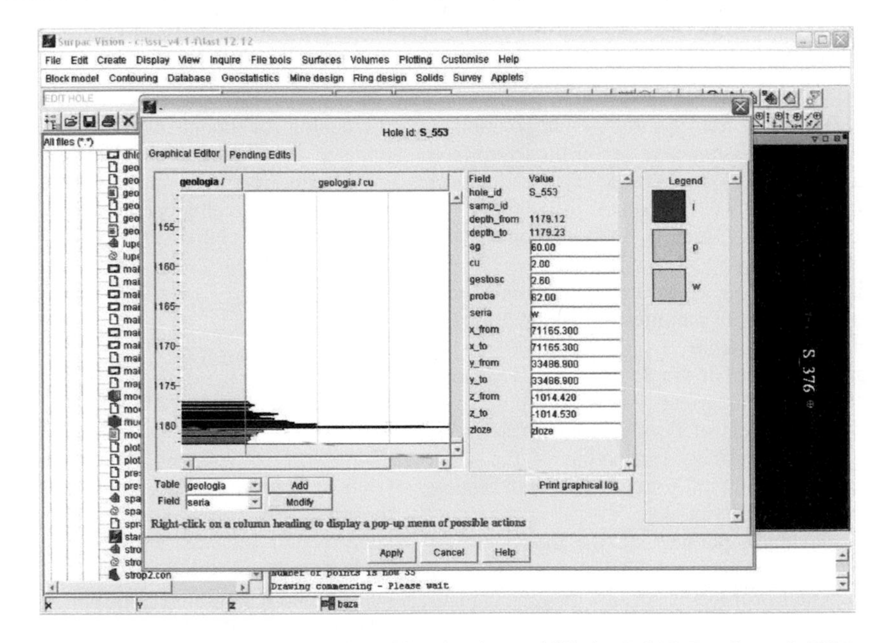

Figure 6. The possibility of visualization and editing of data for chosen drillholes (Kicki, Dyczko et al. 2005).

The first stage of preparation of the digital model involved creation of a geological database containing all information describing the deposit (geological data, quality parameters etc.).

The created database took the information about particular drillholes (x, y coordinates, collar height above sea level) into account together with series of lithology (carbohydrate, shale and sandstone) which were acquired from copper and silver samplings. This data was gathered in a operating spreadsheet in MS Excel. The next step included import of the data to Surpac and storing the data as MS Access database.

The database prepared this way allowed for visualization of geological drillholes in 3D environment (Figure 5) which allowed for both visualization of the geological data and its editing (Figure 6). All implemented changes were saved in the tables of the database.

The prepared database allowed for analysis of gathered geological and quality information.

The quality model was created basing on the block model module. As it was mentioned above, this kind of modelling allows for not only storage of data, but also for searching of necessary information. The difference between traditional databases is the fact that the database of the block model allows for interpolation of the stored data, determination of probability distribution of quality parameters in space and for their calculations.

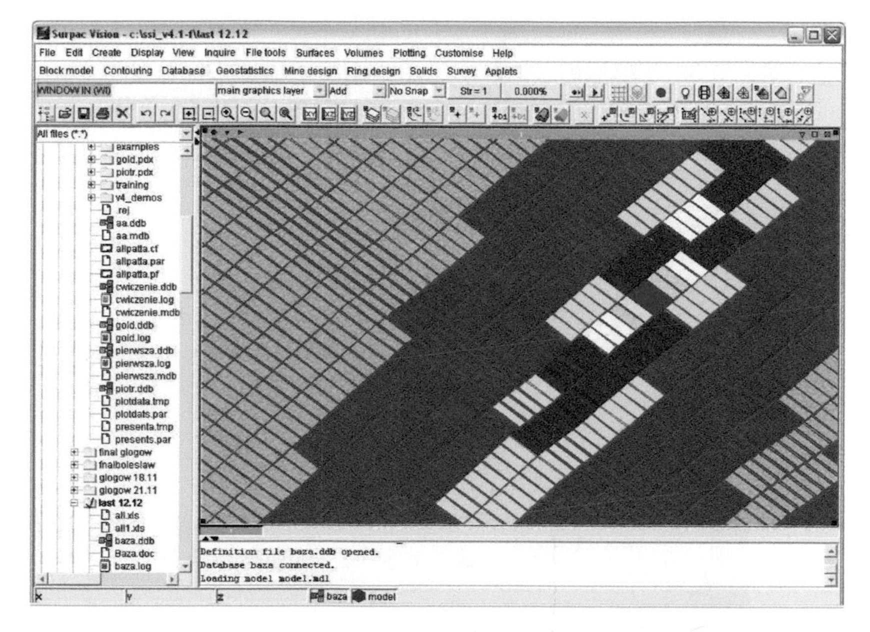

Figure 7. The block model of GG-P allows for visualization of chosen data assigned to particular blocks (Kicki, Dyczko et al. 2005).

The deposit is described by blocks of a given size – as it was in the case of GG-P – the blocks could be rotated to find a best fit to the deposit's strike and dip. In the next step, some constraints were added to the block model. In the analysed case, those constituted for the floor and roof of the deposit.

4 USE OF MODELS OF DEPOSIT IN PRODUCTION SCHEDULING

The final verification of the deposit's model, being either block model or a grid model, is conducted during actual exploitation and processing of the mineral. In order to determine if the model was built and is functioning properly, one must compare such parameters as (Pincock et al. 2006):

– data from blasting drillholes with data from exploratory drillings and data contained in the model of the deposit;

– the amount and quality of the mined mineral with the amount and quality of the mineral predicted by the model;

– the amount and quality of the mined mineral with the amount and quality of the mineral that went into the processing plant;

– the amount, quality and yield of the mineral that

went into the processing plant with the same values predicted by the model.

Gathering of this information will allow for verification of the model and will improve its predicting value about the future exploitation.

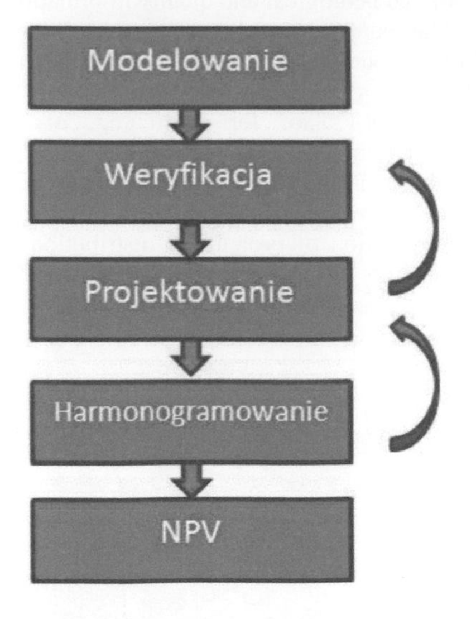

Figure 8. The course of the mine planning process (own study based on Barber et al. 2000).

Model of the deposit is always a first step for development of an exploitation plan and production schedule. At every stage, a close cooperation of mining engineers and geologists is required. Verification of the model involves not only checking its correctness, but also assigning particular blocks with economic parameters, such as costs and revenues related to mining of a particular volume of rock. The design stage includes determination of the mining borders, location of excavation, choice of technology etc.

Creation of schedule, meaning the sequence of exploitation is the next stage on the road to the final step: making a profit, showed in Figure 8 as Net Present Value (NPV). The repetition of steps 2-4 in the figure is necessary for maximization of NPV, meaning achievement of maximal income with minimal exploitation costs. However, instead of maximization, one should rather speak of optimisation (after considering of geological and mining conditions) of NPV in the life-of-mine period. The thing is that, for example, negligence of overburden stripping in an open-pit mine or driving preparatory headings in an underground mine in short-term leads to a drop of costs of the enterprise, but in a few years - when new resources are not accessible – will result in an obvious disaster.

In LW Bogdanka all seams together with their parameters (roof, floor, seam thickness, partings, quality and chemical parameters of coal etc.) were modelled as two-dimensional grid models with the use of Gemcom Minex software. As a result a giant database of information about the deposit was created which allows for a quick and simple updating of the gathered information when new data flows in, for example from exploratory boreholes.

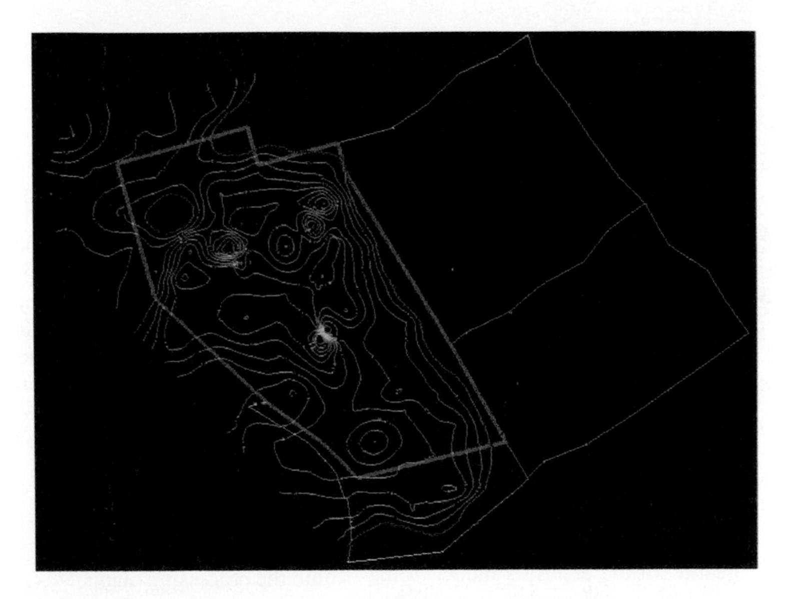

Figure 9. Contour-line model of seam thickness of one of the seams in LW “Bogdanka”.

These models are in turn imported into Surpac and from it to the scheduling tool - MineSched. This tool has broad capabilities of configuration of production parameters or creation of visual results of calculated schedules. Because of the significant improvement in the time the calculations in relation to the methods used earlier (average time of calculation of a single production schedule is a dozen or so minutes) it is possible to create multiple scenarios and their optimisation. Thanks to the macros it is possible to download data from Excel spreadsheets, performing calculations, creating additional results in MineSched and repeated export to Excel in the form users are accustomed to.

5 CONCLUSIONS

This paper is the result of implementation works of Department of Mineral Acquisition operating in Institute of Mineral and Energy Economy of PAS in Krakow concerning computerisation of the work of the mining departments of coal and ores mines. The experiences gathered during realisation of large projects for KGHM Polska Miedź SA, LW Bogdanka SA, JSW SA and Południowy Koncern Węglowy SA allowed the employess of MEERI PAS for development of proprietary approach for the realisation of large IT projects suitable for the mining reality of the domestic mineral industry.

The presented article's goal was to describe the influence of deposit modelling on scheduling of future mining production. As it was proven, it is hard to undermine the advantages resulting from development of a precise model of the deposit, its constant updating and its use in the work of mine planning engineers.

The main advantages of modelling include (Wilkinson 2010):

– reduction of time needed for modelling and preparation of mining plans;

– releasing employees from menial auxiliary activities (import/export of data between different software, preparation of input data);

– reduction of costs, improvement of employees' and the mine's efficiency,

– reduction of exploration costs;

– increase in mineable reserves thanks to a better knowledge of the deposit;

– increase of accuracy and reliability of the prepared mining schedules;

– the possibility of developing alternative productions scenarios thanks to reduction of time for preparation of a single schedule;

– increase of exploitation safety thanks to identification of the parts of the deposits which may endanger the machines and people (i.e. heavily watered running sands).

Mining engineers have determined the structure of deposits, calculated reserves and planned the production using only paper maps and their own experience. The modern IT significantly simplified all of these processes. Practically all of the available software destined for the mining industry offer the modules for creation of three-dimensional models of the deposits, mine design and production scheduling. Integration of these activities in one programme or a set of programmes of a single provider allows for significant acceleration and improvement of the mine planning process.

Every mine which has implemented a modern IT system for planning and scheduling mining production (in the domestic coal market, LW Bogdanka is the single example up to date) felt significant benefits, of which increased efficiency and acceleration of scheduling process are the most important. Because of the trends in the global mining industry leading to increasing automation and computerisation of the mining activity, one may expect that also in Poland, more and more mines will implement specialist mining software.

REFERENCES

Barber Jon, Al-ZoibyBassam, Al-Soi'ody Ziad. 2000. *Recent Advances in Mine Planning at JPMC*. Gemcom Minex White Paper. gemcomsoftware.com

Cox, D.P., Barton, P.B.& Singer, DA. 1986. *USGS Buli.* nr 1693: 1-10.

Janik, T., Nycz, J & Galica, D. 2011. *O potrzebie modelowania złoża w warunkach polskich kopalń węgla kamiennego na przykładzie doświadczeń LW „Bogdanka" S.A. WydIGSMiE PAN.* Monografia Geomatyka górnicza – praktyczne zastosowania.

Kicki, J, Dyczko A. i inni. 2005. *Wielowariantowa ocena udostępnienia i eksploatacji złóż Głogów Głęboki-Przemysłowy (GG-P), z analizą ryzyka przedsięwzięcia.* Praca nie publikowana.

Materiały reklamowe ze stron www firm CAE Mining, Gemcom, Mincom, Maptek.

Model behavior – Mining Magazine. Vol. 194, Nr 2/2006.

Mucha, J., Wasilewska, M. & Sekuła, T. 2010. *Dokładność geostatystycznej prognozy wielkości zasobów węgla we wstępnych etapach rozpoznania złoża.* Mat. XXXIII Sympozjum „Geologia Formacji Węglonośnych Polski", Kraków: AGH: 59-64.

Mucha, J. & Wasilewska-Błaszczyk, M. 2011. *Praktyczne doświadczenia geostatystycznego modelowania i dokumentowania polskich złóż – przegląd wybranych zastosowań. Wyd. IGSMiE PAN.* Monografia Geomatyka górnicza – praktyczne zastosowania.

Nieć, M. 1990. *Komputeryzacja geologiczno złożowych prac dokumentacyjnych. Zakres i możliwości.* Przegl. Geol. nr 7-8 (447 -448): 300-302.

Nieć, M. 1991. *Modelowanie śląski - krakowskich złóż rud cynku i ołowiu; Kierunki metody badań.* Przegl. Geol. nr 3 (55): 129-137.

Nieć, M. i inni. 1999. *Cyfrowe modele złoża i ich wykorzystanie w dokumentowaniu złóż i obsłudze geologicznej kopalń.*

Pincock, Allen and Holt - Model what you mine – Pincock Perspectives No. 67, June 2006.

Reed James. *Volumetric analysis & Three-Dimensional Visualization of Industrial Mineral Deposits.* RockWareIncorporated 2007.

Siata, Ewa. 2008. *Model geologiczny złoża i jego rola w zarządzaniu produkcją.* SzkołaEksploatacjiPodziemnej, MateriałyKonferencyjne.

Wilkinson, W.A. 2010. *Benefits of building efficient mine planning process.* Mining Engineering September.

State Higher Educational Establishment
«National Mining University»

DEPARTMENT OF UNDERGROUND MINING

Underground Mining Department was established in 1900 in order to prepare specialists in underground mining.

There are 54 people working at the department nowadays, among them there are 10 professors, 20 professor associates and 10 assistants.

The department conducts preparation of the future specialists based on such areas and specialities:

Area: 0503 Mining of minerals

Speciality: 7.05030101, 8.05030101

Mining of the deposits and extraction of minerals

Specializations:

- Underground mining of stratified deposits;
- Underground mining of ore deposits;
- Projection of mines and underground structures;
- Underground mining of minerals with profound learning of information technologies;
- Underground mining of minerals with profound learning of profession-oriented English language;
- Underground mining of minerals with profound learning of management in production field.

Address:

49005, Ukraine, Dnipropetrovs'k
19, Karl-Marks ave.
SHEE «National Mining University»
Department of Underground Mining

Tel.: +38 (0562) 47-23-26, 47-14-72
E-mail: v_domna@yahoo.com
olga.malova@yahoo.com

Requisites:

CORRESPONDENT BANK: SWIFT: DEUTDEFF
DEUTSCHE BANK
FRANKFURT \ DEUTDEFF, EUR
BENEFICIARIES BANK: K/acc 10094986271000
PJSC UKRSOTSBANK
STREET KOVPAKA 29, KIEV
UKRAINE SWIFT: UKRSUAUX
BENEFICIARY: №26009000082587
Science research Institute of mining problems AES of Ukraine
49050, DNEPROPETROVSK , GAGARIN av, 105/44
UKRSOTSBANK
DNEPROPETROVSK, MECHNIKOVA, 11
CODE BANK 300023 OKPO 00039019

We invite you to take part in the VII International Scientific-Practical Conference «School Of Underground Mining - 2013»

Address of organizing committee:

49005, Ukraine, Dnipropetrovs'k
19, Karl-Marks ave.
SHEE «National Mining University»
Department of Underground Mining
Tel./fax +38 (056) 374-21-84
+38 (0562) 47-14-72
+38 (0562) 46-90-47
E-mail: olga.malova@yahoo.com
kosganush@yahoo.com
vvlapko@mail.ru
http://www.msu.org.ua

T - #0242 - 251019 - C248 - 246/174/11 - PB - 9780367380915